# SCIENCE IN THE PROVINCES

# SCIENCE IN THE PROVINCES

Scientific Communities and Provincial Leadership
in France, 1860–1930

MARY JO NYE

UNIVERSITY OF CALIFORNIA PRESS
Berkeley   Los Angeles   London

University of California Press
Berkeley and Los Angeles, California

University of California Press, Ltd.
London, England

**Library of Congress Cataloging in Publication Data**

Nye, Mary Jo.
    Science in the provinces.

    Includes index.
    1. Science—France—History. I. Title.
Q127.F8N94  1985     306'.45'0944     85-8503
ISBN 0-520-05561-6

1 2 3 4 5 6 7 8 9

FOR BOB AND LESLEY

# CONTENTS

# ACKNOWLEDGMENTS

It is a pleasure to acknowledge and to thank those who have helped me gather materials, interpret them, and write this book over the course of the last fifteen years. What has become a detailed monograph on French provincial science began as a study of Paul Sabatier and turned into a fascination with Toulouse, the Midi, and provincial France. Is it correct, I asked myself, that Paris should be almost exclusively the center of attention for historians of science concerned with French science? Is the portrayal of French science as Parisian science justified?

The focus of my first research on French science was the physicist and physical chemist Jean Perrin, who spent most of his life in Paris after growing up in Lyon. What kind of education might he have had if he had stayed in Lyon to study and to teach? Who would have been his colleagues? How might his career and his interests have been different?

I am grateful for the warm hospitality and gracious aid extended to me in provincial cities: Jacques Aubry, professor of mineral chemistry at the Université de Nancy I, Dean Depaix and Mme. Viatoux at the Université de Nancy I, and Mme. Ballot at the Rectorat de l'Académie de Nancy-Metz; to F. Mathis, professor of chemistry, and Mme. Périé at the Université Paul Sabatier, Université de Toulouse I; to Mme. C. Chabaneau at the Université de Bordeaux I; to Roger Grignard, chemical engineer and son of Victor Grignard; and to Pierre-Sadi Carnot and Mlle. Lucie Carnot, who opened family archives to my husband Bob and me in Nolay in Burgundy.

Staffs were gracious and helpful at the university archives already mentioned and at the Archives Départementales d'Isère et de l'Ancienne Province de Dauphiné in Grenoble; the Bibliothèque Universitaire de Lyon, Bibliothèque Municipale de Lyon, and the Archives Départementales du Rhône; the Archives Nationales and Bibliothèque

Nationale in Paris. I am also grateful to Mme. Paule Rene-Bazin for help with the use of material then housed in the Archives de l'Université de Paris I (Sorbonne) and to Pierre Berthon at the Archives de l'Académie des Sciences in Paris.

Mlle. Marion Parlange, the granddaughter of Paul Sabatier, spoke with me in Paris and gave me some valuable materials when I first began my studies on Sabatier; and Pierre Laszlo, professor of chemistry at the Université de Liège, provided insights into the organization of modern French chemistry when he was visiting in Norman.

I am grateful to Wilhelm Odelberg and the Swedish Academy of Sciences for permission to look at materials from the Nobel Prize Archives, and I am especially indebted to Elisabeth Crawford for introducing me to and guiding me through these sources. Discussions with her and with Terry Shinn and Harry Paul have been stimulating and important for my work.

John Heilbron and members of the Office for History of Science and Technology at the University of California at Berkeley have extended hospitality and aid to me on several occasions, and Ted Feldman provided graphic interpretations of some of the Berkeley office's valuable data on French science. I have also used university libraries at the University of Lausanne, University of California at San Diego, Princeton University, and, of course, the library and History of Science Collections at the University of Oklahoma.

A very special thanks is due the Institute for Advanced Study for inviting me to be a member of the School for Historical Studies in 1981–82. I am also indebted to generous grants then and on other occasions from the National Endowment for the Humanities, the National Science Foundation, the American Philosophical Society, and the University of Oklahoma Research Council.

I would like to thank *Isis*, *Minerva*, and *Historical Studies in the Physical Sciences* for permissions to reprint all or parts of articles previously published, and Fritz K. Ringer for permission to reprint table 1. For photographs used in this volume, I am indebted to those mentioned in the legends.

Valli Powell has typed and retyped the manuscript with affability and skill. I am grateful to staff members at the Institute for Advanced Study for typing early chapter drafts. I thank Bettyann Kevles for wise counsel, and Sheila Berg for careful copyediting of the manuscript.

I cannot read through the text without feeling appreciation for the advice, encouragement, and friendship in faraway places of Vikki and Michel Lockwood, John Biro, Jim Briscoe, Clarice Fisher, Jim Fisher, Franz Moehn, Kris and Guido Ruggiero, and Carol and Lee Congdon. Without Lesley and Bob, I would not have had the stamina to complete this study or the joy in doing so. My work owes considerable intellectual debts to Bob's knowledge, insights, and advice. I am lucky to have two such companions.

April 1985                                                    Mary Jo Nye

# INTRODUCTION:
# THE PARIS-PROVINCES
# DICHOTOMY

Perhaps the most powerful evocations of the French provinces during the last century are found in the great novels of Balzac and Flaubert. The images and moods created in Balzac's *Scènes de la vie de province* and Flaubert's *Madame Bovary* are enduring ones. Life in the novelists' provincial towns is marked by petty rivalries and squabbles, scandal and malice, self-interest, and self-importance. According to Balzac, provincial life is ordered as in a cloister: houses are like monasteries and drawing rooms are like convent parlors. Virtue can be found, but only as the necessary result of the drab monotony of provincial life.[1] The church, the market, the town hall, and the local pharmacy are the dominant features of the provincial town's physical and mental landscape. In Madame Bovary's Yonville-l'Abbaye,

> what catches the eye most of all is Mr. Homais' pharmacy right across from the Lion d'Or. In the evening especially its lamp is lit up and the red and green jars that embellish his shopfront cast their colored reflection far across the street . . . at the back of the shop, behind the great scales fixed to the counter, the word 'Laboratory' appears on a scroll above a glass door on which, about half-way up, the word Homais is once more repeated in gold letters on a black ground.[2]

Homais describes their new world to Madame Bovary and her husband Charles, who is to be the town doctor:

> Ah! you will find many prejudices to combat, Monsieur Bovary, much obstinacy of routine, with which all the efforts of your science will daily come into collision; for people still have recourse to novenas, to relics, to the priest, rather than come straight to the doctor or pharmacist.[3]

Inevitably, Flaubert describes provincialism through the juxtaposition of province and Paris. Emma Bovary lies awake at Tostes, near Rouen, dreaming of the capital:

> At night, when the carts passed under her windows, carrying fish to Paris to the tune of "La Marjolaine," she awoke, and listened to the noise of the iron-bound wheels, which, as they gained the country road, was soon deadened by the earth. 'They will be there tomorrow!' she said to herself.[4]

Men and women of talent fly to Paris. "Paris écrème la Province," wrote François Mauriac in 1926, adding "C'est vrai pour le talent, non pour la vertu."[5] Using a more parasitical image, Mauriac suggested that Paris drains the provinces of their life and vitality: "La Province recrée sans cesse Paris par une transfusion du sang ininterrompue et dont elle fait tous les frais."[6]

Transforming into a scientific principle the charmed attractions of Paris, Edward Shils has argued that "in every social system, there is a center from which authority emanates and to which deference is granted." He states, "Inequality brings forth the distinction between center and periphery. . . . The metropolis is a center of vitality . . . of creativity." Conceding that "much of what comes from the center . . . might be not better in itself than what originates in the province," he attributes the influence of the ideas emanating from the metropolis to the "special nature" of their place of origin.[7]

Shils offers a center-periphery model for British intellectual life in the 1950s: "If a young man, talking to an educated stranger, referred to his university studies, he was asked 'Oxford or Cambridge?' And if he said Aberystwyth or Nottingham, there was disappointment on the one side and embarrassment on the other. It had always been that way."[8] The cosmopolite's attitude was familiar to the Manchester scientist James Joule in the mid-nineteenth century. Remarking on the Royal Society's rejection of his fundamental paper on electrical heat, Joule said: "I was not surprised. I could imagine those gentlemen in London sitting around a table and saying to each other 'what good can come out of a town where they dine in the middle of the day.' "[9]

As far as the sciences are concerned, traditional wisdom is that the most important scientific work in France has occurred in Paris and that what goes on in the provinces does not count. At its founding, the Institut de France required Paris residency for election to its membership. It is not surprising that as a result the most highly acclaimed scientists, philosophers, writers, and artists came to live in Paris.[10]

In his history of nineteenth-century scientific thought, John Merz described the great scientific institutions of the European continent as the "Paris Institute [the Academy of Sciences], the scientific and medical schools in Paris, and the German universities."[11] Several decades later, the biologist Maurice Caullery wrote in his history of French science: "Rare are the great careers in Science which are made outside Paris."[12] Histories of French science have largely illustrated, by omission, the principle of center-and-periphery. It has been the scientists and institutions of the Academy of Sciences, Paris Observatory, Collège de France, Sorbonne, Museum of Natural History, Paris Faculty of Medicine, the grandes écoles, Conservatoire des Arts et Métiers, and other Parisian establishments that have formed the principal focus of histories of French science.[13]

That focus has begun to change. Terry Shinn's 1979 essay on the French Sciences Faculty system broke ground in surveying budgets, enrollments, curricula, and productivity for the provincial Sciences Faculties at Lyon, Montpellier, and Besançon. Harry W. Paul has looked at French Catholic universities and applied science institutes in provincial cities, with special emphasis on Lille and on the natural sciences. Robert Fox has analyzed the broad range of activities of nineteenth-century sociétés savantes, some 513 of them outside Paris, with particular attention to relations between science and industry in Alsace, especially Mulhouse.[14]

Other historians, including Fritz Ringer, Theodore Zeldin, George Weisz, John Burney, and John Craig, have written accounts of the university system—the latter two with emphasis, respectively, on Toulouse and Strasbourg—establishing the context of university science in the provinces.[15] These recent works are concerned principally with social politics and recruitment of elites. Thus, we have learned a good deal about the social origins, occupations, and rates of productivity of students and their professors. But we still do not really know the place, the people, and the preoccupations that were at the core of scientific work and scientific community in provincial settings in the nineteenth and twentieth centuries.

A new approach is needed which integrates the history of French provincial science into the national context and portrays in depth provincial scientific communities. In fact, science in the provinces enjoyed special vitality at the turn of the century, briefly threatening the hegemony of Parisian institutions. The foundations were laid at this time for the present-day influence of important French scientific centers outside Paris.

In this history of French provincial science, I demonstrate how local economic and cultural interests resulted in important specialties in Nancy, Grenoble, Lyon, Toulouse, and Bordeaux. University science in these cities prospered around 1900, as it prospers today. At the turn of the century, only the university communities at Lille, Montpellier, and Strasbourg (the latter in German hands from 1870 to 1914) were similarly prominent in French provincial science. As will be shown, the peculiar successes of each of these Faculties resulted from administrative, social, and economic circumstances, on the one hand, and initiatives and actions of individual scientists, on the other. A single explanation—social or individual—will not do.

Scientific biographies provide concrete illustrations of provincial scientific careers. The joint award in 1912 of the Nobel Prize in chemistry to Victor Grignard at Nancy and Paul Sabatier at Toulouse became a powerful symbol of provincial achievements. An analysis of their work, as well as that of three other notable provincial scientists, the chemist François Raoult and the physicists René Blondlot and Pierre Duhem, will demonstrate the role of the individual. Specialties developed in the provinces included their fields of research, namely, physical chemistry, organic chemistry, and electrical physics. As we shall see, the engineering sciences also became strengths of provincial university Sciences Faculties.

One intention of my study is to redress conventional emphasis on Parisian science and scientists, diffusing the focus of French science. In addition, I have the more ambitious and problematic objective of answering some vexing questions about the general character of French science and the reciprocal effects of the organization and content of scientific work. These kinds of questions, especially the issue of French scientific "decline," have been addressed in some recent literature on French science.[16] Let me briefly summarize the background for the kinds of questions I have in mind, relating social organization and the intellectual content of science.

The period 1860–1930 was one of attempted reform of the educational system and of scientific research. In the 1860s, it was alleged that French scientific work was in decline, compared to French achievements at the end of the eighteenth century and relative to contemporary German scientific accomplishments. Claude Bernard (1867), Louis Pasteur (1868), and Adolphe Wurtz (1870) wrote detailed reports on the state of French science. The journalist Emile Aglave claimed in the *Revue des Cours Scientifiques* (later, *Revue Scientifique*) that it was decades since France's "scepter" in science had last been held "without contest."[17]

In analyzing this fear that French science was in a state of decline, we may ask whether the French system of education and research changed, or whether it remained fundamentally the same. Under what conditions could, or did, indigenous research traditions spring up and thrive, and to what degree were they independent of Parisian influences and directives? Did there exist non-Parisian networks of importance? Were factors of scientific creativity or prosperity principally cultural and economic, or idiosyncratically individual and personal? How was the content of scientific work determined by these factors? Did local scientific centers and local scientific personalities exert force and prestige beyond their parochial settings? To what extent did provincial scientific centers contribute to a scientific revival, and did the perception of scientific decline alter?

As I shall argue in detail, the post-Napoleonic division between Paris and the provinces concentrated centralization of authority in Paris and decreed an educational hierarchy with legal advantages for Parisians. This system resulted in what its contemporaries, as well as later historians and sociologists, have deemed general weaknesses for French science. The centralized system exacerbated the potentially harmful effects of poor decisions or deeply rooted prejudices regarding the character and organization of scientific education and research. We will see many examples of such problems and difficulties in the chapters that follow. In particular, scientific education, like all education, was geared to preparation for state examinations, the most prestigious of which qualified secondary students for the Parisian grandes écoles. Thus, the national examination system guaranteed uniformity in scientific studies and ensured that the best students gravitated to Paris, just as the Paris residency requirement by the Academy of Sciences ensured that the most ambitious scientists eventually moved there. This system made attending lectures and classes or working with professors on individual projects less important than cramming with tutors and reading books geared to the examinations. Mathematics was the most important subject on the examinations. Other scientific subjects, especially chemistry and natural history, were considered intellectually inferior to the "queen of sciences."

The salary system for Faculty professors was fixed for each rank and did not vary from city to city, except for Paris where salaries legally were higher. A penurious salary scale at lower ranks and an expectation of bourgeois living at the higher ranks perpetuated the practice of *cumul*, the holding of multiple positions by a single scientist, which could most easily be done in Paris. Furthermore, cumul and a de facto seniority system only rarely allowed a younger scientist's election to a chair over the head of an elder. The fixing by the education ministry of a limited number of professorial chairs throughout France placed an artificial ceiling on the number of scientists living comfortably in university communities.

The system was inborn and inbred. An influence network emanated from the grandes écoles and the education ministry, determining the position and prestige attained by any French scientist. The requirement that all professors be French citizens blocked influence or innovation from beyond French borders. I illustrate in detail how this system worked and how it affected the lives and contributions of individual scientists in different scientific communities.

This centralized system changed dramatically in the 1880s and 1890s, evolving more gradually in the next decades into an infrastructure that resulted in two striking innovations: the establishment in 1939 of the Centre Nationale des Recherches Scientifiques (CNRS) and the creation after the war of the Ecoles Nationales Supérieures d'Ingénieurs (ENSI). The education of traditional scientific elites remained the prerogative of Parisian institutions, but new scientific and engineering elites, many of them foreign students who later returned to their own countries, invigorated provincial university settings. Both institutionally and intellectually, provincial innovations and developments exerted force and influence on Paris and the state, while provincial scientists and administrators often provided the lead in developing research and organizational initiatives.

The history of French science is not simply a history of centralized planning but of reciprocal tensions and relations between the established Parisian network and the competitive and more open networks of the provinces. This study demonstrates that the strengths of regionalism consistently provoked fears among Parisian administrators that further decentralization might endanger their own power and values.

The ideological values motivating the governing state elite included the building of republicanism, the combating of organized Catholicism, and the preparation for revenge, whether economic, political, or military, against Germany. Science was an essential tool for attaining these ends, and the governing elite paid much homage to the role of science in training the republican citizenry for rational thinking, civic virtue, and secular progress. Not surprisingly, various parts of the republican program sometimes rested in uneasy equilibrium with one another. For example, reformers weighed the success of the German model of scientific pluralism and institutional competition against the fear of Catholic and monarchist political strength in provincial enclaves. Even on the political right, Charles Maurras expressed fear, in 1908, that newly powerful French provinces might reclaim the recruitment of the army, debilitating the power of a unified French state for the sacred duty of revenge against Germany.[18]

By 1908 there was a great deal of vigor in the provinces, which gave pause to both republicans and royalists. Provincial liveliness was partly the result of enthusiastic local support for the sciences, which could add luster to regional prestige and prosperity to local industries. In his work on the provincial scientific societies, Fox has shown the breadth of local support for science. It was the Sciences Faculties, united into universities, not scientific societies, which were the recipients of regional funds and loyalties. The great successes of scientific centers like Nancy, Grenoble, Toulouse, and Lyon at the turn of the century—and the lesser success of Bordeaux—were due to agricultural, commercial, and industrial interests that favored sciences aiding regional development. Thus, at Nancy, in the industrial northeast near the German border, programs in chemistry, brewing, electricity, and metallurgy thrived. In the Pyrenees and in the French Alps, where the potential for development of new kinds of energy through hydroelectric power was recognized locally, electrical physics and industrial chemistry experienced extraordinary growth at the end of the nineteenth century. Lyon, which along with Mulhouse was a traditional textile center in France, supported chemical programs. Wine growers around Toulouse and Bordeaux were favorable to supporting agricultural sciences, but without an industrial base Bordeaux fell behind her sister scientific centers in physics and chemistry in the early twentieth century.

Regional settings affected the choice and content of scientific research by favoring or ignoring certain orientations in scientific fields. For example, organic chemistry and agricultural chemistry became Sabatier's interests after his arrival in Toulouse. And Grignard was persuaded to follow chemistry, which he previously thought boring and intellectually inferior to mathematics, after he became a student at Lyon.

Another example of the influence of environment on science is the possible relation between Grenoble's relative isolation and the lack of specialization there and the evolution of Raoult's new approach in chemistry. Robert Root-Bernstein and John Servos have observed that modern physical chemistry emerged as a boundary or hybrid discipline in out-of-the-way or peripheral places like Utrecht (J. H. van't Hoff), Uppsala (Svante Arrhenius), and Riga (Wilhelm Ostwald). If this is a meaningful generalization, perhaps it is not coincidental that Raoult, in out-of-the-way Grenoble, was a founder of the experimental basis for an ionist physical chemistry that was anathema to mainstream Parisian chemists of his generation.[19]

In some cases, we see that local interests caused the sacrifice of basic science to engineering and applied science programs. In other cases, like that of Blondlot's alleged discovery of N-rays at Nancy, we find that local pressures, influence networks, and individual psychology resulted in damage not just to the scientific prestige of a single scientific center but to French science as a whole. Petty bickering and personality differences could damage the scientific enterprise among a university faculty, as we shall see in the case of Duhem and Bordeaux. The First World War posed another set of problems. Studying scientific endeavor in its social context aids us in understanding the mutual effects of social and intellectual factors, as well as group dynamics and individual personality, in scientific work.

The strength of French science lay in the diversity rooted in the provinces. Provincial scientists were often leaders and innovators, not followers of Parisian scientific culture and values. Provincial university science frequently was more open and innovative than Parisian science. What is lacking in the Shilsian dichotomy of center-periphery is an understanding of the dialectical, symbiotic relationship between center and periphery, and a sufficient appreciation of the diverse forces working within and across provincial communities. There is a richer variety in French styles and traditions than may be contained within the clichés based solely on profiles of Parisian science.

Still, what is striking about the organization of French institutions is the effectiveness of the Parisian elite in incorporating innovation and diversity into new centralized structures. As Mauriac wrote in 1926:

> The Provinces cultivate differences: no one there dreams of blushing at his accent, his manners. Paris imposes on us uniformity; it puts us, like its houses, in alignment; it blurs features, and reduces us all to a common type.[20]

While Madame Bovary's Yonville-l'Abbaye may have changed relatively little during the course of the nineteenth century, our five provincial centers entered a new era of science and technology. The church, the market, and the town hall still marked the *centre ville*, but more imposing than the pharmacy, lying a little on the outskirts of the oldest part of town, were the palatial buildings of the university and the edifices of the institutes of science. The aim, as Sabatier put it, was that "light should come not only from Paris but also from the provinces."[21] This study will aid in assessing whether that aim was effectively realized.

# 1

# FRENCH UNIVERSITY SCIENCE BEFORE THE FIRST WORLD WAR

In fall 1904, the American professor Barrett Wendell arrived at the Sorbonne as the first representative of Harvard University to lecture in French universities about America. He recalled later that after an initial cordial conversation with his French host at the Sorbonne he was asked to pass into the professor's study.

> This proved to be a snug library full of books and papers, and remarkable chiefly for the blackboard on which was sketched a somewhat complicated diagram, resembling the plans of Hell, Purgatory and Paradise to be found in most editions of the "Divine Comedy." Indeed this likeness was so marked that, unaware of what my friend's special branch of learning might be, I was disposed to take for granted that he was occupied with some minute study of Dante. In fact, it presently appeared, this impressive diagram had been ingeniously devised for my personal benefit. Rightly assuming that I could not find my way in France without a clear knowledge of where I belonged there, he had prepared it to illustrate a concise little discourse on the present structure and constitution of French universities.[1]

The most important laws establishing the university structure in modern France were those of May 10, 1806, and March 17, 1808. Faculties were suppressed and reestablished, governing councils received more or less autonomy, and curriculum revisions took place, but substantial elements of the system remained intact until 1968, when reforms were instituted following the student May Days.[2]

What changed most significantly in the course of these years was the definition of the aims and functions of the university in society. This trend became apparent in the 1870s, perhaps even a little earlier. It became clear that the university Faculties would become centers of scientific research, taking their place alongside traditional research institutions like the Museum of Natural History and the Collège de France, and superseding in importance local scientific societies and academies.[3]

The fastest growing area of university study was in the curriculum related to "technique" and engineering, which was previously a preserve of the grandes écoles. Theoretical and applied science became ever more closely related in the activities of university scientists toward the end of the nineteenth century, and many of them came to think that the success of their work depended on industrial, agricultural, and public support at the local level. Connections between science and technology were hardly new, nor was it unprecedented for the public to make practical demands on science. But, through the universities, a trend was beginning which led to modern scientific communities that were different from their predecessors.[4]

New means of transportation, as well as economic and administrative measures, brought Paris and province into greater interdependence in the twentieth century. But it must be remembered that the Parisian and provincial university system was fashioned from the debris of institutions that had thrived during the seventeenth and eighteenth centuries. The strengths and weaknesses of the modern system cannot be understood without some sense of the deep roots of French education and culture in older local institutions, some of them founded in the Middle Ages. Regional pride and local prejudices were stronger by the end of the nineteenth century than at the beginning, when most Frenchmen and Frenchwomen reveled in the glories of Revolution and Empire.

## STRUCTURE AND REFORMS IN THE EARLY THIRD REPUBLIC

There were twenty-two universities in France before the Revolution, more than in the nineteenth century until the establishment of Catholic universities in 1875.[5] Many, but not all, eighteenth-century universities had four Faculties. They had grown up separately and had very different histories. While Paris was the largest of the old universities, with approximately 6,000 students in the late 1780s, the eighteenth-century

Faculty of Medicine at Montpellier gave more certificates in medicine than did the Paris Faculty. Toulouse, a medieval university like Montpellier and Paris, was the largest university next to Paris, with an enrollment of almost 1,000 students. Other provincial universities were attended by several hundred or fewer students on the eve of the Revolution.[6]

After the universities were dissolved in 1793, for almost a decade advanced scientific and mathematical training took place in the professional schools that had predated the Revolution, such as the Ecole des Mines and the Ecole des Ponts et Chaussées, and in the new creations of the Convention, namely, the Ecole Polytechnique and the Ecole Normale Supérieure. These *grandes écoles parisiennes*, unlike the secondary-level *écoles centrales*, survived the Revolution.[7]

In 1802, Napoleon's new law for secondary education established a system of lycées, modeled on the Parisian "Prytanée français," formerly the Collège Louis le Grand, the only secondary school of the Old Regime to survive the Revolution.[8] Napoleon's choice of the Louis le Grand model reflected the value he placed on discipline and order as well as the influence of the scientist Antoine François de Fourcroy, who counseled him on the advantages of Jesuit education: reasoned exegesis of texts, learning by heart, translations, imitation and disputation, exemplary punishment, the absolute authority of the teacher, the detailed timetable for the pupils. This format was to dominate lycée education for the next hundred years.[9] Further, performance on the postlycée baccalaureate examination was to establish each year the small elite who might go on to advanced training in higher education or to the upper ranks of the civil service.

In 1808, France was divided into educational regions called "academies," similar only in name and locale to the provincial academies of belles lettres, arts, and sciences which were sprinkled throughout France in the seventeenth and eighteenth centuries. In principle, each academy included at least one lycée and one or more Faculties of Sciences, Letters, Medicine, Law, and Theology.[10] From 1808 to 1810, the government created fifteen Sciences Faculties, ten of them in the territory that remained French after Napoleon's defeat: Paris, Besançon, Caen, Dijon, Grenoble, Lyon, Metz, Montpellier, Strasbourg, and Toulouse. Three of these (Besançon, Lyon, and Metz) were temporarily suppressed in 1815.[11]

Each Sciences Faculty was required to maintain four professors, one

each for differential and integral calculus; rational mechanics and astronomy; physical sciences; and natural history. Candidates for these
positions had to be French citizens and at least thirty years old; the
Sciences Faculty submitted two names to the administrative council in
Paris, and the education minister made the final choice. Professors
wishing to transfer from one Sciences Faculty to another could do so
only with the approval of the Faculties, deans, and academy rectors
involved, and with permission from the Ministry of Public Instruction
in Paris. There was a fixed salary scale for Parisian Faculties and a
separate one, with lower salaries, for the provincial Faculties.

A March 17, 1808, decree required professors to administer and grade
examinations given in the spring and fall to students seeking the baccalaureate diploma. In 1808 the *licence* degree required only one step
beyond the baccalaureate: a short oral test in an area of specialization.
The doctorate required two written dissertations, one in the area of
specialization of the licence, and an oral defense of the dissertation.
The most important postbaccalaureate degree was the *agrégation* diploma, the result, like the baccalaureate, of high-ranking performance
on a competitive examination testing specialized and memorized
knowledge. The *agrégés* received the credential that made it possible to
teach in the lycées, and some went on to write doctoral dissertations,
usually while teaching in lycées.

Professors were to lecture three times a week for a total of four and a
half hours each week during the academic year, which began in late
November and ended in June. They often gave these lectures in the
evening in order to reach a larger audience, which included lycée students, Faculty students, and the general public. Many professors in the
Sciences Faculties also taught in the local academy's lycée and during
the period of the baccalaureate examinations, they were busy traveling
from town to town to preside over examinations.[12] By the 1840s a publishing industry had sprung up offering manuals to prepare students
cramming for the "bac," and courses offered at the lycée and the Faculty became less crucial to success. The art of preparing students for examinations was substituted for the art of teaching science.[13]

After the July Revolution of 1830, there was interest in restructuring
the university system, with Victor Cousin and François Guizot arguing
for the notion of instituting several large universities as regional counterweights to Paris. Guizot suggested Strasbourg (the only university
center other than Paris with five Faculties), Rennes, Toulouse, and

Montpellier.[14] But instead of concentrating the number of Faculties in a few cities, the French educational ministry under Louis-Phillipe increased them, adding, for example, three new Sciences Faculties.[15] Under the educational minister Hippolyte Fortoul (1851–1856), a system of fifteen academies or university centers was established in 1852, to which after considerable lobbying by city notables, Nancy was added in 1854.[16]

A turning point occurred in French scientific education in the 1860s which is rightly identified with the policies of the third educational minister of the Second Empire, Victor Duruy (1863–1869). In support of secular and scientific education, Duruy enthusiastically took up a scheme already under development by his predecessor, Gustave Rouland (1856–1863), and the chemist Jean-Baptiste Dumas. This plan came to be called "special secondary education." In 1866, Duruy founded a normal school at Cluny to train teachers for a "modern" education aimed at the agricultural, industrial, and commercial middle classes who wished to send their sons to lycées.[17] The Cluny curriculum included very little of the classics, rhetoric, and philosophy learned by most boys in the sciences track of traditional lycées.[18]

The Cluny school's director was Ferdinand Roux, who had been principal of a small college at Castres which independently developed nonclassical secondary education with the financial support of the municipality and the special interest of the Tarn parliamentary deputy, the Saint-Simonian banker Eugène Péreire.[19] The Cluny normal school graduates, preparing to be lycée teachers, competed in the regular agrégation examinations in the physical sciences, mathematics, the economic sciences, or modern languages. But, despite some apparent success, the Cluny school was closed in the educational reforms of 1891, and a modern track became part of the ordinary lycée baccalaureate preparation. Of twenty students in the Cluny school's last two scientific classes, four completed the agrégation in classical studies (for which the Cluny school had not prepared them), seven eventually obtained the state-conferred doctoral degree in the sciences, and one became a chief engineer in Ponts et Chaussées. Victor Grignard, who was later to win the Nobel Prize, was among the last group of students.[20]

Duruy also "invited" the Sciences Faculties to teach applied science. His predecessor had established a nonbaccalaureate, special degree in the applied sciences, but was ambivalent about local provincial professors' initiatives and enthusiasm for offering public courses in the mechanical arts, industrial and agricultural chemistry, and public hygiene. In 1855 the ministry reproached the Sciences Faculties at Grenoble, Nancy, and Lyon for offering unauthorized evening lectures in applied sciences and for wanting to charge fees to attending students. Rouland dismissed Nancy's dean of the Sciences Faculty in 1857 for designing basic courses in applied science.[21]

Perhaps the most significant change under Duruy's administration was the redefinition of the function of university professors to encompass a stronger emphasis on research. This change took place largely in response to concern with recent German advances in science and technology, as demonstrated at the International Exhibition at Paris in 1867. Germany's victory over the Austrians in 1866 also contributed to the perception that it was moving ahead of its European and British neighbors. But it should be noted that even before 1866–67, Duruy lobbied for educational reform, in keeping with demands appearing regularly in literary and scientific journals like the *Revue des Cours Scientifiques*.[22]

Like English educators, who were similarly concerned with evidence of German advances in science and technology,[23] Duruy sent envoys abroad to study rival systems of education. He arranged for the Cluny normal school students studying modern languages to spend their last year abroad under exchange arrangements with schools in England and Germany (an arrangement that collapsed when war broke out in 1870).[24] Young agrégés, including the philosophy student Emile Boutroux, were sent to Germany to attend university classes and to report on the content of German philosophical and scientific studies. Duruy also sponsored visits abroad by Faculty professors. The chemist Adolphe Wurtz, for example, spent a couple of months visiting Griefswald, Bonn, and Leipzig.

What the envoys found in Germany was a system of twenty-two universities, half of them in Prussia. There was a sharp distinction, which had existed since the early 1800s, between secondary and higher education. The *Abitur* degree was conferred after a nine-year education at a German classical gymnasium. Like the baccalaureate, this degree allowed a student to enroll at a university, take higher state examinations, or enter directly into the middle ranks of the civil service. Unlike their Faculty counterparts in France, German universities conferred two degrees, a doctoral degree calling for independent research and a thesis, and the *Habilitation*, a second dissertation permitting one to teach at a university as a Privatdozent.[25] The diplomas were conferred by individual universities, not by the state. In addition, the Technische Hochschulen in Germany were modern, university-level institutions that conferred doctorates after 1899.[26]

What especially drew plaudits from both English and French visitors was the emphasis in German universities on research, exemplified not only in the financing of research laboratories for university professors but also in the institution of seminars oriented toward research methods and topics.[27] German professors lectured on their current research as well as on well-established texts, and German scientists stressed practical laboratory work for students in the scientific curriculum.[28]

Reports sent to Duruy emphasized these innovations. They were similar in theme to the views of Matthew Arnold, who counseled English university reformers to learn from the Germans: "The French university has no liberty, and the English universities have no science; the German universities have both."[29] Wurtz argued the comparative advantages for German science of government subsidies, regional industrial support, and the building of fully equipped, modern laboratory facilities. He recommended greater autonomy for the French Faculties, echoing Arnold's theme of greater liberty, especially in the recruitment of professors, developing curriculum, and emphasis on research.[30] Boutroux admired the interdisciplinary seminars in German universities, and the stress on scholarship and research rather than on eloquent lecturing. In his report he expressed a preference for the German examination system, which graded students by an "absolute" set of standards for achievement rather than by the relative values of ranking in the French *concours*.[31]

With Duruy's encouragement and financial support, a stream of reports and monographs on the state of French science and higher learning inundated Emperor Louis-Napoleon and the public.[32] Duruy was unwilling, however, to challenge the traditional baccalaureate and agrégation system, which demanded full-time lecture and grading duties from university professors. Rather, he recommended the founding of a separate institution devoted primarily to scholarship and research, the Ecole Pratique des Hautes Etudes (EPHE). This "school," established in 1868, was in fact an *arrangement* for organizing scientific teaching and laboratories. Instruction was to occur in small conferences or seminars, modeled on the existing conferences for students at the Ecole Normale Supérieure and the German university seminars. Older laboratories and observatories were to be modernized and attached to the EPHE, and new ones were to be completed, including laboratories for Pasteur, Bernard, and Marcellin Berthelot, all of whom were vocal educational reformers.[33]

In terms of the structure of French higher education, the EPHE offered little real innovation, and it had immediate critics, who had hoped for more radical reform and especially for decentralization and increased provincial development. The professors in the EPHE were those already teaching at the Sorbonne, Museum of Natural History, Collège de France, and other institutions of higher learning. Essentially a Parisian institution, critics have suggested that it had no influence whatsoever on scientific research outside Paris.[34] However, by integrating into this research organization Faculty professors who were charged primarily with duties relating to the French teaching and examination system, the establishment of the Ecole Pratique des Hautes Etudes signaled a new orientation for all Faculty professors. It shifted the basis for prestige.

An indication of this shift is that demands on the quality of doctoral research stiffened considerably. For example, a jury which rejected a thesis by the abbot P. A. Issaly in 1886 said that many theses in the past had been simply works of erudition, demonstrating and presenting new points of view rather than establishing real discoveries. Issaly's work was good, said the jury; it would have earned him a doctorate in 1870, but it did not fulfill the requirements of the 1880s.[35]

The defeat of French forces by the Germans in 1870 quickened demands for educational reform in the 1870s and 1880s. To be sure, defeat was not blamed on educational mediocrity alone, but on military inferiority, economic backwardness, and even biological "decadence."[36] The old interest in general reform and in modern or scientific and engineering education understandably rekindled. An internal threat, as well as the exterior German menace, hastened university reform when the republicans gained a majority of seats in the national assembly in 1876. The previous year, the bill for "liberty of higher education" had become law, and five Catholic universities were established. Republicans now in power immediately enacted the greatest relative annual increase in the financial history of French education.[37] In 1877, 300 university scholarships (bourses d'études) were created to attract candidates for the licence certificates, 160 of them for the licence ès sciences.[38] In 1882, 200 additional scholarships were instituted for agrégation candidates, and these scholarships enabled many of the original licence scholarship holders to continue their studies. Closed courses (cours fermés), which could not be attended by the general public, were established for serious university students, and renewable one-year appointments for lecturers (maîtres de conférences) were instituted for agrégés who wished to remain in the university while doing doctoral research.[39]

At first the lecturers were young and their functions were modest, with the expectation that they would, in turn, become Faculty professors. As we shall see, however, the positions became long-term ones because of a lack of mobility in the university system—a major problem for French university science. Decrees of 1885 also established the new category of adjunct professors (professeurs adjoints) which was governed by the rule that a Faculty might request one adjoint for each three full professorships on its staff. A new policy in 1920 allowed the ratio of adjoint to full professors to be as high as 1:2.[40]

Republicans in the 1870s and 1880s were ambivalent about the best course to take for fundamental structural reform. On the one hand, they found in the German example of competing regional universities a model for creating provincial educational and scientific centers as rivals to Paris. It was said, for example, that a strength of German universities lay in their struggle against each other for the allegiance of scholars, scientists, and students. On the other hand, there was a long-standing identification by French republicans of provincial interests with monarchist, Catholic political parties and anti-Jacobin sentiments, with greatest strength in the rural countryside. As a consequence, republicans took a middle course in educational decentralization.[41]

The philologist and educational minister William Waddington drew up, in 1876, a project recommending the creation of seven universities.[42] Faculties in the provinces and in Paris were asked to make recommendations on the reorganization of higher education, and some, like those at Nancy, submitted detailed and carefully thought out proposals. Xavier Bach at Nancy's Sciences Faculty pressed the ministry, too, for radical modifications in the statutes governing the Parisian grandes écoles in order to encourage greater recruitment of good students to the provincial Faculties.[43] The issues were discussed regularly in a popular new periodical, the Revue Internationale de l'Enseignement, founded by a group of reformers including Ernest Renan, Ernest Lavisse, Paul Bert, Pasteur, and Berthelot.[44]

Under the guidance of educational minister Jules Ferry from 1879 to

1883, the Chamber of Deputies enacted laws making primary education free and compulsory for all children aged six through thirteen, establishing lycées for girls, and affirming the religious neutrality of public education. At the university level, two decrees in 1885, prepared by Louis Liard, gave each Faculty in the academies a civil personality, enabling it to receive gifts and subsidies independently from the state. Each academy's Faculties were organized into a hierarchical administrative structure with a general council at the top made up of representatives from all the Faculties, along with the rector. In addition, each Faculty was to have regular meetings of a general assembly of all Faculty staff members and of a Faculty council made up of all titled professors.[45]

Liard, director of higher education from 1884 to 1902, complained in a book published in 1890 that the existence of separate faculties had led to sterility and rigidity, and the creation of unified, separate universities would lead to rivalry of the provincial Faculties with each other and with Paris, reversing the traditional flow of the best talents and minds into the capital city. Provincial commerce and industry would flourish, he predicted, as competition between the new universities led to different regional specializations and vocational expertises.[46]

In early summer 1890, the Chamber of Deputies began debating Liard's project for establishing regional universities from the existing Faculties. What was envisioned was the kind of pruning that the Prussians had done in the early part of the century. After regional leaders collared their friends and representatives in Paris, however, the government renounced its plan of five or six universities. The chamber instead created, in 1896, seventeen "universities" corresponding to the sites of existing Faculty enclaves. Since government funding was now declining relative to the heady and generous period of the mid-1870s and the 1880s, the new universities found themselves increasingly dependent on local financial resources, although still ruled (or overruled) in disposition of their funds by the Paris ministry. The hierarchical structure of authority within the university system still ultimately flowed to Paris, when the Parisian ministry chose to exert vested authority.

The general thrust of reforms from 1875 to 1900 was to foster the stronger orientation toward research that was begun by Duruy in the 1860s. Before that time, the Faculty had as their primary responsibility the grading of baccalaureate examinations. In the early 1880s, they saw their primary pedagogical responsibility as "the preparation of students for the examinations of licences and agrégations." But by the late 1880s, the emphasis in Faculty reports and meetings was the triple function of research, teaching, and service to society. An example is Dean Benjamin Baillaud's demand at the Toulouse Sciences Faculty for a faculty which would do more than supply teachers for secondary education. He wanted a faculty which "by their research [would play a role] in the remarkable movement which . . . will restore to our country the rank in science which twenty years of neglect and feebleness have made it lose."[47] Reforms of the baccalaureate introduced junior Faculty members (1887) and then senior lycée professors (1902) onto the examination juries, relieving and distributing the burden of grading duties. Grading licence and agrégation exams was not onerous in the next few decades because of the small numbers of students involved.[48]

In keeping with the emphasis on research, the government educational ministry now emphasized original research, rather than longevity in lycée service or eloquence of speech, in appointing new Faculty professors. As I have noted, the doctoral thesis was to be an exemplar of original research rather than of erudition. It no longer sufficed for the level of teaching at the Faculty to merely surpass that of the requirements for the lycée baccalaureate, as Baillaud claimed was the case when he arrived at Toulouse in 1879.[49] New appointees were to be trained in the contemporary practice of their disciplines, spend time in the laboratory, and teach seriously. This was a policy of faculty appointment that had been practiced by the Prussian Kultusministerium in the period 1817–1840 under its minister, Karl Altenstein, and his advisor, Johannes Schulze, in building the German research universities.[50]

A consequence in France of the hiring of those who were research-oriented in the Faculties was new specialization within the curriculum for the licence and agrégation. The program for the licence was revamped so that, in 1896, the Sciences licence degree was awarded after the completion of three specialized certificates (certificats d'études supérieures) centered around a program in mathematics, the physical sciences, or the natural sciences.[51] There were recommendations that the licence should place a stronger emphasis on individual work and specialization, with the last semester of a three-year program of study requiring an individual project. This idea may have been inspired by the German seminar model. It was pushed by Sabatier at Toulouse, but made no headway with the ministry.[52]

The university system expanded dramatically in enrollments, teaching positions, and budgets. Between 1876 and 1877 the state's budget for higher education rose by nearly half, the biggest annual increase in the history of the university. From 1875 to 1900, the Parisian authorities spent 13 million francs on Sciences Faculty construction and renovation, over 9 million of which was directed to the Sorbonne, where 6 new laboratories were built, bringing the total to 14, with separate laboratory space for professors and doctoral candidates.[53] It has been estimated that after the 1885 law making it easier for Faculties to accept private gifts, more than 30 million francs were forthcoming from the local level between 1885 and 1900 for construction and renovation, or over 75 percent of the total sum of monies spent on construction in the provinces during the period. The Sciences Faculties of Lyon, Bordeaux, Nancy, and Lille alone received almost 20 million francs in local gifts and subsidies. But state support for higher education increased only 10 percent from 1887 to 1900, and matériel funds provided by the state to the Faculties for ordinary equipment expenses actually decreased by 30 percent, dropping another 6 percent during the period 1900–1910.[54]

There was an almost exponential increase in the number of staff members in the Sciences Faculties during the period 1850–1890. In 1900, the Sciences Faculties staff included approximately 200 teachers, a doubling in twenty-five years. Over 100 of these teachers were teaching at the rank of *maître de conférences* (lecturer), rather than in chaired professorships, and half of the staff positions were laboratory assistants and technical personnel.[55] Many of the new positions were funded through local resources and university investments rather than by the state. The increase in staffing corresponded to an even more dramatic rise in student enrollments, in large part the result of a greater demand for scientific teachers for secondary education. In addition the "PCN" program introduced in 1893 substantially increased science enrollments. The PCN was a new prerequisite for medical school, a one-year "certificate of physical, chemical, and natural sciences." Overall, during the period 1875 to 1891, the number of university students in France increased from 9,963 to 19,821, doubling again to 39,890 from 1891 to 1908.[56] In the sciences, the numbers of students increased from only about 300 in 1876 to 700 in 1881, 1,200 in 1886, 3,900 in 1901, and 6,100 in 1911.[57] (See table 1.)

## TABLE 1
### FACULTY DEGREES, ENROLLMENTS, AND FRENCH NATIONAL POPULATION, 1851–1961

(Degrees and Enrollments in Thousands, Population in Millions)
Degrees (in parentheses) and students in:

| Year | Law | Medicine | Pharmacy | Letters (& Theol.) | Sciences | All faculties | Total population | Students per 1000 pop. | Pop. aged 19–22 | Students per 1000 aged 19–22 |
|---|---|---|---|---|---|---|---|---|---|---|
| 1851 | (1.0) | (0.4) | (0.2) | (0.1) | (0.1) | | 35.8 | | | |
| 1856 | (0.8) | (0.4) | (0.2) | (0.1) | (0.1) | | 36.0 | | | |
| 1861 | (0.8) | (0.4) | (0.2) | (0.1) | (0.1) | | 37.4 | | | |
| 1866 | (1.1) | (0.5) | (0.3) | (0.1) | (0.1) | | 38.1 | | | |
| 1876 | (1.0) 5.2 | (0.6) 4.0 | (0.4) 1.4 | (0.1) 0.3 | (0.1) 0.3 | 11.2 | 36.9 | 0.3 | 2.4 | 4.7 |
| 1881 | (1.3) 5.2 | (0.7) 4.1 | (0.4) 1.1 | (0.2) 0.9 | (0.2) 0.7 | 12.0 | 37.7 | 0.3 | 2.6 | 4.6 |
| 1886 | (1.4) 5.7 | (0.5) 5.7 | (0.4) 1.6 | (0.3) 2.1 | (0.4) 1.2 | 16.3 | 38.2 | 0.4 | 2.7 | 6.0 |
| 1891 | (1.2) 7.7 | (0.6) 6.2 | (0.6) 2.5 | (0.3) 2.7 | (0.3) 1.6 | 20.7 | 38.3 | 0.5 | 2.6 | 8.0 |
| 1896 | (1.2) 8.8 | (1.1) 8.5 | (0.8) 3.1 | (0.4) 3.5 | (0.3) 3.1 | 26.9 | 38.5 | 0.7 | 2.7 | 10.0 |
| 1901 | (1.5) 10.2 | (1.2) 8.6 | (0.6) 3.3 | (0.5) 3.9 | (0.3) 3.9 | 29.9 | 39.0 | 0.8 | 2.5 | 12.0 |
| 1906 | (1.7) 14.3 | (1.1) 8.1 | (0.6) 2.7 | (0.5) 5.0 | (0.3) 5.6 | 35.7 | 39.3 | 0.9 | 2.5 | 14.3 |
| 1911 | (2.0) 17.3 | (1.0) 9.9 | (0.3) 1.6 | (0.5) 6.2 | (0.5) 6.1 | 41.2 | 39.6 | 1.0 | 2.5 | 16.5 |
| 1921 | (2.3) 17.4 | (1.4) 11.3 | (0.4) 2.2 | (0.8) 8.1 | (0.8) 10.9 | 49.9 | 39.2 | 1.3 | 2.5 | 20.0 |
| 1926 | (1.5) 17.4 | (1.5) 12.3 | (0.4) 3.7 | (0.6) 12.5 | (0.7) 12.6 | 58.5 | 40.7 | 1.4 | 2.7 | 21.7 |
| 1931 | (2.1) 20.7 | (1.1) 18.1 | (0.8) 5.5 | (1.2) 18.7 | (0.8) 15.5 | 78.7 | 41.8 | 1.9 | 2.7 | 29.1 |
| 1936 | (2.9) 21.6 | (1.4) 17.7 | (0.8) 5.7 | (1.6) 17.5 | (0.8) 11.3 | 73.8 | 41.9 | 1.8 | 2.0 | 36.9 |
| 1946 | (6.3) 42.3 | (1.5) 20.5 | (1.1) 8.5 | (2.0) 29.0 | (1.8) 22.9 | 123.3 | 40.5 | 3.0 | 2.5 | 49.3 |
| 1951 | (3.0) 38.7 | (2.3) 30.2 | (1.1) 7.1 | (1.8) 36.6 | (1.2) 27.0 | 139.6 | 42.0 | 3.3 | 2.6 | 53.7 |
| 1956 | (3.1) 37.0 | (2.2) 30.0 | (0.8) 7.9 | (2.4) 43.2 | (1.8) 39.3 | 157.5 | 43.6 | 3.6 | 2.4 | 65.6 |
| 1961 | (1.9) 34.3 | (2.3) 31.7 | (0.9) 9.2 | (4.0) 64.4 | (6.3) 70.2 | 210.9 | 45.9 | 4.6 | 2.2 | 95.9 |

SOURCE: Fritz Ringer, *Education and Society in Modern Europe* (Bloomington: Indiana University Press, 1979), Table XI, p. 335. (Reproduced with permission.)

## THE APPLIED SCIENCES AND SPECIALIZATION

The provincial Faculties of Sciences used their new legal status as civil bodies within a local university to accelerate prosperous development, building an infrastructure of specialized scientific institutes attached to the Faculties. The focus of most of these institutes was directed toward the applied sciences, for which there was increasing demand by students and generous support from local industrial and agricultural interests. Public and government support for these developments reached a high point around 1900.

It was said that France had relied for too long on the engineering schools founded in the eighteenth century. The state-run schools trained small elites who oversaw the country's mines, bridges and highways, military installations, and state monopolies.[58] In contrast, the Technische Hochschulen in Germany were an impressive example of the possibilities for a university-level education that included the applied sciences, while promoting basic science as well.

Another reform important to German research got under way following the International Electrical Congress in Paris in fall 1881. The Prussian minister of culture was concerned that the French exerted too much authority in establishing international electrical and metrological standards.[59] An appointed committee engaged in deliberations that led to the founding in 1887 of the Physikalisch-Technische Reichsanstalt (PTR) with the support of the industrialist Werner Siemens, whose son-in-law Hermann von Helmholtz became its first director. That Siemens's motivations were patriotic and chauvinistic is apparent in his Darwinian remarks:

> Recently England, France and America, those countries which are our most dangerous enemies in the struggle for survival, have recognized the great meaning of scientific superiority for the material interests and have zealously striven to improve natural scientific education through the improvement of their instruction and to create institutions that promote scientific progress. . . . In the present and vigorously conducted struggles of peoples, the country that opens new paths and newly creates or enlivens important branches of industry has a decisive superiority.[60]

As is shown by David Cahan in an analysis of the early history of the PTR, basic science, applied science, and industrial demands were very closely knit. On the one hand, the institution had ties with academic physics through the requirement that its director teach two to four hours each week at the University of Berlin. But, simultaneously, much of the work of the Sciences Section, as well as the Technical Section, was directed toward accurate analyses and determinations for German industry. A striking example of these interconnections is that work in radiation physics, one of the great triumphs of German physics in the late nineteenth century, was done in connection with photometric studies requested by the German Union of Gas and Water Specialists. Thus, we have Otto Lummer and Wilhelm Wien's statement in an 1895 paper on black-body radiation that their study was "as equally important for technology as for science."[61]

In 1895, the mathematician Felix Klein helped establish Walther Nernst's Institute for Physical Chemistry at Göttingen, again with monies from industrialists. A division for technical physics was established in 1897 at the Institute,[62] a facility cited by the French as yet another example of Germany's accelerating support for science and technology. This strengthened the French complaint that in physics, and particularly in electrical technology, there was a large gap separating France and Germany. The Germans were said, too, to have achieved superiority in industrial and agricultural chemistry. Ferdinand Lot, like others, concluded that these fields should be developed vigorously in reconstituted French provincial universities.[63] Lot's demand for this program in 1906 can accurately be seen, however, as a stamp of approval on a *fait accompli*.

For decades, the Ministry of Public Instruction had suppressed or expressed ambivalence about the development of applied science in the university Faculties. On the face of it, France appeared already to have one of the finest systems of scientific and technical education in the world, with the Ecole Polytechnique in a star role, turning out at this time some 200 graduates annually. But, as discussed by Shinn in his monographic study of the Ecole Polytechnique, *polytechniciens* did not begin to enter industry until around 1900, and they were found inadequately prepared since physical mechanics was rarely taught at the school and the teaching of electrical theory and its applications was neglected. The polytechniciens were trained primarily for leadership, not for professions.[64]

For the requirements of industry and industrial innovation, the French system included several systems of secondary-level technical schools. The most successful were the *écoles d'arts et métiers*, whose graduates had high success rates in competing for places in the higher-level technical schools, the best of which were the Ecole Centrale des Arts et Manufactures (founded in 1829), the Ecole de Physique et de Chimie Industrielle de Paris (founded in 1883), and the Ecole Supérieure d'Electricité (founded in 1893), all in Paris.[65]

But among regional industrialists, particularly in the northeast of France and in the Rhône Valley, there was strong interest in having local expertise available in scientific technology and scientific agriculture. Lycées with strong science programs thrived especially in northeastern towns with military traditions, like Metz and Nancy.[66] On the whole, it was not highly specialized knowledge or education that was demanded but a certain familiarity with scientific and technical terms and methods that might be useful. As in England, local interest in science was sometimes fashionable, as well as practical, fulfilling the desire of an entrepreneurial civic elite to identify with the rise of scientific culture. As David Cardwell has pointed out in his work on the organization of science and technology in England, industrial or practical demand for real specialists probably did not exist until the universities produced them.[67]

The movement toward university specialization in applied science began after 1870, when provincial Faculties in France faced less opposition from the ministry against their perennial attempts to teach applied science. Unable to persuade the ministry to favor recruitment of highly ranked students away from the Ecole Normale Supérieure and the Ecole Polytechnique, the provincial Faculties accelerated their interest in the applied sciences at times when it was allowed or encouraged by the state.[68] When the 1885 law allowed Faculties to accept private and municipal gifts, courses in applied science were turned into professorial chairs. At Nancy, industrial chemistry became a chaired curriculum in 1886, thanks to funds from the Henri Giffard legacy administered through the state.[69] At Lyon, Dean Raulin organized, in 1883, an Ecole de Chimie Industrielle staffed by Sciences Faculty members who taught general and experimental chemistry, chemistry applied to industry and agriculture, industrial physics, and applied mineralogy in the school. This arrangement had been impossible in earlier years because of the opposition of the ministry. As a result, private industrialists had independently set up their own Ecole Centrale Lyonnaise for chemistry in 1857.[70]

Courses and staff positions in applied science proliferated in the 1880s and 1890s, and by 1905 there existed in France some dozen chairs and lectureships in agricultural and industrial chemistry and in applied physics. These were associated principally with the provincial Faculties, and expenses were borne mainly by the municipalities and by university sources independent from the state, including donations and trusts. As local revenues directed to applied science improved provincial facilities and raised the general financial prosperity of provincial Faculties, Parisian scientists began to experience a sense of competition with their provincial colleagues.

Albin Haller, who left Nancy in 1899 for the chair of organic chemistry at Paris, told the Faculty council that he was surprised to find the situation at Paris the poorer of the two Faculties. Edmond Bouty, whose brother-in-law, Baillaud, was dean at Toulouse, told his Parisian colleagues that Toulouse professors were better off despite the fixed salary differential between Parisian and provincial Faculty members because Parisians had to deal with larger numbers of students.[71] Yet the Faculty in Paris would not make elaborate commitments to the applied sciences, which were bringing in large sums of money to the provinces. As George Weisz notes in his recent book on the French universities, around 1900 financial arrangements still favored Parisian Faculties over the provincial ones, and the Sorbonne Sciences Faculty did not experience so much pressure for increasing enrollments in order to enhance revenues.[72]

In 1899 the Paris Sciences Faculty council discussed transforming Charles Friedel's chair of organic chemistry into one of applied chemistry. The inorganic chemist Henri Moissan subsequently cautioned that "it is important not to allow the teaching of applied chemistry to go beyond its true path by giving it a character so extensive that we become a second Ecole Centrale for chemistry." Still, he favored the retention of a course (not a chair) in applied chemistry, and later persuaded the Faculty to offer a chemical engineering diploma on the grounds that it would make students' entry into industry easier.[73]

Tension over the issue of applied science surfaced again in a 1907 meeting of the Paris Faculty when Paul Janet, who had taught industrial chemistry and physics at Grenoble, proposed granting official recognition to a course of industrial design that students had already organized among themselves. This type of course had been proposed and organized elsewhere, for example, at the University of Toulouse in 1887–88.[74] However, at Paris, the mathematicians Gaston Darboux and Emile Picard expressed dismay with Janet's proposal, saying that the Faculty was in danger of becoming a prep school for industrial schools.

Haller sided with Darboux and Picard, a somewhat surprising move given his strong support for applied science at Nancy. But then he had not wanted a chair of applied chemistry for himself at the Sorbonne.[75] Once at Paris, Haller seemed to share the view that applied science should be taught primarily at institutions outside the Paris Sciences Faculty like the provincial Faculties and the Ecole de Physique et de Chimie, for which he was director.[76]

By the early 1900s, a clear consensus had emerged in the Ministry of Public Instruction and in Parisian academic circles that the provincial Faculties and the Parisian Faculty were to have different functions. There was an effort to encourage foreign students to study at provincial universities rather than in Paris. An 1895 decree, soon reversed, actually forbade foreign students to enroll at the Paris Faculty of Medicine.[77] One assumes that the ministry preferred to reserve Parisian places for the French elite. The Collège de France philologist Michel Bréal seems to express this view in an article published in the *Journal des Débats* and reprinted in an American student guide to French universities. He supported the founding in 1895 of the Franco-American Committee and the Alliance Française.

> But we counsel you in your own interest not to crowd yourselves into a city already overflowing, on to benches already overfilled . . . choose one of our provincial universities . . . Nowhere will you find scientific institutions that are larger or more convenient than those of Lyons, nowhere equipment more extensive than at Lille. . . . Would you revel in a rare climate, a rich and unforward nature? Go to Dijon, to Toulouse, to Bordeaux or to Montpellier. For my own part, were I to begin my life again, I would not be a student elsewhere than at Grenoble, within sight of the Alps, beside the swift waters of the Isère. . . . Keep Paris, if you like, for the end, when having completed your studies, you want to become acquainted with the furnace in which all these diversities are united and smelted down.[78]

The establishment of a new set of degrees, conferred by the individual universities rather than by the state, further encouraged studies in France by foreign students who wanted a diploma but had not earned the baccalaureate and did not intend to seek public employment in France. These were the *diplômes de l'Université*, which included a *doctorat de l'Université* that did not entitle its holder to the benefits or prestige of the *doctorat d'Etat*.

The university doctoral degree was established in 1898 on the initiative of the Paris Faculty of Sciences precisely to attract foreign students. The physicist Gabriel Lippmann suggested at a meeting of the Faculty of Sciences council in 1896 that France should have two kinds of doctorate, as was the case in German universities. Once the degree was instituted, there was an effort made to convince foreign university centers that the university degree differed from the state degree not in scientific caliber but only in not requiring the "encyclopedic knowledge" valued by French education.[79]

The university doctoral degrees transformed provincial universities, especially the Sciences Faculties. Finding their programs attractive to foreign students, especially students from Eastern Europe, provincial deans set about adapting facilities in local communities to a new clientele. Comparisons with German, English, and American universities resulted in a conviction that foreign students required dormitories, sports fields, and meeting rooms—a perception that focused attention on building a university with an identifiable physical presence. As Lot wrote in 1906, "Very often in France you do not even see the university. . . . You will find three or four old buildings dispersed in the far corners of the city."[80] Gradually, provincial universities came to have attractive new "campuses."

Whereas in 1894–95, Paris had 80 percent of those foreign students enrolled in the sciences (or Faculties of Sciences), by 1929 the Sciences Faculty in Paris taught 25 percent of science students in the universities, Toulouse 21 percent, and Grenoble 13 percent. At Toulouse, there had been only 47 foreigners in the entire university in 1895, and most of them studied law. In contrast, in 1909–10, there were 223 foreign students in Toulouse, with 174 of them at the Faculty of Sciences. By 1929 the proportion of foreign students enrolled in the Toulouse Faculty of Sciences was roughly 55 percent of those inscribed, that is, 952 of the 1,479 students enrolled in the Sciences Faculty; most of these students were in the technical institutes.[81] A substantial proportion of the foreign students were women. For example, over 20 percent of foreign university students in France were women on the eve of the First World War. In the Sciences Faculties, 25 percent of the students were foreign, and 2 percent of all science students were foreign women.[82] (See table 2.)

Catering to a clientele of foreign students was not the original intention of most provincial Sciences Faculties. Nor would the strong provincial focus on applied science and foreign students necessarily have occurred had the provincial universities enjoyed a mechanism for

## TABLE 2
### Numbers of Students Enrolled in Faculties in Six University Centers

| Year | City | Faculty of Law | | | | Faculty of Sciences | | | | Faculty of Letters | | | | Faculty of Medicine | | | |
|---|---|---|---|---|---|---|---|---|---|---|---|---|---|---|---|---|---|
| | | All Students | | | | All Students | | | | All Students | | | | All Students | | | |
| | | French Males | French Females | Foreign Males | Foreign Females | French Males | French Females | Foreign Males | Foreign Females | French Males | French Females | Foreign Males | Foreign Females | French Males | French Females | Foreign Males | Foreign Females |
| 1877–78 | Paris | 7,156 | | | | 259 | | | | 6,536 | | | | 6,216 | | | |
| | Bordeaux | 1,419 | | | | 89 | | | | 1,140 | | | | 859 | | | |
| | Grenoble | 423 | | | | 184 | | | | 357 | | | | 964 | | | |
| | Lyon | 560 | | | | 30 | | | | 503 | | | | 762 | | | |
| | Nancy | 562 | | | | 44 | | | | 489 | | | | 443 | | | |
| | Toulouse | 1,447 | | | | 28 | | | | 1,253 | | | | 2,728 | | | |
| Jan. 1894 | Paris | 2,663 | 1 | 228 | 3 | 409 | 21 | 35 | 9 | 1,153 | 180 | 94 | 25 | 3,417 | 22 | 627 | 139 |
| | Bordeaux | 635 | 0 | 4 | 0 | 133 | 1 | 0 | 0 | 195 | 15 | 0 | 0 | 809 | 7 | 21 | 0 |
| | Grenoble | 252 | 0 | 2 | 0 | 63 | 3 | 0 | 0 | 63 | 6 | 0 | 0 | – | – | – | – |
| | Lyon | 466 | 0 | 7 | 0 | 158 | 3 | 0 | 0 | 185 | 17 | 3 | 2 | 915 | 0 | 49 | 0 |
| | Nancy | 222 | 0 | 16 | 0 | 122 | 0 | 3 | 0 | 89 | 3 | 0 | 0 | 201 | 0 | 76 | 0 |
| | Toulouse | 732 | 0 | 14 | 0 | 81 | 0 | 5 | 0 | 179 | 6 | 0 | 0 | 346 | 1 | 7 | 2 |
| Jan. 1900 | Paris | 3,681 | 0 | 331 | 0 | 1,118 | 23 | 115 | 32 | 1,261 | 53 | 69 | 65 | 3,366 | 81 | 380 | 98 |
| | Bordeaux | 770 | 1 | 2 | 0 | 245 | 4 | 6 | 0 | 190 | 15 | 0 | 0 | 625 | 21 | 15 | 2 |
| | Grenoble | 236 | 0 | 16 | 0 | 86 | 1 | 0 | 0 | 67 | 18 | 23 | 5 | – | – | – | – |
| | Lyon | 436 | 0 | 2 | 0 | 409 | 2 | 15 | 8 | 208 | 30 | 2 | 1 | 1,018 | 2 | 49 | 19 |
| | Nancy | 345 | 0 | 9 | 0 | 270 | 1 | 21 | 6 | 92 | 1 | 2 | 1 | 206 | 0 | 40 | 10 |
| | Toulouse | 802 | 1 | 19 | 0 | 280 | 0 | 2 | 1 | 131 | 4 | 2 | 0 | 450 | 12 | 29 | 1 |
| Jan. 1910 | Paris | 6,751 | 37 | 813 | 87 | 1,285 | 99 | 331 | 130 | 1,354 | 553 | 461 | 747 | 3,124 | 191 | 448 | 317 |
| | Bordeaux | 972 | 1 | 1 | 0 | 260 | 10 | 11 | 0 | 194 | 77 | 2 | 0 | 842 | 23 | 14 | 10 |
| | Grenoble | 290 | 0 | 68 | 1 | 302 | 7 | 39 | 5 | 128 | 42 | 80 | 99 | – | – | – | – |
| | Lyon | 804 | 2 | 47 | 0 | 455 | 21 | 31 | 4 | 317 | 92 | 18 | 8 | 927 | 13 | 29 | 6 |
| | Nancy | 457 | 0 | 25 | 1 | 415 | 9 | 335 | 22 | 103 | 7 | 20 | 57 | 315 | 5 | 47 | 40 |
| | Toulouse | 1,308 | 1 | 13 | 3 | 464 | 14 | 155 | 8 | 235 | 70 | 4 | 4 | 444 | 1 | 9 | 8 |
| Jan. 1930 | Paris | 6,968 | 1,085 | 1,757 | 161 | 2,243 | 1,080 | 834 | 148 | 2,681 | 2,562 | 1,296 | 1,193 | 3,271 | 824 | 1,732 | 245 |
| | Bordeaux | 594 | 88 | 45 | 2 | 464 | 111 | 96 | 3 | 296 | 222 | 32 | 15 | 1,217 | 172 | 129 | 31 |
| | Grenoble | 210 | 42 | 285 | 59 | 458 | 58 | 460 | 19 | 155 | 259 | 316 | 505 | 40 | 18 | 8 | 3 |
| | Lyon | 711 | 66 | 178 | 9 | 801 | 145 | 129 | 14 | 373 | 290 | 74 | 26 | 1,154 | 152 | 138 | 8 |
| | Nancy | 405 | 61 | 249 | 51 | 536 | 74 | 492 | 52 | 214 | 222 | 31 | 23 | 504 | 230 | 375 | 183 |
| | Toulouse | 644 | 81 | 74 | 5 | 529 | 130 | 667 | 34 | 352 | 325 | 23 | 49 | 532 | 94 | 116 | 8 |

SOURCE: France: *Annuaire Statistique*, 3 (1880), 250; 15 (1892–1894), 266–267; 20 (1900), 88–89; 29 (1909), 56–57; and 46 (1930), 39.

recruiting first-rate French science students to their programs. Provincial Faculties had long demanded reforms in recruitment, which tracked the best students to the Ecole Polytechnique and the Ecole Normale Supérieure through the yearly concours and distribution of scholarships. In 1903, following long debates in private chambers and public journals, the Ecole Normale Supérieure became part of the University of Paris in what appeared at first to be a major reform in the structure of French higher education, a victory for those who had argued for decentralization, and an abrogation of the special status and privileges of the *normaliens*.

But this result was not achieved. At Toulouse, the academy rector Claude Perroud, in his annual report for 1903–04, remarked on

> the anxieties that the recent reform of the Ecole Normale have raised among provincial Universities . . . [there are] conditions which appear to menace the number and quality of scholarship students who will be distributed to us . . . and these students are like a leaven of vitality in the greater mass of students. It is important to us, then, that they should not be rare or of inferior quality.[83]

What became clear was that the reform had not transformed the elite normalien students into ordinary university scholarship students; instead, it had a completely different effect. The scholarship students at the University of Paris now became a privileged group of Faculty of Sciences students with new access to the teachers and facilities of the Ecole Normale Supérieure because of the legal affiliation of the two institutions. The school's library, from which students could check out books, and the conferences that prepared students for the agrégation were open to the scholarship students of the Sorbonne. This made it even more certain that high-ranking concours students would choose to study in the Paris Sciences Faculty rather than in the provinces since they received highly individualized preparation for the national agrégation examination, which took place in Paris.[84]

The normaliens had always had high rates of success in the agrégation, and they traditionally received the best positions in higher education. Of twenty-eight major chairs of physics in Paris around 1900, 52 percent were held by graduates of the Ecole Normale Supérieure and 31 percent by graduates of the Ecole Polytechnique. Students at the Paris Sciences Faculty welcomed the opportunity to study alongside the normaliens.[85]

When a provincial student did outperform the Parisian students, he was said to be at a disadvantage in job placement. The story was told of the provincial student who scored first in the agrégation above the Paris students, but who—unlike them—did not get placed. "Ah! si vous étiez normalien!" the minister said to him, according to a staunch critic of the system.[86] The sociologist Celestin Bouglé, who taught at Toulouse until 1910, summed up the provincials' attitude to the reform in saying that far from giving new blood to provincial Faculties, the reform would "aspirer vers le centre le peu qui en reste."[87]

Resolutions were passed at the Universities of Bordeaux, Lyon, Grenoble, and elsewhere demanding a reduction in the number of non-normalien scholarship holders at the University of Paris and favoring the posting of these students to the provinces. It was demanded that all scholarship students selected by concours be called normaliens.[88] Some provincials also argued in favor of a disciplinary specialization that the Ecole Normale continued to resist. "General culture" was defended by normaliens:

> general French culture, the taste for ideas which are clear, objective and coherent, the sense for what matters and what can be proved, [has] been maintained until this day in the University against the excesses of specialization, which so often have produced in Germany only a vain heap of vain science.[89]

The course of action open to provincial Sciences Faculties was clear. The reform of the Ecole Normale was at odds with real decentralization. Attempts to restructure access to the Ecole Polytechnique likewise were stymied by that school's refusal to abide by the regulations of the 1902 Educational Reform Act on the equivalence of several baccalaureate options.[90] As a consequence, acknowledging what now seemed inevitable, the Grenoble Sciences Faculty in 1907 requested that its students preparing for the agrégation in mathematics and physics be permitted to transfer with their scholarships intact to the Ecole Normale Supérieure before their final year. The Grenoble Faculty also passed a resolution in favor of concentrating the agrégation scholarships at Paris, while increasing the number of licence scholarships and first-year agrégation scholarships in the provinces. The Nancy Faculty voted to no longer accept as students agrégation candidates.[91] What this meant was that fewer students were encouraged to stay in the provinces to do preparation for the state agrégé degree, and consequently, these students were more likely to do research for the state doctoral degree in Paris than in the provinces.

In fact, these recommendations were in keeping with a trend already begun in physics, mathematics, and astronomy in the late nineteenth century. In every scientific field except chemistry, the percentage of French theses (state doctoral degrees) done in the provinces was lower in the period 1850–1890 than in 1810–1850. The percentages in the natural and veterinary sciences remained about the same.[92] Generalized, "polyvalent" education demanded for the elite remained the prerogative of Parisian institutions, while specialized and technical education was the preserve of the provinces. In most scientific fields, the French student who was oriented toward research and university teaching gravitated to Paris to complete his or her doctoral thesis.

There were a few exceptions to this rule (notably, Victor Grignard, whose career we shall study in detail). Although decentralization of elite French scientific education generally failed, and the outlines of that failure became evident by 1914, vigorous scientific activity was the rule not just in Paris, but in a handful of university cities and towns on the eve of the First World War. In particular, the "university" degree program provided a framework for high-quality education and research in basic and applied science. Some distinguished provincial scientists, motivated by loyalty to their regions and satisfaction with conditions for research, chose to remain in provincial cities. They contributed in important measure to the general achievements identified with French science, and they helped advance local industry, agriculture, and commerce.

We turn now in detail to the history and character of local scientific communities, analyzing individual scientific careers and research activities as well as the social, economic, and institutional settings in which these took place. We begin with Nancy, where in a 1934 celebration of the university's past, historian Charles Bruneau spoke of the transformation of the city by a new collaboration between science and industry:

> The old industries, based on ancestral routines, the commerce of yesterday, founded on the shopkeeper's nose for things, is making way for modern structures, established on scientific laws. Because of this, the role of the universities has expanded in a singular manner. . . . What would Guizot have thought, what would Victor Cousin have said to the proposition to add an Institute of Milk Science to the Institut de France![93]

# 2

# NANCY: THE GERMAN
# CONNECTION, SCIENTIFIC
# RIVALRY, AND N-RAYS

The city of Nancy was described at the height of the Belle Epoque as "a city, the refinement of which recalls, on a small scale, that of Athens."[1] In 1900, it was resplendent with new decorations and designs by the "Ecole de Nancy," which dominated the year's International Exhibition at Paris with glassware, ceramics, pottery, wood sculpture, cabinets, and furniture in the distinctive art nouveau style.[2] In Nancy, art nouveau flowers, plants, ferns, trees, and insects curved around balconies and windows of the newly built houses near the fashionable eighteenth-century Place Stanislas. Art nouveau shops adjoined the square of the Hôtel de Ville, the episcopal palace, municipal theater, and Grand Hotel. At the square's edges gleamed magnificent iron grillworks, embellished in gold, one of them leading to the public promenades of the Pépinière, with its English-style gardens and lawns. Just beyond the Pépinière, hidden behind the Renaissance palace of the dukes of Lorraine, lay the site of the new Chemical Institute for engineering science.

Nancy had not long been part of France; it was only in the eighteenth century, shortly after the Place Stanislas was built, that Lorraine lost its independence. The dukes of Lorraine were the hereditary rulers of the region of the Meuse and Moselle from the eleventh century until 1736, when François III married Marie-Thérèse of Austria and exchanged the duchies of Bar and Lorraine for the grand duchy of Tuscany. Lorraine was given to Stanislas Leczinski, the dethroned king of Poland and father-in-law of Louis XV. It was Stanislas who initiated an ambitious building program in the 1750s, founded the Société Royale des Sciences et Belles Lettres de Nancy, instituted chairs and prizes for this academy, and established a library.[3]

When Stanislas died in 1766, Lorraine became part of France. It is paradoxical, perhaps, that in another hundred years Lorraine was among the most patriotic and nationalistic of all the French administrative regions (départements). Nancy is separated from Strasbourg, about 150 kilometers to the east, by the soft blue lines of the Vosges mountains. Low hills cut off Lorraine from Belgium, Luxembourg, and Germany to the north; the Rhine River separates Strasbourg from Germany to Strasbourg's east. During the Franco-Prussian War, the Germans entered Nancy on August 12, 1870, and did not leave until May 1, 1873. They bombarded Strasbourg on August 24, 1870. Only then, according to Emil Alfred Weber, an Alsatian professor, did French cultural influence finally triumph in Alsace.[4]

Alsace had been under French rule longer than Lorraine, since 1681, but Protestantism and the Alsatian and German languages continued to be influential in French Alsace, especially among the urban bourgeoisie of Strasbourg.[5] The treaty of Frankfort gave three départements to Germany, including Alsace and that part of Lorraine corresponding to the current département of the Moselle, encompassing Metz but not Nancy. Only 18 kilometers from Nancy, the imperial eagle signaled an international frontier.[6] As Paul Appell remarked, the most pessimistic initially thought the treaty would last no more than three years.[7] But events were to take a different course, and a substantial exodus, from Strasbourg toward the west, soon took place.

In 1870, the population of Nancy was 50,000, but it more than doubled within a couple of decades, as many Alsatians fled their homeland, bringing with them their industrial, commercial, and cultural traditions. Each year at Mars-la-Tour, a few steps from the frontier, the combat of 1870 was commemorated by the citizens of Lorraine and refugees from Alsace. The mention of the fort of Metz, the blue line of the Vosges, or the cathedral of Strasbourg had an emotive function now almost forgotten.[8] The theme of Alsace-Lorraine is found in artistic work of the Ecole de Nancy as well as in literature and political debate. Emile Gallé depicted in porcelain the forget-me-not of Alsace withering, nailed to the gallows post of Germany beneath the gaze of an owl and the spires of Strasbourg. The thistle, emblem of the city of Nancy,

became a bristling symbol of hostility in wood marquetry or on glassware when coupled with the motto "qui s'y frotte s'y pique." A large
table, exhibited in 1889, bears the shield of Lorraine and, carved in full
relief across the base, thistles intertwine with the legend "Je Tiens au
Coeur de France, Plus me Poignent, plus j'y Tiens."[9]

This intense French patriotism and anti-German hostility fused with
traditional regional pride after 1870. The chauvinism took many forms,
among others, military, economic, and cultural; of particular interest is
a strong support for education in the late nineteenth century, especially
for scientific and technical education that would aid economic development and military security. In the aftermath of defeat at Sedan, the
university Faculties of Nancy became a primary vehicle for intellectual,
scientific, and technological competition with Germany.


## SCIENCE AND THE AFTERMATH OF THE GERMAN
## VICTORY AT SEDAN

A university had been founded in the sixteenth century at Pont-à-
Mousson, about 30 kilometers north of Nancy on the Moselle River. In
1768, Louis XV transferred the university to Nancy as a result of an edict
against the Jesuits in Lorraine. In 1788, buildings were completed for the
university on the rue Stanislas to the west of the Place Stanislas. Here
were housed Schools of Law and Medicine, while Arts and Theology,
along with a municipal college, remained near the porte Saint-Nicolas.[10]
Under Napoleonic educational reforms, in 1809, Nancy became seat of
an academy that included Lorraine, Meurthe, Meuse, and the Vosges.
However, Nancy was given only one Faculty, that of Letters, which the
city then lost in 1815. The nearest Sciences Faculty was at Metz and the
nearest Law Faculty at Strasbourg, which had five Faculties, one more
than Paris.

It was not until the 1850s that a Faculty was reestablished at Nancy.
Nancy's Ecole Libre de Médecine continued to exist from 1797 until
1872, when it became a Faculty.[11] The intervention of the Nancy
municipal council and Baron P. G. de Dumast prevented the omission
of Nancy from the list of academy seats drawn up in Fortoul's educational reform law of 1854. Dumast, a philologist who wrote songs and
poetry, was committed to a regionalist program against Parisian centralization.[12]

Sciences and Letters Faculties were created under the rectorship of the Parisian astronomer Hervé Faye, and a Law Faculty was added in 1864. The Faculties were housed in a new *palais*, completed in 1862, decorated with statues of Napoleon III and the dukes of Lorraine. As was true at other provincial universities, university Faculty professors were busier grading lycée students' baccalaureate exams than teaching higher-level courses to postbaccalaureate students. During the years 1854–1871, the Nancy Sciences Faculty awarded 66 licence degrees and the Letters Faculty, 104.[13]

It was the War of 1870 that decisively transformed the university Faculties and directed the attention of Nancéiens and French administrators to the quality and character of higher education in Lorraine. When the Germans took Alsace, the Reichstag debated the reorganization of what had been France's Académie de Strasbourg. The Wehrenpfennig bill, which passed on May 24, 1871, directed three lines of criticism at French higher education, as exemplified in conditions they found in Strasbourg: (1) the existence of separate Faculties in the city contradicted the principle of the unity of all knowledge; (2) Strasbourg's Faculties possessed neither academic freedom nor administrative autonomy under the centralized Napoleonic system; and (3) the academy's dominant concern was not the advancement of learning but "bread and butter" studies.[14] The historian Heinrich von Treitschke ridiculed French institutions for what he said was their utilitarian orientation, characterizing French Faculties as a "collection of trade schools" with buildings scattered all over French cities. The French did not take to such criticism, but neither did they ignore it: educational reforms of the late nineteenth century addressed exactly these points.[15]

On May 30, 1871, the French National Assembly resolved that Strasbourg's Faculties should be transferred to Nancy, making Nancy the only French academy other than Paris with four Faculties. "It will be . . . the living fount where the youth of Alsace and Lorraine—the youth of the cities of Metz, Strasbourg, Colmar, Mulhouse—would come to reinvigorate its love for the French *patrie*. There is a political interest in this whose importance it is unnecessary to emphasize."[16]

Among scientists, most of the Medical Faculty and some of the Sciences Faculty members moved to Nancy in 1872.[17] At this time the Sciences Faculty at Nancy included five staff members: D. A. Godron in natural history, Nicolas-Aimé Renard in pure and applied mathematics, Jules Chautard in physics, Camille Forthomme in chemistry, and Louis-

Nicolas Grandeau in chemistry and physiology applied to agriculture. Godron had been with the Nancy Faculty since its founding in 1854, when he and three other men were named: J. M. Seguin in physics, J. Nicklès in natural history, and Faye in astronomy and mathematics.[18] There was some turnover of faculty in the years 1854–1870, and, as noted earlier, there were only a few students—an average of a dozen per year. In addition, some 250–300 baccalaureate examinations were graded annually in the 1860s.[19]

The most famous pupils taught by the Faculty members were lycée students. Gallé, whom Godron taught at the Nancy lycée, immortalized Godron's *La flore française* and *La flore lorraine* in the style of art nouveau.[20] Another pupil whose fame far excelled that of his teachers was the lycée student Henri Poincaré, whose family lived midway between the Place Stanislas and the university palais. Poincaré and Appell took special mathematics classes together at Nancy while preparing for the entrance examinations of the Parisian grandes écoles. They wrote their Ecole Polytechnique examinations in Nancy at precisely the time the German troops left the city in August 1873. Appell later recalled that Poincaré

> made very nervous by emotion, had been particularly unsuccessful with his drawing, an exercise in which he hardly excelled anyway; he was in a hurry to join his family at the Hôtel de Ville, Place Stanislas, where they awaited the arrival of the French troops.[21]

The Nancy lycée was an important source of recruitment for Parisian Sciences Faculty students and for engineering students. Later named for Poincaré, the lycée was strongly oriented toward sciences and mathematics. In 1831, it started "industrial classes"; in 1836, it had a "modern" laboratory; and in 1838, an "electromagnetic apparatus." In 1860, 200 of the 261 boys in the top class were in sciences. Many, like Poincaré, went on to the Ecole Polytechnique; during the early years of the Third Republic, as many as fifteen were admitted in one year.[22]

Alsatian emigrés were integrated into the Nancy Faculty in the early 1870s. Xavier Bach of Strasbourg was named professor of pure mathematics in September 1871, as Renard's chair became one only of applied mathematics. Godron retired so that his chair might be doubled into zoology and botany, to which a new chair of mineralogy and geology was added. Alexis Millardet of Strasbourg was named chargé de cours in botany; Jules Baudelot of Strasbourg, professor of zoology and animal physiology; and Joseph Delbos of Mulhouse's Ecole Préparatoire, professor of geology and mineralogy.[23] The number of chairs now stood at nine, making Nancy's Sciences Faculty one of the largest of provincial Faculties. In the academic year 1879–80, the Faculty enrolled 44 students, 7 of whom held scholarships established in the 1877 national scholarship program. Of the 44 students, 19 were enrolled in mathematical sciences, 11 in physical sciences, 6 in natural sciences, and 8 in sciences applied to agriculture.[24]

The general programs inaugurated by the French education ministry in the 1870s and 1880s for scholarships, faculty positions, new budgetary funds, and research added strength to the growing vitality of the Nancy group. The Nancy Faculty heartily approved the request by scholarship students that they be allowed to organize periodic conferences at which students would give papers.[25] In response to a ministerial inquiry of 1882, the Sciences Faculty assembly approved the teaching of open lectures (cours libres) in the Faculties, observing that it was good to introduce into France this kind of teaching, which was analogous to the German system of Privatdozenten. The faculty showed themselves amenable to encouraging mobility among university students, recommending that students not be required to obtain the three certificates for the licence degree from the same Faculty. And when the Paris Sciences Faculty proposed the "university" doctoral degree in 1897, the Nancy Faculty quickly adopted the idea.[26] In general, the Nancy Sciences Faculty was open to change, especially one that emulated or was analogous to the German university system. It is interesting to note, too, that there was less dissension and argument in Nancy Faculty meetings than in those of some other French Faculties (notably, Toulouse and Bordeaux), a circumstance explained by present-day Nancy science professors as perhaps characteristic of a "northern or German mentality," in contrast to a different (southern) mentality at Toulouse.[27]

Like other provincial faculties, however, Nancy had a basic problem of recruitment. Bach, who served briefly as dean from 1871 until 1873, sought to address this dilemma by asking for radical modifications in the statutes of the grandes écoles.[28] Stymied in this effort, it was through the actions of two young scientists hired in 1876 and 1879 that the Faculty's fortunes changed to a new tack that influenced not only Nancy but all of French science. These new professors were Ernest Bichat, a physicist, and Albin Haller, a chemist. Both were native sons of Alsace-Lorraine and firm believers in the reciprocal relationship of science and industry. Just as practical aims and craftsmanship traditions nourished the art of the Ecole de Nancy, so industrial problems and engineering concerns provided new vigor to scientific development there. Nancy-based institutes of applied sciences became a significant factor in a renaissance in French science and engineering at the end of the century.

This orientation was not new; from the establishment of the Nancy Sciences Faculty, it had a strong commitment to applied and engineering sciences, in keeping with 1854 directives from the educational ministry encouraging a "new education" that would demonstrate applications of scientific theory. This education was for "young people who do not have the taste for the strong and complete studies of the Faculty, or who lack the secondary instruction required to study there."[29] According to Fortoul, it was for a "class" of society which needs an education different from the preparation for the "academic grades" ordinarily directed by the Faculties.[30]

Nancy's first dean, Godron, characterized applied science as a special and important orientation for regional Faculties:

> The Sciences Faculties today no longer have as their exclusive aim the development of theoretical knowledge, but also teaching with care the applications of this knowledge for diverse regional industries; they have as their mission, not only forming educated men, but further, giving to the country useful citizens.[31]

This was the view of the academy rector as well, who went so far as to recommend in 1855: "It is up to us to do nothing less than create at Nancy, with only the resources of our provisory organization, a third Faculty, a Faculty of Industry."[32]

The Sciences Faculty organized a two-year program in applied science, crowned by a state "certificate of capacity." The program included mechanics, architecture, physics, chemistry oriented toward printing and dyeing, natural history, industrial and commercial geography, and hygiene. Some of these courses were taught by men who were not regular Faculty members, and Faculty professors themselves taught both pure and applied courses.[33] The applied science courses were open to the public in the evenings, and the lectures moved toward elementary popularization. When the Ministry of Public Instruction sent out "invitations" to regional Faculties in 1864 to add applied science to their formal teaching duties, the Nancy Sciences Faculty seemed a little miffed. "It would be rather difficult for the professors of our Faculty to add anything else to the teaching they already are doing."[34] But the number of inscriptions dropped, and the program was discontinued in 1871.[35] The only applied science course that survived past 1870 was a course established in 1868 which was directed toward agriculture rather than industry. In 1872, this curriculum became, for Grandeau, a chair of chemistry and physiology applied to agriculture.[36]

Within the next decade a decisive turn took place at the Sciences Faculty, largely through the efforts of Bichat and Haller. Bichat arrived in Nancy in October 1876 as chargé de cours in physics, replacing Chautard. He was named professor in 1877 and became dean in 1888. Born at Lunéville, near Nancy, Bichat was a normalien who passed the agrégation and completed his doctoral thesis at Paris in electricity. Civic minded, Bichat became a member of the Nancy municipal council and of the départemental general council.[37] While defending the cultivation of science "for its own sake, without being preoccupied too much with its applications," he emphasized the difficulties of drawing clear distinctions between pure science and applied science. A monument erected in Bichat's honor shows figures symbolizing "the union of science and industry."[38]

Haller shared Bichat's conception of the relationship between science and engineering, but from a rather different educational experience. Haller was an Alsatian who served as apprentice to a carpenter and then to a pharmacist in French Munster. The pharmacist, Achille Gault, saw to it that Haller had lessons in French and Latin, and the young apprentice passed the baccalaureate in 1870 when he was twenty-one years old. After fighting against the Germans at Belfort, Haller was demobilized and joined Gault and the staff of the previous Ecole de Pharmacie de Strasbourg at Nancy. In the next years, Haller prepared a licence and a doctorate, as well as the agrégation in pharmacy, at the Nancy Sciences Faculty. In 1879, he became lecturer in chemistry at the Faculty and then, in 1881, succeeded Forthomme in the chemistry chair.

Haller was appointed to the University of Paris in 1899 as Friedel's successor in organic chemistry. So respected was Haller internationally that he was nominated later for the Nobel Prize in chemistry, by Wilhelm Ostwald and William Ramsay, among others. In 1905, he became director of the Parisian Ecole Municipale de Physique et de Chimie Industrielle. Haller was the third Alsatian who headed the school; the others were Charles Schutzenberger and Charles Lauth.[39]

From his appointment at Nancy in 1879, Haller began pushing for a program at Nancy to train chemical engineers in competition with Germany. His doctoral thesis on camphor dealt with a field in which the Germans seemed to be taking a clear lead in both scientific research and industrial production.[40] Indeed, from 1867 to 1875, 90 percent of the papers on aromatic compounds published in the *Bulletin de la Société Chimique de Paris* were written by Germans. Among French scientists, it was primarily Alsatians who took up William Perkins's synthesis of the aniline dye "mauve," for example, those associated with the Société Industrielle de Mulhouse. French production of aniline dyes (after 1856) and alizarin dyes (after 1869) languished, partly because of unfavorable French patent laws, and partly because of a more rigid separation in France than in Germany of academic science and commerce. After 1870, many Alsatian dye manufacturers emigrated to Switzerland, around Basle, or to Lorraine. As an Alsatian pharmacist, Haller had a background that predisposed him toward developing practical organic chemistry at Nancy and forging links with manufacturers, on the German and Alsatian models.[41]

When Albert Dumont, director of French higher education, passed through Nancy after a visit to German universities in 1883, he talked with the Sciences Faculty about founding specialized institutes. Permission was forthcoming from Liard, minister of public instruction, on the condition that the city and the département would raise substantial funds for the project. The ministry eventually contributed 500,000 francs for the construction of an Institut Chimique and an Institut Anatomique.[42]

With the law of 1885 giving Faculties legal civil "personality," the task of raising and holding funds was made easier for the Sciences Faculty. Haller, with the assistance of Bichat, intitially raised 300,000 francs for the construction of the Chemical Institute. By 1890, the General Council of Meurthe-et-Moselle contributed 100,000 francs, the Vosges, 10,000 francs, and the city of Nancy, 390,000 francs.[43] In addition, the Société des Amis de l'Université de Nancy, a new fund-raising organization comparable to other such societies springing up at French university centers, was established for the university in 1891. It was presided over by a local industrialist.[44] Bichat told the municipal council that scientific laboratories and researchers were essential to the well-being of labor and industry:

The industry of stearic candles, the soda industry, iron, steel, aluminum, illuminating gas and all oil products, the astonishing applications of electricity, and so many other industries which require the
manual labor of millions of workers: all this has resulted from the
laboratory of our workers who work with their minds.[45]

## TOWN AND GOWN: THE RECIPROCAL RELATIONSHIP

The moment was propitious for the establishment of programs of applied science at Nancy because the Ministry of Public Instruction had
become convinced of the need for technical education, for increased
funding of scientific education, and for some measure of decentralization, in emulation of German successes. In addition, regional industries
in Lorraine were undergoing significant new developments in the last
quarter of the nineteenth century.

Until 1870, the Lorraine economy was principally agricultural, with
some industry in the Moselle valley, especially metallurgy in the area
between Metz and Thionville, where waterfalls could be used for
sources of power. The oldest industry was glass-making, which had
originated in a fifteenth-century decree of René d'Anjou. In the sixteenth century virtually every village in Lorraine boasted its own glass-
works. It was this tradition that had given rise most recently to the
designs and innovations of the Ecole de Nancy.[46] Textiles had been
manufactured at Sedan since 1646 and became increasingly important
south of Nancy in the nineteenth century. Other traditional and expanding industries included papermaking, tanneries, printing, and
breweries.[47]

Lorraine is a region rich in deposits of iron, salt, soda, and coal. But
the iron ore in the basins of Longwy, Briey, and Nancy had not been
useful because of their high phosphorus content, which makes steel
brittle. This changed with the development by Sidney Thomas and his
cousin Percy Gilchrist of a process for the removal of phosphorus in the
Bessemer converter, a principle extended in 1884 to the open hearth
process.[48] By the 1930s, Lorraine was producing 97 percent of French
steel and 90 percent of exploited iron; the importance of its coal deposits was second only to the Département du Nord.[49]

Another major new development in the Lorraine economy occurred
in 1872, when Ernest Solvay built a factory at Varangéville-Dombasle
for the manufacture of soda.[50] His factory near Nancy was Solvay's first

in France. By 1909, it produced 200,000 tons of soda and employed approximately 2,500 workers, for whom Solvay took pride in providing medical and pharmaceutical benefits, housing, exercise facilities, a library, and adult education courses.[51] He took a strong personal interest in the Sciences Faculty at Nancy, hoping to draw on its scientific expertise for production and management skills in his enterprises. Disinclined to make gifts and subsidies to his hometown university, the University of Brussels, the Belgian industrialist became instead a principal benefactor of Nancy.[52]

Permission for establishing a chemical institute at Nancy came from the ministry in fall 1885.[53] The municipal council of Nancy, in addition to contributing building funds, underwrote the costs of a new chair in industrial chemistry (1890) and a lectureship in the chemistry of dyes (1899). The first position went to Georges Arth, who was director of the Chemical Institute from 1899 to 1909, when he was succeeded by Antoine Guntz (1909–1929), who was succeeded by Alexandre Travers (1929–1940). Alfred Guyot, who was appointed to the chemistry lectureship, was one of many Nancy Faculty members who later left the university for industry. Guyot became scientific director of the firm Alais, Froges and Camargues after the First World War.[54] Guntz was an Alsatian who settled with his family in Nancy after 1870. Having completed a thesis with Berthelot in 1884, he returned to Nancy and was named professor of mineral chemistry in 1884, when Forthomme's chemistry chair was doubled into organic chemistry (Haller) and mineral chemistry (Guntz).[55] There were now four chairs of chemistry and an additional program in the chemistry of dyes at Nancy.

First opened in 1890, the Chemical Institute stood to the northwest of the Pépinière, not far from the old ducal palace. Its formal inauguration in 1892 brought the president of the Republic, Sadi Carnot, to the city. From 1892 to 1933, 14 state doctoral theses and 95 university doctoral theses were completed in the Institute's laboratories. In 1900, it enrolled an average of 100 students each year, and 145 in the 1930s. From 1901 to 1932, 806 chemical engineers received engineering certificates, many with scholarship aids not only from the government but also from the Société Michelin, the Société Rhône-Poulenc, the Comité de l'Union des Industries Chimiques, and other organizations.[56]

Beginning in 1890, three lectures a week at the Chemical Institute were devoted to physical chemistry, taught by Paul-Thiébaud Müller, another refugee from Alsace who completed a Paris doctoral degree in

1893.[57] At the time, Nancy's was the only systematic program in physical chemistry in France. In 1896, Haller proposed that the program in physical chemistry suffice as the certificate of physical studies required for the licence degree. When he left Nancy for the Sorbonne in 1899, his chair of organic chemistry was turned into a chair of physical chemistry, the first in France.[58] The claim later was made that it was at Nancy that the ionization theory was first taught in the normal chemistry curriculum.[59]

From 1899, a subsidy of 6,500 francs from the Nancy municipal council was used for teaching electrochemistry at the Faculty.[60] Most students in the Chemical Institute took formal physics courses, as well as physical chemistry, in addition to traditional general, organic, and mineral chemistry courses. The emphasis on physics and physical chemistry has been viewed as a strength of the Nancy chemical program, in comparison with most other chemical schools in France, including Lyon and Paris.[61]

After the founding of the Chemical Institute, Bichat, a physicist and the Faculty's dean, gave high priority to establishing an Ecole de Brasserie et de Malterie, "for reasons of a patriotic sort."[62] A laboratory of brewing opened in 1893 and the school in 1896, at a time when the French still routinely went to Germany to study brewing. Paul Petit, who succeeded Grandeau in the chair now called agricultural chemistry, organized the new school. Over the next decades, from 1893 to 1934, the school trained some 650 students, contributing to a tripling of French exports of beer and a precipitous decline in the importation of German beer, which had stood at about 100,000 hectoliters in 1893.[63]

Simultaneous with the establishing of the Ecole de Brasserie, Bichat began laying plans for the Institute of Electrotechnology, a project in which Solvay was interested and to which the Solvay Company initially contributed 100,000 francs. It later contributed an additional 400,000 francs.[64] The new Institute opened in 1900 in a building facing the Chemical Institute. Its curriculum was modeled on the Ecole Supérieure d'Electricité in Paris and took three years. After 1905, entering students were required to have completed the mathematics baccalaureate; in 1920, an entrance examination required special work in mathematics beyond the baccalaureate; and in 1927, a concours was instituted at an even higher level.[65]

In 1900, several staff members were carrying out fundamental research in electricity and radiations. Blondlot, who succeeded Bichat in 1906, had been teaching in a physics lectureship since 1882 and was an expert on electromagnetic radiations. Camille Gutton, who succeeded Blondlot as lecturer in physics, did comparative studies of the velocities of radio and light waves and became director of the National Laboratory of Radio-Electricity in 1934.[66] Alexandre Mauduit, a graduate of the Ecole Supérieure d'Electricité, was appointed lecturer in electrotechnology in 1901, a position that became a chair in 1913, a year after he completed a doctoral degree at Nancy.[67]

With the founding in 1905 of the Institut de Mathématiques et de Physique, all the physical scientists and mathematicians had a base for laboratory research and special programs in "institutes." Other fields were not neglected: in 1903, Edmond Gain established the Institut Agricole et Colonial; in 1906, Pol Bouin founded the Ecole de Laiterie and the Institut de Zoologie Appliquée; during the period 1908–1911, Nicklès established the Institut de Géologie; and, in 1919, another engineering institute, the Ecole Supérieure de la Métallurgie et de l'Industrie des Mines (ESMIM), was founded by Petit.[68]

The 1903 student handbook lists a staff at the Nancy Sciences Faculty of 21 men who were professors, adjoint professors, chargés de cours, or lecturers, as well as 6 laboratory supervisors (*chefs des travaux*) and 3 laboratory assistants (*chargés des travaux*).[69] In the 1904–05 academic year, 598 students were inscribed in the Sciences Faculty, compared with 69 in 1889–90 and 44 in 1879–80. Of these, 128 were licence candidates, 74 PCN, 112 Institut Chimique, 206 Institut Electrotechnique, 32 Ecole de Brasserie, 11 Institut Agricole, 9 Institut Colonial, and 11 were not following programs. Three students were candidates for the state doctorate and 12 for the university doctorate.[70]

In the early 1900s, as happened at other provincial university centers after the granting of university status and the founding of institutes, foreign students began to make a conspicuous appearance. They were welcomed at Nancy, with the hope that foreign students exposed to French culture and ideas would help spread sympathy and influence for France and for the "European spirit."[71] A 1903 Nancy student handbook proudly listed the amenities of the Nancy student association, housed in a building on the rue de la Pépinière: a large basement for fencing and athletics, a concert hall, a billiard room, a library and study hall.[72]

Of 769 students enrolled in the Sciences Faculty in 1910–11, 336 were foreign. Of these, 257 were Russian, 20 Bulgarian, and 10 Rumanian. The number of foreign sciences students rose to 624 in 1913–14 on the eve of the war. Their nationalities differed dramatically from foreign students in the Letters Faculty, where 157 students were German (and 88 Russian). Not surprisingly, in 1920–21, in the wake of the war and the Bolshevik Revolution, only 187 of 979 Nancy sciences students were foreign, and the entire university enrolled only 308 foreign students. But by 1929–30, the sciences foreign student enrollment was 544 of 1,154 enrolled. The distribution now included 122 Poles, 60 Russians, 57 Rumanians, and 29 Bulgarians.[73]

As elsewhere, Nancy Sciences Faculty enrollments became dominated by Institute enrollments. This does not mean that pure research ceased among Faculty members, most of whom taught both in the Faculty (state) and the Institute (university) programs. Most notable at Nancy was the chemical work of Victor Grignard, who taught at Nancy in the chair of organic chemistry from 1909 until 1919, and was awarded the Nobel Prize jointly with Paul Sabatier of Toulouse in 1912. Details of Grignard's work and his experiences at Nancy are discussed later.

The Nancy Sciences Faculty was home for more than two decades to Jules Molk, a Strasbourg native who had studied in the mobile German manner at Zurich Polytechnik, the Sorbonne, and Berlin, doing work with Weierstrass and Helmholtz, among others. Molk came to Nancy in 1890 in a chair of applied mathematics, renamed rational mechanics in 1898. He published, with Jules Tannery, the four-volume *Traité des fonctions elliptiques* and then undertook in 1902 the direction of what turned out to be a forty-volume *Encyclopédie des sciences mathématiques pures et appliquées.*[74]

Elie Cartan taught in a chair of differential and integral calculus at Nancy from 1904 until 1909 when he was named lecturer at the Sorbonne, a not uncommon example of a provincial scientist taking a step down in formal rank when moving to Paris.[75] Georges Darmois, who taught in one of the mathematics chairs from 1919 to 1936, had a strong reputation in the field of general relativity; he left for the Sorbonne in 1936. Jean Delsarte, the mathematician, played a role in the foundation of the "Bourbaki" group. Among physicists, Eugene Darmois, who taught at Nancy from 1919 to 1926, did work in optical activity and electrolysis, and also finished his career at the Sorbonne, as did F. Croze, who did physics research in spectroscopy and relativity theory. Eugene Darmois's pupil, René de Mallemann, who remained at Nancy, became a nonresident member of the Academy.[76]

Among those in the natural sciences, perhaps the most distinguished was Lucien Cuénot, who became a member of the Academy of Sciences after it established a class of nonresident members in 1913.[77] Cuénot worked on the application of Mendel's laws to animals. Under his direction the Faculty offered in 1920–21 the prelicence certificate "evolution of organized beings," until then offered only at Paris and Strasbourg. From 1896 to 1934, fourteen state doctoral theses were completed in zoology at Nancy, as well as some university theses on topics in applied zoology such as agricultural parasitology.[78]

In 1912, the year that Grignard was awarded the Nobel Prize in chemistry, Cuénot received the Prix Cuvier for his work on the origin of species, André Wahl (later named to the Conservatoire des Arts et Métiers) won the Prix Berthelot for his work in chemical synthesis, and Nicklès won the Prix Joseph Labbé for his work on the geology of Lorraine. It was largely because of laboratory facilities and funding inspired by utilitarian motives that much of this work was possible. When the unified University of Nancy was established in 1896, the city of Nancy had already spent more than 2.3 million francs during the period 1870–1890 to install and develop its four Faculties and new institutes.[79]

Research facilities at Nancy compared favorably with those anywhere in France in 1900. As we have noted earlier, when Haller arrived at the Sorbonne to teach organic chemistry, he told the Paris Sciences Faculty council that he was surprised to find the laboratories there the poorer of the two. When Henri-Gustave Vogt arrived at Nancy from Zurich Polytechnik, he said that he had seen no teaching laboratories comparable to Nancy's anywhere in France.[80] At the Universal Exposition of 1900 in Paris, the University of Nancy as well as art nouveau carried away several Grand Prix, and Louis Liard, the director of higher education, delighted Nancéiens and piqued Parisians when he said, "Allez à Nancy, voir ce qui c'y passe, et faites-en-autant."[81]

By the end of the decade, however, Nancy suffered a major scandal because of spurious researches carried out in the Sciences and Medical Faculties. We will examine this work, by Blondlot and his colleagues, in detail. But before turning to that episode and its implications, we will look ahead briefly to the years immediately afterward. It is difficult to estimate the emotional impact of the scandal on morale at the university. It is tempting to link to it a slight later decline in enrollment in the Sciences Faculty, but there are good reasons to think there is no link whatsoever.

As is demonstrated in table 3, the number of students enrolled in the Sciences Faculty rose exponentially during the period 1902–1909 (from 353 to 850), fell slightly from 1909 to 1911 (to 782 in 1910–11), and then rose again. Enrollments in many Faculties declined briefly after 1906, when exemptions to three-year military service were ended for students, but enrollments continued to increase at Nancy (precisely after the Blondlot exposé). However, it is not surprising that the Blondlot scandal failed to affect enrollments among licence and engineering students, and there were never many agrégation and doctoral students in pure physics. Grignard's arrival in 1909 was a boost, too, to the Chemical Institute.[82]

The end of the decade 1900–1910 was a period of economic boom and mild inflation, a period in which the central government and industrialists, to a large extent, decreased funding to provincial universities, on the grounds that new facilities had been built and there were more graduates than jobs. This was a period, too, when Charles Maurras argued that decentralization was becoming a danger to the French government.[83]

The building program at Nancy was not over, however, and the Solvay Company continued to fund additional construction at the Electrotechnical Institute. Still bent on expanding the engineering program, the Sciences Faculty assembly voted a double transformation in 1913: the chair of mineralogy and geology was to become one of electrotechnology (for Mauduit), and a lectureship in electrotechnology would become one in crystallography and mineralogy (for the successor to Julien Thoulet). In a letter, Appell asked the Faculty to reconsider. A decision from the ministry maintained the chair of mineralogy. The Faculty got its way, however, when the Solvay Company underwrote the costs of the electrotechnology chair as part of a 500,000 franc donation to the university.[84]

Some Nancy Sciences Faculty members were ready to give up entirely doctoral-level programs in research and the preparation of students for the state agrégation. To a large degree, this attitude was the result of the 1906 centralization of the agrégation examination at Paris, followed by the ministry's encouragement of provincial faculties to specialize in only a few fields for postlicence study. Guntz, a professor in mineral chemistry, proposed at a Sciences council meeting in spring 1909 that Nancy no longer enroll candidates for the agrégation since

## TABLE 3
### Number of Students Inscribed at Nancy Faculties, 1901–1912

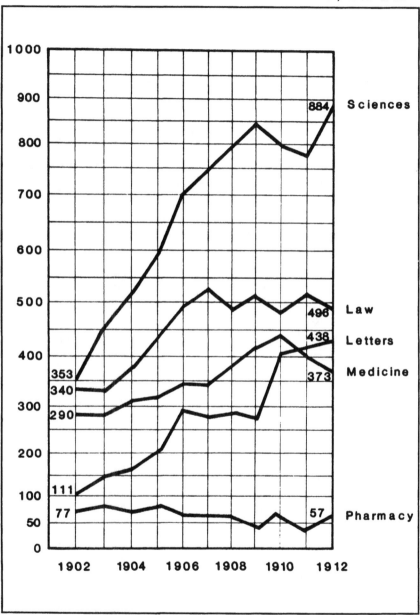

University of Nancy. Number of Students by Faculty during the Period 1901–1902 to 1911–1912.

their small number was so disproportionate to the special preparation required by professors working with them. The physicist Edmond Rothé opposed Guntz in this matter, arguing that it was absolutely necessary for the good of the Physics Institute that a small nucleus of students and aides (préparateurs) continue to do research and work on doctoral theses.[85]

The war years were hard on the city and the university. Faculty members kept teaching programs going until close to the end of the war, often on their own initiative. Vogt met in Paris with students of the Institut Electrotechnique. Grignard and other staff members carried out war work in chemical and physics laboratories in Paris. The city was often bombarded, not only by ordinary cannon but also by airplanes and by long-distance cannon ("Big Bertha") installed near Château-Salins. The Nancy University library was bombed shortly before the Armistice, and two-thirds of the books were lost, to be replaced at the war's end in part by donations from the Poincaré and Boutroux families.[86]

The war over, the university lost some Alsatian Faculty members who chose to return to Strasbourg, Mulhouse, and Colmar. Müller, Hackspill, and Rothé left almost immediately for Strasbourg, Wahl moved to Paris, and Grignard returned to Lyon. Because of the city's losses during the war, Nancy received the special attention of the Ministry of Public Instruction, and plans were made for a new natural sciences institute, an aerodynamic institute, and additions at the Institut Electrotechnique. Lucien Cuénot, recently returned from the United States, reported the magnificence of American laboratories: American professors and laboratory directors were blessed with aides who did all the busy work so that the scientists could concentrate on innovative thinking; budgets were enormous; and the researcher need not even take his own notes; he could dictate into a phonograph and find them recopied and classified in a file when he needed them. In France, alas, the reality was far different—at the Institut Chimique, for example, in 1921 only 3 aides were available to assist 7 professors and 177 students.[87]

Conditions were not to improve a great deal in the next few years, for Nancy, like other European universities, suffered an erosion of real funding in a period of rapid inflation. The Sciences Faculty began to use its reserve funds in 1925. And, like other Sciences Faculties, Nancy found it difficult to pay for subscriptions to German periodicals and had the additional problem of replacing periodicals and books destroyed in the war. In June 1925, the Faculty council discussed a debt of

56,000 francs to German bookstores. They appointed a committee to decide which periodicals would be canceled, hoping that laboratory fees would help pay the current periodical costs.[88]

If the Faculty did not receive increased funding from the state, it would have to raise fees,[89] partly to pay young scientists and especially engineers as aides, when they might receive higher salaries in industry.[90] But while more students, including significant numbers of women (about 12 percent of the Sciences Faculty enrollment in 1930), might be enrolled to increase receipts from fees, the additional students could not be well taught, nor could their teachers be active researchers when the student/teacher ratio was so high. Since the state was cutting back on the higher education budget in general (nationwide, firing 56 university Faculty members in 1922), the situation was bleak indeed.[91]

In addition, Faculty members became very sensitive to the charge that the Nancy Sciences Faculty and institutes were principally an "école des arts et métiers," which the rector, Charles Adam, explicitly denied in 1934.[92] Only from the 1920s is there a record of disagreement among Faculty members about the balance between fundamental science and engineering science, or between academic experience and practical experience among new appointees. One of the few occasions of clear division among Faculty council members during the entire period 1854–1930 was a vote in 1924 for the Faculty's nomination to the chair of chemistry, recently vacated by J. Minguin. The chemists favored Raymond Cornubert, a chemical engineer, specializing in stereochemistry, who three years earlier had completed a doctorate. Other Faculty members favored Henri Pariselle, who was *professeur sans chaire* at Lille. Gain preferred Pariselle because his had been purely a university career, although he also was said to have links with industry in Lille. The council, following a vote of eight to eight, agreed to nominate Pariselle because he was older than Cornubert. This decision was overturned by the ministry on the grounds that it was illegal. The next ballot resulted in ten votes for Cornubert, one for Pariselle, and five abstentions.[93]

A few years later, the council more clearly showed a preference for practical experience in a vote for the new director of the Chemical Institute, replacing Guntz. A decision was to be made between Charles Courtot, one of Grignard's students at Nancy, and Alexandre Travers, who had been teaching for at least a decade in industrial chemistry, specializing in ionization, oxidation potentials, and the chemistry of cements and corrosion. Three chemists spoke for Courtot, but Gustave Vavon, who taught organic and applied chemistry, argued that Travers was preferable because he had personal relations with industrialists and with teachers in secondary education, as well as more practical experience and comprehension of engineering mathematics and physics. The vote went in favor of Travers by ten to six, with one abstention. Travers left Nancy permanently in 1942 to work in industry.[94]

By 1930, most of the laboratory supervisors and aides at the Institut Electrotechnique were alumni with industrial experience. The school regularly enrolled graduates of the Ecole Polytechnique, as well as candidates who ranked high on the school's own concours, which was inaugurated in 1927.[95] The Ecole Supérieure de la Métallurgie et de l'Industrie des Mines, founded in 1919, was now a significant force within the Faculty and the university. Its students, well schooled in hydraulics, dynamos, turbines, and electrical installations, always found jobs. Included among alumni were several directors of electrical companies. In 1932, the ESMIM had 142 candidates for its concours, or six candidates for every available place. Between 1919 and 1934, thirty-nine graduates of the Ecole Polytechnique studied there, in a curriculum taught by mining engineers as well as Faculty professors.[96]

In a recent article comparing the chemical schools at Lyon and Nancy, Bauer and Cohen have praised the Nancy school as progressive and untypical of France in its mixture of physics and chemistry education, theoretical and practical curriculum, and administrative organization linking the university with industry and local interests without relinquishing Faculty control.[97] When the national grouping of thirty Ecoles Nationales Supérieures d'Ingénieurs was established in 1948, with a national concours for admission, Nancy had six schools in the plan.[98] Admission to the schools had become so competitive and the future of its graduates so assured that they came to consider themselves an elite in comparison to their comrades in the Faculty courses, as has been pointed out by Delsarte, the Nancy mathematician who was a firm advocate of decentralization in the 1950s.[99]

As will be seen in other cases, too, so successful was the regional program of applied/fundamental science at Nancy—as an idea, if not as a fact—that it was appropriated by the Ministry of Public Instruction. Haller had argued that the superiority of nineteenth-century German industry lay in collaboration among science, technology, and laboratory research. Supported by strong patriotic and regional sentiments and by an expanding, largely new economy, the city of Nancy and the residents of Lorraine devoted great effort to building a university that would serve the region and the nation, both in emulation of and in competition with their German neighbors. As Adam commented, "Le motif fut d'abord une pensée patriotique, comme il convenait alors à l'Université-frontière."[100]

A symbol of French patriotism and "revanchisme" against Germany, Nancy was a border town. When Paul Doumer, president of the National Assembly, visited Nancy in 1905, he talked with Faculty and municipal officials about the request of 300,000 francs for the university. The story was told by Adam that Doumer then took a tour by automobile of all the *forts d'arrêt* from Verdun to Epinal, to examine "s'ils etaient en état de barrer la route à l'envahisseur." The tour completed, he concluded, "On ne passe pas." After that the credits were forthcoming.[101]

Proud of her heritage as the duchy of Lorraine, protective of prerogatives against the ever-increasing centralization of Paris bureaucracies, inhabited by refugees from occupied Lorraine and Alsace, Nancy was a cosmopolitan city that merited the metaphor of a small Athens in the Belle Epoque. A dominant emotional motif was competitiveness, directed both west, toward Paris, and east, toward the enemies who had bombarded the cathedral of Strasbourg and occupied the city of the dukes of Lorraine and Stanislas Leczinsky. Unfortunately, as we shall see, the vehemence of emotions seems, in one outstanding case, to have affected the quality of scientific work, leading to scandal for French science.

## SCANDAL AND THE SCIENTIFIC COMMUNITY: THE N-RAY EPISODE

In spring 1903, Nancy's most illustrious physicist, René Blondlot, announced the discovery of a new radiation, which he named "N": a symbolic letter identified with his birthplace, his university, and the coat of arms of Nicolas François, the seventeenth-century regent of Lorraine.[102] Blondlot said that the rays emanated from both inert and living bodies and that they increased in strength with the "psychic activity" of the source. Their effects were observed by at least forty people and analyzed in some 300 papers by one hundred scientists and medical doctors between 1903 and 1906. Since N-rays do not exist, the episode has become an exemplar of "pathological science" among some of its historians and commentators.[103] Typically, accounts of N-rays have cautioned scientists not to abandon the scientific method, implying that the N-ray case is an example of the intrusion into the scientists' laboratory of prejudices and motives that are abnormal in the scientific community.[104]

But the history of N-rays and the prejudices and motives influencing the work of Blondlot and his colleagues are not so anomalous psychologically or historically that we cannot learn a good deal from the episode. Indeed, this history aids us not only in better understanding the scientific enterprise in general but, more particularly, the social context and internal organization that characterize French science, both provincial and Parisian, at the beginning of the twentieth century. An important element in the N-ray story is the effect on scientific work of scientists' sense of competition with peers. This took the form of competition of the Nancy group with other French scientists, particularly Parisians, and of French competition with foreign scientists, especially the Germans. The N-ray history demonstrates the mechanisms and the perils of intellectual rivalry among communities of scientists; it also illuminates the tightly knit structure of the French scientific community and the loyalties within this elite.

In late 1902, Blondlot claimed that he had demonstrated that X rays are propagated with a velocity equal to that of light. His work, although later rejected as inadequate, excited considerable interest.[105] In February 1903, he reported an investigation based on the widely discussed idea that experimental failures to polarize X rays might be due to their already being plane-polarized at emission because of the unidirectionality of the parent cathode rays.[106] To detect the supposed polarization, Blondlot used a Hertzian-type analyzer and two sharpened wires communicating at A and A' with the terminals of a Ruhmkorff coil B and B' (figs. 1a, 1b). An aluminum sheet 0.4 m on a side shielded the cathode-ray tube from the electric spark in order to prevent any direct electrical influence on the spark from the electrodes.

As X rays were emitted by the cathode-ray tube, a spark simultaneously played between the points of the spark gap, since the points cc' were connected to the same coil as the tube. With the gap parallel to the cathode rays, Blondlot said that he could diminish the spark's intensity by placing a lead or glass screen between the tube and the gap;

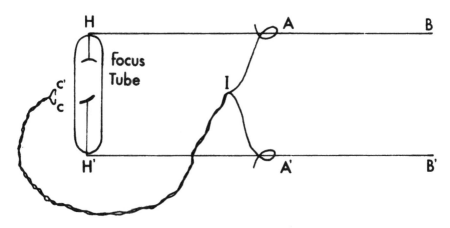

Fig. 1a. Blondlot's diagram as it appeared in his 1903 paper on the polarization of X rays. Like many of his descriptions of his apparatus and results, the
diagram lacks detail.

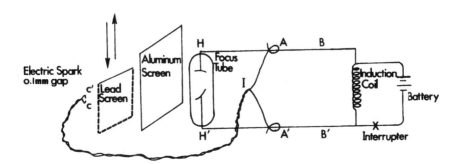

Fig. 1b. A reconstruction of Blondlot's apparatus for demonstrating the polarization of X rays, later said to be N-rays. The focus tube, emitting X rays, is
linked to an induction coil by the insulated wires BH and B'H'; two other
insulated wires AIc and A'Ic' terminate at A and A' in two rings around BH
and B'H'. AI and A'I are rolled up on each other so that the distance between
c and c' can be regulated easily. The inductive surge at each interruption of the
coil current produces a small spark at cc' at the same time that an X ray pulse
comes from the tube. Because of the flexibility of AIC and A'Ic', the orientation
of the spark gap can be changed. A square sheet of aluminum, 0.4 meter on a
side, was placed between the tube and the spark to prevent direct influence on
cc' of the focus tube electrodes.

with the spark turned normal to the direction of the cathode rays, the screen had no effect on the intensity of the spark. The results seemed to confirm the supposition that X rays might be plane-polarized. Blondlot next collimated the X rays through a slit in a lead plate and reported that the direction for maximum brightness of the spark changed when quartz and sugar pieces were introduced between the slit and the spark; he suggested that these crystals rotated the plane of polarization of the X rays.[107]

Blondlot noted the changing brightness of the spark, not, like Heinrich Hertz, its length. The action of X rays on the sparking distance of an electrical discharge was well known. However, Blondlot claimed that the change in spark brightness was a different effect than the increase in spark distance, although he ascribed both to a decrease in air resistance to the passage of electricity. He argued that a short, weak spark (apparently less than 0.1 mm in length) was necessary to detect the effect on spark brightness; in a strong, hot spark the decrease in air resistance owing to the X rays would not be appreciable.[108] Blondlot's February 1903 paper was unique in stressing brightness as a key property of the spark, in insisting that the sparks must be short and weak, and in arguing that the changing brightness of the spark, as it was moved about in space, demonstrated polarization of X rays.

Within a couple of months Blondlot had concluded that since the radiation he was studying could be refracted by a quartz prism, it could not be X rays. The brilliance of the spark detector changed when a quartz prism was placed between the aluminum-screened cathode-ray tube and the spark gap; the sparking device had to be moved laterally from its original position in order to regain its point of maximum luminosity. As detected by the spark, the radiation was bent toward the base of the quartz prism; its index of refraction in quartz was approximately 2.

When Blondlot decreased the cathode tube voltage so that no fluorescence occurred in the tube walls (and therefore, according to him, no X rays were being produced), he still obtained changes in brightness of the spark. He deduced the existence of a radiation different from X rays, a radiation that traversed black paper and wood as well as aluminum, that was polarized on emission, and that could be reflected, refracted, and diffracted. All these properties were detected by means of the spark analyzer; the rays produced no fluorescence on bromide or cyanide screens and no direct photographic effects.[109] While a cautious experimenter might have examined the hypothesis that photoelectric effects were responsible for many of his results, Blondlot rushed to christen his findings a new radiation.[110]

When he announced these results in spring 1903, Blondlot was a distinguished professor at Nancy, a corresponding member of the Academy of Sciences, and the recipient of the Academy's Gaston Planté Prize (1893) and La Caze Prize (1899).[111] His 1881 doctoral thesis under Jules Jamin at Paris had been called "brilliant."[112] His next researches, inspired by Hertz's, centered on measurement of the velocity of electric waves. By one method, which depended on the calculation of the period of the waves, he got an average value of $2.976 \cdot 10^{10}$cm/sec. By a second method, photographing electric sparks reflected by a rotating mirror, he found values very close to the figure $3 \cdot 10^{10}$cm/sec calculated for light from both optical experiments and the ratio of electrodynamic and electrostatic charges.[113] Eleuthère Mascart praised Blondlot's "originality," Poincaré lauded his results as an "experimentum crucis" for Maxwell's theory, and J. J. Thomson cited his investigations as exemplary measurements of the velocity of electric waves.[114]

After Roentgen's discovery in 1895 of radiations emitted from a cathode-ray tube, Blondlot set out to determine whether Roentgen's rays, like electric waves produced by oscillatory discharges, have a velocity near that of light. He again used an electrical spark as an analyzer, this time stressing the delicacy of the experiments and the dependence of success on determining a maximum *brightness* of the spark. He did not, as in the earlier work on electric waves, photograph spark discharges but merely described the changing brightness as he saw it. Now, in spring 1903, he concluded that this set of experiments had in fact measured the velocity of N-rays and not of X rays.[115]

The discovery of a new radiation was not a surprising event in the decade 1895–1905. Radiations old and new, demonstrated and refuted, interpreted and reinterpreted abounded: cathode rays; Becquerel rays, including alpha, beta, and gamma rays; "black light"; and now N-rays. Properties of X rays were at this time hotly disputed, and different methods of investigation yielded different conclusions. C. G. Barkla, for example, was to report in 1905 what historians regard as the first successful detection of X-ray polarization by the "electrical" or ionization method. H. Haga tried a similar experiment using a photographic method and failed to confirm any polarization effects.[116]

Blondlot's method for studying polarization by visually detecting changes in brightness of a spark was inferior to his foreign colleagues' photographic effects and ionization measurements. However, electrical sparks had worked for him in the past, and the method of visual detection of changes in luminosity was one of the war-horses of French physical optics. Oblivious or indifferent to the very complicated physics of variations in intensity of an electric spark in the vicinity of a cathode tube, Blondlot pursued his announced discovery. The perils of using as his primary data the extremely subjective variations in brightness of a feeble electric spark did not seem to daunt him in the least.

On the grounds that the refractive index of N-rays in quartz was similar to that for red light, Blondlot began looking for N-ray sources different from a cathode-ray tube. Drawing on Heinrich Rubens's experimental work in the far infrared, Blondlot wondered if his rays might be produced by a gas-burning Auer lamp. Using the changing brightness of a spark as a detector for the precise focusing of his radiations by a quartz lens (fig. 2), he deduced a series of indices of refraction in quartz greater than 2.0, the principal one of which was 2.93. These indexes, he said, correspond to separate bundles of rays in different parts of the N-ray spectrum emitted by an Auer lamp. For the experiment, the lamp was encased in corrugated iron and equipped with an aluminum window, 0.1 mm thick, through which the N-rays passed before being focused by the quartz lens.[117]

Rubens found Blondlot's results impossible. The assertion that electromagnetic radiation of long wavelength could traverse an aluminum window of 0.1 mm thickness contradicted Maxwell's theory and Rubens's own observations. With the help of Otto Lummer and Ernst Hagen, Rubens failed to confirm Blondlot's results.[118] Quartz is opaque to infrared radiation longer than 4.20 microns.[119] According to calculations made by Georges Sagnac using Blondlot's figures for refraction index, the N-rays should have wavelengths four times greater than Rubens's longest wavelengths.[120] It seemed unlikely that they could make their way through quartz. Blondlot conceded that his rays were not exactly like Rubens's infrared radiations, and he suggested they might be in the "five octaves" of radiations still unexplored between Rubens's rays and shorter ones.[121]

Many of the properties that Blondlot reported for N-rays were similar to ones established for radioactive rays. Pierre Curie had noted, for example, that under certain conditions Becquerel rays facilitated the

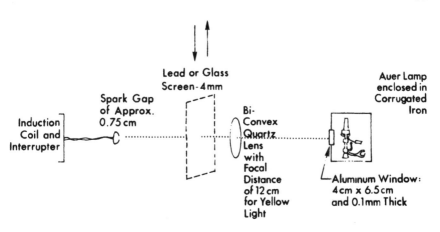

Fig. 2. A reconstruction from Blondlot's text, showing his first systematic attempt to estimate wavelengths for N-rays. An Auer gas lamp within a lantern of corrugated iron sends its rays through a rectangular aluminum window and a biconvex quartz lens to fall on a sparking device operated from a very weak induction coil furnished with an interrupter. A lead or glass screen is used occasionally to block the N-rays' action on the spark.

passage of sparks between conductors placed in air.[122] N-rays did not directly excite fluorescence like Becquerel rays, but Blondlot reported that phosphorescent bodies after exposure to the sun appeared to increase in brightness on exposure to N-rays. Edmond Becquerel had demonstrated this particular effect for infrared radiations, but, again, Blondlot denied that his radiations were heat rays. A thermocouple, he said, did not respond to N-rays.[123]

A new arrangement did respond: luminous calcium-sulfide spots. Blondlot recommended diluting calcium-sulfide powder in a collodion spread out with ether, and then painting spots of this mixture, each spot several millimeters in diameter, on black cardboard. According to Blondlot, the luminous spots became distinct and more luminous when an ironclad Auer lamp with its aluminum N-ray window was brought near. The N-rays acted more powerfully when focused on the calcium-sulfide screen with a quartz lens. In making these observations, Blondlot noted, it was indispensable to avoid all constraint of the eye, all effort at vision. The observer should not look directly at the screen; "the observer should play an exclusively passive role, under penalty of seeing nothing."[124] However, these instructions did not reassure doubters.[125]

Using the screen as detector, Blondlot found that N-rays were emitted by the sun, a flame, or any incandescent body, including the incandescent electric lamp. Next, he claimed that a quartz lens or sheets of quartz, feldspar, and Iceland spar, once exposed to N-rays, reemitted them. He was led to this discovery of secondary radiation, he said, by the fact that his calcium-sulfide screen continued to increase in brightness after the Auer lamp was extinguished. Hence, he concluded, the quartz focusing lens, still in place, must be reemitting radiation stored up in it.[126] Soon he added the humoral fluid of the eye to the list of secondary emitters, and he concluded that the N-rays are "not without influence on certain phenomena of animal life or vegetable life."[127] To the list of primary sources, he added tempered steel and other materials in a state of strain, suggesting that the energy of emitted rays is borrowed from the potential energy corresponding to an inner strained state of matter. This was an explanation not unlike hypotheses of the source of the energy of radioactivity.[128] Blondlot did not propose means to measure quantities of energy associated with his N-rays.

Perhaps the most exciting properties of N-rays were those announced by one of Blondlot's Nancy colleagues in winter 1903. Several Sciences Faculty and Medical Faculty professors took up the subject, and the most successful, Auguste Charpentier, found N-rays and other "physiological" radiations to be disbursed by animal muscles and by the human body. Jules Meyer, Blondlot's aide, corroborated these findings in Blondlot's laboratory. In the Medical Faculty, M. Lambert published on the emission of N-rays by organic ferments, especially the digestive ferments of the albumoids, and Edouard Meyer investigated the emission of N-rays by plants in the dark.[129]

Charpentier's papers created a sensation when they were delivered to the Academy of Sciences from December 1903 to July 1904. At first he claimed that N-rays were emitted by both the muscles and the nervous system; then he settled on the nervous system alone. As described enthusiastically by the Sorbonne physicist and physiologist Arsène d'Arsonval, these radiations increased in a noticeable way when an individual talked or exerted intellectual effort. It was from "Broca's center" in the brain that the rays emanated during speech. "You can see an increase in brilliance of the phosphorescent body. There is a sort of quantitative relationship between the degree of psychic activity and the phosphorescence of the screen." Similar effects were claimed for sensation; for if nervous sensation were eliminated by a dose of curare and a muscle made to contract electrically, the N-ray activity disappeared.[130]

Two members of a psychophysics study group in the Institut Général Psychologique, Courtier and Youriévitch, described their visit to the Nancy laboratories where they witnessed N-ray phenomena. The detector was either a cardboard screen covered with a thin film of calcium sulfide or a little lead tube closed at one end with a wooden cork, the outer surface of which was painted with calcium sulfide. An experimenter held the open end of the tube against part of his body, and he or another observer noted the increase in brilliance of the phosphorescent cork. Using a diaphragm one could orient a thin stream of physiological N-rays toward lenses or prisms, concentrating or refracting the rays in order to better study their effect on a calcium-sulfide screen.[131]

Experiments were carried out, too, on hyperacuity, the heightened visual perception of an individual after a bundle of N-rays fell on his eyes. After exposure of the experimenter's eyes to N-rays, a feebly illuminated page of writing or the face of a clock that he previously could not see became decipherable.[132] Some members of the Institut Général Psychologique, which included in its directing committee old-time positivists like d'Arsonval, Théodule Ribot, and Liard, the director of higher education, proposed that the emanation of N-rays might help explain medical and parapsychological phenomena like the visual hyperacuity of hysterics.[133] D'Arsonval ventured the hypothesis that "there exists around individuals a kind of atmosphere of an altogether special nature, a radiation for which one can determine physical constants absolutely, as for all other manifestations of energy."[134] Spiritualistic auras were to be explained away by physics. Spiritualists like Carl Huter claimed priority over Charpentier in the discovery of these properties of the new radiation; Huter's claims were rebuffed at the Academy of Sciences in d'Arsonval's report, which declared that the priority of "facts" belonged to Charpentier.[135]

Although confirmatory studies appeared for several years, trouble had surfaced early; already in his paper of November 2, 1903, Blondlot remarked that the aptitude for recognizing the feeble variations of intensity varied from one person to another. Certain persons saw at first glance the anticipated variations, while for others they lay "almost at the limits of what can be distinguished."[136] In letters to *Nature*, as well as in reports, many scientists, including the young physicist Jean Perrin, criticized the subjective nature of the methods of detection. They concluded that N-ray effects on phosphorescent surfaces were caused by the heat of the human body or by the heat of infrared radiation emitted by an Auer or Nernst lamp.[137]

Physicists critical of Blondlot's conclusions also argued that his observations were psychophysiological effects rather than physical ones. Lummer was the first to present this interpretation systematically in a paper read to the German Physical Society in November 1903. He and Rubens had found that a set of Blondlot's experiments could be almost exactly imitated without employing any source of illumination whatsoever and could be attributed to the interplay of the rods and cones of the retina in seeing in the dark. Making use of accepted theories of retinal vision, Lummer explained that the color-blind, peripheral vision of rods in the retina comes into action in the dark. When a lead plate or the hand is interposed between the source of illumination and phosphorescent screen, the observer fixes his gaze on the screen, thereby inactivating the rods. If he gazes continually at the screen, he senses no changes in brightness; but as he looks slightly away, the rods come into action and he perceives a slight white luminosity.[138] Indeed, the trick of peripheral vision had long been employed by astronomers to "look" at faint stars, and the Czech physiologist Jan Purkinje had explained as early as 1825 changes in the apparent relative luminosity of objects in dim light.[139]

Burke at the Cavendish Laboratory concurred with Lummer's explanation, as did C. C. Schenck of McGill University, who wrote that much of Blondlot's work could be explained through indirect vision, the sensitivity of the eye in a dark room, the natural decay of phosphorescence, and temperature effects. Enrico Salvioni and M. Munoyer suggested, too, that psychological factors played a role in observations of N-rays.[140]

The sources of error with phosphorescent screens or corks were noted in the generally sympathetic meetings of the Institut Général Psychologique. D'Arsonval commented already in December 1903 that heating the screen increased its phosphorescence. He said that as an experimental control, N-rays had been studied at Nancy as they emanated from cold-blooded animals.[141] Later, Courtier discussed his visit to Nancy and the fact that some experimenters saw an increase in light intensity, while others did not. He reported that Charpentier, an expert on the physiology of the eye, was not unaware of these problems. The retina had been the topic of his doctoral thesis as well as other work, including a contribution to the International Congress of Physics in 1900.[142]

According to Courtier, Charpentier reported that the observer who barely perceived light at the beginning of prolonged confinement in a dark room later could detect light one thousand times less intense than light that originally excited sensation. Moreover, this differential sensibility varied during the course of experiments in detecting N-rays. Looking directly at the calcium-sulfide screen, one thought that it brightened and then lost its luminosity; but after closing the eyes for a moment, one sensed immediately on reopening them an increase of luminosity on the screen.[143] Blondlot and Charpentier seemed to think that awareness of these effects was sufficient guard against their becoming significant sources of error.

There were strong objections to the method of visual detection of variations in the brightness of an electric spark, a technique decidedly more arbitrary than measuring spark distances. In addition, John Burke noted that the diminished brightness of Blondlot's spark when a lead plate was placed between the spark and an N-ray source might be due to the lead plate itself damping the oscillations of the spark. Paul Langevin, who visited Nancy to see Blondlot's experiments and attempted to verify them in Mascart's laboratory at the Collège de France, failed to get positive results. To produce a stable, unvarying spark he used a small electrostatic machine run by a motor in uniform rotation; a small condenser, charged by the machine, was discharged periodically through the length of a high-resistance wire. No effect on the spark was registered by presumed N-rays from Auer lamps or other sources. Blondlot countered Langevin's disconfirming evidence by saying that it was necessary for the electric spark to be unstable ("short and feeble") in order to get the effect of increased brilliance. But how, wondered Langevin, can one distinguish meaningful minute variations in the brightness of an already unstable spark?[144]

Rather than depending only on subjective visual observations of variations in an electric spark or variations in calcium-sulfide spots, Blondlot and others attempted to obtain systematic photographs of N-ray effects. Rothé, then lecturer in physics at Grenoble, described in June 1904 a photographic method for following variations in phosphorescence under the influence of N-rays; and H. Bordier, an agrégé

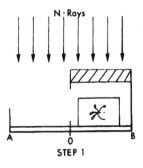
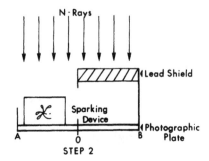

Fig. 3. The spark is enclosed in a cardboard box that is open only on the side of the photographic plate. A lead screen covered with moist paper shields the sparking device and half of the photographic plate from N-rays during half of the experiment. The spark moves back and forth from AO to OB at five-second intervals to the beat of a metronome.

in physics at the Medical Faculty of Lyon, did the same in a paper presented to the Academy. Those present at the December Academy session could not agree whether there was any difference between Bordier's photographs of calcium-sulfide spots illuminated by N-rays and spots not so illuminated.[145]

Blondlot had designed an experiment to register on photographic plates the variations in brightness of an electric spark. An electrical sparking device was moved back and forth manually from one side of the photographic plate, where it was exposed to N-rays, to another side, where it was protected by a lead shield (fig. 3). The photographic plate registered the spark more brightly on one side than on the other.[146] Those critical of Blondlot suggested that since the back-and-forth movements were carried out manually rather than automatically, his laboratory assistant, L. Virtz, may have unconsciously allowed the electrical sparking device to remain on the unshielded side of the plate slightly longer than on the other side. The American physicist R. W. Wood ungenerously referred to Blondlot's assistant in these experiments as "a sort of high-class laboratory janitor."[147] Later Virtz was to become a scapegoat for those who chose to award Blondlot's errors to his overzealous assistant. In April 1903, well before the public scandal over N-rays broke out, Blondlot said of Virtz:

> If I have accomplished a lot, it is by grace of the assistance of M[on-sieur] L. Virtz, mechanic attached to our laboratory; not only has my

apparatus been constructed by him, but I owe him more than one idea for its arrangement. He has repeated all the measurements, a necessary control in such delicate researches.[148]

In a letter to the editor of *Revue Scientifique* published in November 1904, Blondlot maintained that the "mechanic who moves the sparking device back and forth is disinterested in the results and at first he had no idea of the significance of what he was doing. . . . [Further], I accept exclusively the responsibility for the experiments that I have published myself or at least verified."[149] After Blondlot retired from the Faculty at Nancy in 1909, he loaned the Faculty 20,000 francs, with the stipulation that the interest payments be earmarked for the salary of Virtz as aide, mechanic, or curator of collections.[150] Blondlot was not prepared to see his assistant made a scapegoat.

In January 1904, Blondlot assigned new, precise values for wavelengths to the N-rays. Using an aluminum prism with a refracting angle of 27° 15', he detected deviations with a spot of calcium sulfide 1 mm wide and 1 cm high mounted on a piece of cardboard. Again, he reported that there were different groups of N-rays with different refractive indexes. But the newest wavelength estimates showed Blondlot's earlier measurements with a quartz prism to have been gravely in error, apparently because of quartz' then unknown property of storing up and reemitting N-rays. They did not fall in the infrared after all, but in the far ultraviolet. He paid no attention to Victor Schumann's recent work on the ultraviolet which indicated that such high-frequency N-rays should be absorbed by the air.[151]

As the scientific community was becoming increasingly skeptical about all these claims, they were made known to the wider public. The Paris daily newspaper *Figaro* proclaimed that the electromagnetic spectrum, extending from Hertzian waves through the infrared and visible light to the shorter ultraviolet, had been completed with the discovery of N-rays.[152] D'Arsonval delivered a similar message in a popular lecture in March 1904 at the Hôtel des Sociétés Savantes. Blondlot's radiations were of wavelengths as small as 0.008 micron, said d'Arsonval, placing them in the neighborhood of the ultraviolet.[153]

That same month, Schenck again expressed doubts about Blondlot's work, this time about the refraction and diffraction experiments. Schenck argued that according to Blondlot's description of his apparatus, the N-ray beam would be 1 cm wide at a distance of 14 cm from the slit, and the groups said to be separable by their different refractive indexes must in fact be overlapping. Also, Schenck estimated that the collimated N-ray beams used in the diffraction experiments could be no more than 2 percent as intense as the original beam. Blondlot had measured a central image and twenty diffracted images, the intensity of which could have been no more than a thousandth of that of the original beam. All this was done with a radiation so feeble that no observer outside of France has been able to detect it at all.[154]

In Paris, Jean Perrin objected that the reported precision of Blondlot's experiments was impossible given the width of the slit used in the experiments. When R. W. Wood also made this criticism, Blondlot countered that this was merely one of the inexplicable properties of N-rays.[155] Blondlot seemed to his critics to be irrational, even perverse.

Salvioni at first failed to observe refractions of the N-rays, but once he "decided to follow exactly the directions of Blondlot by using a luminous calcium sulfide furrow in cardboard rather than an extended screen," he observed maxima and minima with a diffraction grating. Three months later, however, he could no longer obtain these results. Salvioni concluded that visual and physiological effects accounted for all his observations.[156]

At Lyon, Georges Gouy failed to get Blondlot's results, and in Paris, in addition to Langevin and Perrin, Victor Henri, Henri Pellat, and Henri Abraham reported negative experiments. Jules Violle, Marcel Brillouin, Paul Janet, Georges Sagnac, and Pierre Curie all said they had carried out superficial experiments and failed to get Blondlot's results, but each said his results were not conclusive. This was the opinion, too, of Charles Camichel at Toulouse. At the Zurich Polytechnik, Pierre Weiss defended Blondlot, but later he and Walther Ritz could not successfully duplicate Blondlot's work. Among those who visited the Nancy laboratory and failed to verify Blondlot's results were Langevin, L. P. Cailletet, Lucien Poincaré, and Henri Poincaré (who averred that, after all, his eyes were not very good).[157] Those who journeyed to Nancy and verified the experiments included Jean Becquerel, who returned to Paris and carried out independent investigations. Later, Edmond Bouty said that he believed he had seen at Nancy most of the reported N-ray effects. D'Arsonval and Mascart both distinguished diffraction maxima and minima with Blondlot at his laboratory.[158]

Meanwhile, Lummer spread his interpretation at the British Association meeting in Cambridge and the Deutsche Naturforscher Versammlung at Breslau: Blondlot's observations, he said, were not to be attributed to physical phenomena but to physiological and psychological causes.[159] At the International Congress of Physiology in Brussels, where M. Lambert, agrégé in physiology at the Medical Faculty of Nancy, presented a paper confirming the emission of N-rays by the human body, the nervous system, and vegetable fermentations, Victor Henri and Henri Piéron reported failure to find anything. When Piéron suggested to Lambert that he, Piéron, control the emission of N-rays while Lambert viewed the calcium-sulfide screen and called out the appearance of disappearance of the N-rays, Lambert declined. The Belgian physiologist Augustin Waller remarked that the rays of Nancy should be called "rays of suggestion," alluding to the well-known studies on suggestion by the psychiatrists of the Nancy Medical Faculty.[160]

It was at the British association meeting that Rubens proposed to R. W. Wood of Johns Hopkins University that he go to Nancy to test the experiments. "You are an American," he is reported to have said, "and you Americans can do *anything*." Rubens "felt particularly aggrieved because the kaiser had commanded him to come to Potsdam and demonstrate the rays. After wasting two weeks in vain efforts to duplicate the Frenchman's experiments, he was greatly embarrassed by having to confess to the kaiser his failure." This anecdote, recounted in France by A. Berget, fed the scandal, although Rubens publicly denied the story.[161]

Wood arrived at Nancy in late September. He spoke German, he said, so that Blondlot would feel free to speak in French to his assistant. Wood later wrote of the incident:

> He [Blondlot] first showed me a card on which some circles had been painted in luminous paint. He turned down the gas light and called my attention to their increased luminosity when the N-ray was turned on. I said that I saw no change. He said that was because my eyes were not sensitive enough, so that proved nothing. I asked him if I could move an opaque lead screen in and out of the path of the rays while he called out the fluctuations of the screen. He was almost 100 percent wrong and called out fluctuations when I made no movement at all, and that proved a lot, but I held my tongue.[162]

The crucial test for Wood involved Blondlot's spectroscopic observations, where a vertical thread coated with luminous paint was moved through the region of the N-ray spectrum by turning a wheel. The thread, according to Blondlot, brightened as it crossed the invisible lines of the N-ray spectrum. After Blondlot had run through the spectrum once,

> I asked him to repeat his measurement, and reached over in the dark and lifted the aluminum prism from the spectroscope. He turned the wheel again, reading off the same numbers as before. I put the prism back before the lights were turned up, and Blondlot told his assistant that his eyes were tired. The assistant had evidently become suspicious, and asked Blondlot to let him repeat the reading for me. Before he turned down the light I had noticed that he placed the prism very exactly on its little round support, with two of its corners exactly on the rim of the metal disk. As soon as the light was lowered, I moved over toward the prism, with audible footsteps, but *I did not touch the prism.* The assistant commenced to turn the wheel, and suddenly said hurriedly to Blondlot in French, "I see nothing; there is no spectrum. I think the American has made some dérangement." Whereupon he immediately turned up the gas and went over and examined the prism carefully. He glared at me, but I gave no indication of my reaction. This ended the séance and I caught the night train to Paris.[163]

In a letter sent to *Nature* and translated in *Revue Scientifique* and *Physikalische Zeitschrift,* Wood announced his discovery that "the removal of the prism . . . didn't appear to present any obstacle to the localization of the maxima and minima of the bundle of deviated rays."[164] The *Revue Scientifique* immediately launched a three-month poll of French chemists, physicists, physiologists, and psychologists. The chemist Henri Moissan demanded to know if scientific questions were to be resolved by plebiscites.[165] While most scientists who previously had verified or extended Blondlot's results reiterated their findings despite Wood's account,[166] very few new confirmations were published. In 1905, only one confirmatory account of N-rays was presented at the Academy. Still, the rays did not disappear at Nancy. Gutton photographed what he said were changes in intensity of a secondary spark of an electric oscillator influenced by N-rays falling on a primary spark. Mascart reported new readings with Gutton, Virtz, and Blondlot of deviations of N-rays by an aluminum prism.[167] In March, Cotton and Raveau visited Nancy to see Gutton's new experiments, but they were unimpressed and added their objections to remarks by Perrin decrying the reappearance of N-rays at the Academy.[168]

Albert Turpain of Poitiers wrote angrily to the editors of the *Revue Scientifique* that his paper submitted to the Academy of Sciences reporting negative results for Blondlot's and Gutton's recent experiments had been returned by Mascart with the explanation, "Your results can be explained simply by supposing that your eyes do not have sufficient sensibility for appreciating the phenomena." Turpain now published his reply to Mascart: "If N-rays can only be observed by privileged rarities, then they no longer belong to the domain of experimentation."[169] This attitude did not help Turpain's chances for a chair at Nancy in 1906. The successful candidate turned out to be Gutton, Bichat's godson and Blondlot's collaborator, who was then lecturer in physics.[170]

Blondlot and others who had investigated N-rays hardly made convincing replies to their critics. Blondlot refused control experiments on the grounds that the observer must regulate the emission of N-rays and their detection in order to avoid fatigue. When experiments went awry, Blondlot and others pleaded tiredness or the difficulties of the observations. To Langevin's failure to register N-rays with a steady electric spark, Blondlot replied that the spark must be unstable. When observers failed to see a difference in luminosity between two spots of calcium sulfide, Becquerel explained that N-rays bounding off one spot overexcite the retina so that differences cannot be detected. Despite his critics' willingness to collaborate, Blondlot never agreed to any cooperative venture with them, nor did he develop any incontrovertible techniques to study N-rays. In addition, he failed to meet demands that his papers include more exact details for his apparatus and investigations.[171]

If Blondlot extended too far the flexibility of unwritten rules regarding allowable subjectivity of data, reasonable degrees of precision, and observational reproducibility, so did Lambert, Becquerel, and others. So, while Blondlot's "discovery" set the N-ray affair in motion, one cannot neatly explain the episode as his responsibility alone. An explanation of the N-ray controversy may best be achieved by examining not the structure of Blondlot's psyche but rather the structure of Blondlot's scientific community, its organization, aims, and aspirations around 1900. Here we may uncover the communal scaffolding, if not the personal foundations, of the N-ray debate.

Four aspects of the French scientific community seem especially to have influenced the course of events: (1) the renewed scientific interest in psychiatry and spiritualism around 1900; (2) the perception by French scientists that their international reputation was in decline, particularly vis-à-vis the Germans; (3) the inbred and hierarchical structure of the French scientific community; and (4) the local pride and spirit that dominated the Sciences Faculty at Nancy.

1. The renewed interest in spiritualism was in part a result of traditional Cartesian and Comtian aims in France to explain and unify natural phenomena through the scientific method. By the 1860s, France had become a center for "scientific" approaches in medical studies of mental pathologies, with a strong materialistic bias. A long-term aim of this reductionist research program was a physical explanation for pathological, psychological, and spiritualistic phenomena. Here N-rays seemed to offer a guide. Blondlot's special pleading and insistence that different observers have different sensitivities to N-ray effects might have been motivated by the widespread hope of advancing medical studies of psychological phenomena. The sensitivity to N-rays varied with the individual; just as some people are susceptible to hypnotic suggestion, some people are sensitive to the visual effects of N-rays.

D'Arsonval was keenly interested in psychics and spiritualism, an almost respectable study among fin-de-siècle scientists.[172] Like many others who attended table-rapping and levitation séances around 1900, d'Arsonval sought a scientific or physical explanation for mental suggestion, telepathy, and the like. One spiritualist phenomenon that was a candidate for physical explanation was the "odic" force described by Charles de Reichenbach; it was said to be emitted by the human body and transmitted along an iron wire.[173] Joseph Maxwell, the French medical doctor and popularizer who twice entertained the famous medium Eusapio Palladino at his home near Bordeaux, approached psychic phenomena from a reductionist point of view:

> It is not even necessary [in order to understand the possible mechanism of raps and movements of objects] to suppose that the nervous force acts outside of the limits of the body if we suppose that the experimenters create around them a sort of field analogous to a magnetic field. Nervous force would attain a maximum potential [in the experimenters] or in the "medium"; objects placed in the field would have a different potential: according to conditions, one would have phenomena of attraction or repulsion.[174]

Maxwell thought table raps analogous to noises of an electric spark, the rapping lines of force being determined by the action of the nervous centers. The glows and auras seen by psychics around the fingers and bodies of their subjects were natural phenomena at the limits of physical sensibility, just as Blondlot's N-ray effects were at the limits of perception.

Some spiritualists saw in Charpentier's discovery a confirmation of their views on the future of mental telepathy. Following Becquerel's comparison of the three subbundles of N-rays to the constituents of uranium radiation, one spiritualist related them to the three principal fluids of the *magnétiseurs*, including Mesmer's healing fluid. A military officer named Darget published photographs in *Vie Illustrée* of N-ray effects as well as of the colored auras emitted by persons as each touched a photographic plate; the color of the photographic impression varied according to the kind of fluid possessed by the individual.[175]

Blondlot did not repudiate spiritualist applications of N-rays and he employed the term *émission pesante,* used by spiritualists, for radiations that tend to descend perpendicularly. They associated this property particularly with the *effluves* that flow from hypnotist to subject.[176] Given the reductionist aims of French medical psychology, N-rays had great promise as a material explanation of spiritualist phenomena and certain mental or pathological states. The Medical Faculty at Nancy was distinguished in psychiatric research, and in the decades before 1900, medical colleagues and students, including Sigmund Freud, had come there from Germany, Holland, and Belgium to study the clinical approach to hypnotism of the physician and psychiatrist Hippolyte Bernheim and his colleagues.[177] For those who accepted the idea that nervous sensation and nervous energy were fundamentally electrical, a collaboration between Charpentier in the Medical Faculty and Blondlot in physics augured well for physical explanations of biological and psychological functions. This hope may have generated indulgence in the rash claims for N-rays.

2. When Blondlot's results were first reported in spring 1903, some French scientists were worried that their international prestige was slipping. Even the year's award of the Nobel Prize in physics to three Parisian scientists was not reassuring, for Pierre Curie had not been educated within the traditional academic hierarchy and Marie Curie was from Poland. In certain fields where French scientists had long excelled, such as optical and electrical measurements, their expertise was coming into question, notably in the Pender-Crémieu collaboration.

This collaboration originated in the work of the American physicist

Henry Rowland, who had demonstrated that the motion of a charged body creates a magnetic field. Roentgen (1888) and Franz Himstedt (1889) repeated and confirmed these experiments, but in 1900, Victor Crémieu, an aide in Bouty's laboratory, reported negative results. When Harold Pender, a recent Ph.D. at Johns Hopkins, confirmed the Rowland effect, Henri Poincaré suggested a collaboration between Pender and Crémieu. Lord Kelvin proposed Paris as the site, and Bouty put his Sorbonne laboratory at their disposal. Since the collaboration received a great deal of attention, its outcome—a confirmation of the Rowland effect—caused the French some embarrassment. It was at just this time that Blondlot first announced his discovery. Later, R. W. Wood cited the Pender-Crémieu collaboration as a precedent that Blondlot should follow.[178] The implication was that French work was unreliable.

The editors of *Revue Scientifique* warned that N-rays must not become a subject of ridicule by foreigners. The apocryphal story of Rubens and the kaiser suggested that French science was a laughingstock in the German court. Cailletet said that "for the sake of French science" he hoped that Blondlot and his colleagues were not mistaken, and, after Wood's letter to *Nature*, the chemist Joseph Achille Le Bel exclaimed, "What a spectacle for French science when one of its distinguished savants measures the position of the spectrum lines, while the prism reposes in the pocket of his American colleague!"[179]

French concern about their scientific standing is further shown by the appointment in 1901 of a Sorbonne committee to determine why American scientists sent American postgraduate students to Germany rather than France.[180] That French fears were not groundless is demonstrated in F. Giesel's remark that German scientists paid little attention at first to Becquerel's claims about radioactivity because of recent spurious radiations reported to the Academy of Sciences, particularly by the Frenchman Gustave Le Bon.[181]

3.   Besides the faculty and staff at Nancy, two other groups supplied supporters of N-ray research: young physicists or physiologists, usually at the rank of agrégé or lecturer and usually associated with provincial faculties, and Blondlot's contemporaries and personal friends, such as the Parisian physicists Bouty and Mascart and the physiologist d'Arsonval. Henri Poincaré and his family were illustrious members of the Nancy elite; like Blondlot's father, Poincaré's had taught at Nancy's

Faculty of Medicine. Brillouin, who refused publicly to conclude that Blondlot was wrong despite his failures to verify Blondlot's results, had taught at Nancy. Pierre Curie, who behaved similarly, had written a book with Blondlot.[182] Some younger, distinguished Parisian scientists like Perrin, Langevin, Cotton, and Henri were among the few French physicists who said outright that Blondlot and other N-ray enthusiasts were in error. The reception and extension of N-ray investigations may be explained in part by power and influence networks within the French scientific community.

The peer pressures that operated within the Nancy scientific community stifled criticism and created a sense of camaraderie through native loyalties and shared goals. The clannish forces working within the broader French scientific community were analogous to those at Nancy. Blondlot's contemporaries in the physics community were former classmates; his colleagues among physiologists and medical doctors were friends or acquaintances since student days in the Latin Quarter. Twelve to twenty candidates entered the sciences class of the Ecole Normale Supérieure every year, and most left together. They and other Paris-educated scientists constituted an elite corps that spread out through the Faculties of France, most of them aspiring to the few chairs at Paris and aware that successful competition for these prizes depended on patronage, promotions, and publications on the route from the provinces back to the capital.

Ambitious young scientists found themselves compelled to establish their reputations quickly, so that their credentials would be ready whenever the coveted Paris vacancies fell open. Many of the French scientists outside Nancy who invested effort in spectacular new N-ray investigations undoubtedly were hoping to make their mark in a new field and were doing their first independent research work.

The best known of the young N-ray researchers was the Parisian, Jean Becquerel. His example is instructive, although his formidable heritage keeps him from being entirely representative. Like his father, Henri, his grandfather, and his great-grandfather, Jean Becquerel was eventually to hold the physics chair at the Museum of Natural History. In 1903, he had just become an assistant at the museum after graduating from the Ecole Polytechnique and the Ecole des Ponts et Chaussées. He visited Blondlot's laboratory, where he believed that he saw the effects of N-rays on a calcium-sulfide screen. Blondlot encouraged young Becquerel to pursue the subject, as did André Broca, another family friend and professor of medical physics at the Paris Medical Faculty. In collaboration with Broca, Becquerel published reports on the effects of anesthetics in suppressing the emission of N-rays from living bodies, and he studied the deviability of N-rays in a magnetic field. He ceased his investigations in 1904, published nothing at all the following year in the *Comptes Rendus* of the Academy of Sciences, and afterward turned to new topics.

Becquerel later explained his early mistakes as the result of inexperience and of powerful pressures from the establishment. Speaking of himself in the third person, he wrote:

> Since then [1903–04] the author has acquired too great a habit of experimentation to attribute today the least value to his works on the pretended N-rays. The purely subjective method employed for testing the effects is antiscientific. It is easy to understand the illusion that deceived observers: with such a method you always see the expected effect when you are persuaded that the rays exist, and you have, *à priori*, the idea that such-and-such an effect can be produced. If you ask another observer to control the observation, he sees it equally (provided he is convinced); and if an observer (who is not convinced) sees nothing, you conclude that he does not have a sensitive eye.
>
> The author means to say that he then had various reasons to believe in the existence of N-rays: as we have said, he had been engaged in this direction by a scientist whose previous work inspired only admiration; more, another physicist declared that he had seen the effects, and some of Blondlot's students (who since then have become illustrious) published notes on N-rays. Before such authoritative powers, a young man, just graduated from the Ecole des Ponts et Chaussées and never yet having done any research, can be excused somewhat for getting carried away. Finally, not one note of the author was published without having been submitted to Blondlot and without receiving his approval.[183]

We could hardly have a more effective statement of the corporate pressures operating within the French scientific network. Becquerel might well have been following the advice given by the normalien Ernest Lavisse to a group of Sorbonne history students in 1881: adopt a "corporative spirit" and "deference" toward your teachers.[184]

Becquerel may have been all the more easily enmeshed in the N-ray web precisely because he was an exemplary insider in France's scientific hierarchy. Perrin, Langevin, and Cotton were not direct heirs from birth of the ancien régime of the French scientific community, and they differed significantly from Blondlot's generation and heirs both in their approach to physics and their social and political philosophy. Separated by age, by social and family ties, and by immersion in the new language and theory of molecular physics, Perrin, Langevin, and most members of their circle were independent, as well as outspoken, critics of the Blondlot camp.[185]

The official spokesman of the French scientific community was the Academy of Sciences, an institution to which Blondlot's young critics did not yet belong in the early 1900s. The Academy heard the first formal reports of N-rays, continued to discuss them for several years, and made an explicit choice to defend and protect Blondlot in the face of almost unanimous criticism from abroad and substantial doubts at home. This choice was first made in summer 1904, when an eleven-member committee met to discuss candidates for the Academy's Le Conte Prize. The prestige of the prize can be measured accurately in terms of its value, 50,000 francs, five times the current annual salary of Blondlot, who had then been teaching more than twenty years.

The prize committee members included such old friends and teachers as Henri Poincaré, Mascart, Berthelot, Moissan, L.-J. Troost, and Henri Becquerel. According to a later account, Becquerel prepared a report on Blondlot's work, including N-rays. Some committee members proposed to award the prize to Pierre Curie, but Blondlot won out in the end. Henri Poincaré rewrote Becquerel's original report, reducing to only a few lines the initial exposition of Blondlot's work on N-rays.[186]

This report was published in December 1904, when the Academy formally bestowed the prize on Blondlot and when the *Revue Scientifique* had influenced public opinion against the legitimacy of Blondlot's claims. Poincaré announced that the prize went to Blondlot for the "whole of his work." As for N-rays, Poincaré hedged:

> Circumstances have not allowed all the members of the committee to acquire the conviction on these questions that only personal observation can give. . . . Without yet prejudging the significance and the scope of these new discoveries, the committee has not believed it should defer longer the reward that this scientist has merited for a long time. It has wanted at the same time to affirm its confidence in the experimenter and to give him support in the midst of some of the greatest difficulties that physicists have ever encountered.[187]

Blondlot's sense of debt to the native Nancéien Henri Poincaré is apparent from his later statement that he felt an obligation to Poincaré to make public any doubt on his part about the validity of the N-ray work.[188]

Henri Becquerel lectured on N-rays in his course of 1904–05, and Bouty discussed them in the third supplement to Jamin's *Cours de physique*, published in 1906.[189] Mascart continued to defend Blondlot at the Academy. The next public step by Blondlot was a request for retirement, at age sixty, from higher education. He referred to N-rays as a phenomenon requiring delicate and difficult experimental observations, which many eminent scientists, he said, had not succeeded in confirming.[190] As is evident from sealed notes opened at the Academy on his death in 1930, Blondlot worked on N-rays long after his retirement. He was remembered by Husson of the Sciences Faculty at Nancy in a striking statement of collegial solidarity:

> His ideas were often in advance of his time, and in his important fields of electricity, his studies have become classics. With the measurement of the velocity of propagation of electricity, he put in evidence experimentally the electromagnetic nature of light, and is classed in the highest rank of experimentalists. It seems probable that his last researches on a new radiation could be mended with more powerful means and more extended and corroborated general views.[191]

4.   That Husson and his colleagues refused to make a harsher judgment on Blondlot is perhaps not surprising, given the role of the whole Nancy school of physics in the N-ray episode and the reciprocal allegiances of Blondlot and his colleagues in the Sciences Faculty and the institutes. Blondlot's allegiance to Nancy was deeply rooted. His father, born in the Vosges, had taught at Nancy's Preparatory School of Medicine and Pharmacy, which became the Faculty of Medicine in 1872.[192] At the conclusion of Blondlot's study for the doctorate in Paris in 1882, Berthelot highly recommended him for lecturer in physics at Nancy. Although he was thought "bien supérieur" to the situation, Blondlot lobbied to be sent to Nancy because of his "many attachments of family and interests there."[193]

The year after his appointment to Nancy, Blondlot received an opportunity to make a move within the French academic hierarchy, when he was recommended by the director of higher education as substitute for Jules Violle at Lyon. Conceding that his resolution not to accept this opportunity might appear "strange," Blondlot pleaded delicate health, along with family and material interests at Nancy, and added that he hesitated to relinquish his already productive research with Bichat at

Nancy.[194] Blondlot often expressed his desire to remain at Nancy and to remain free of administrative duties so that he could devote all his time to his teaching and research.

As we have seen, Bichat and Gutton, like Blondlot, were natives of Nancy. Four out of five of their supporters in the N-ray fiasco had taught or studied there. No critical papers originated at Nancy.[195] An explanation of the hasty publication at Nancy of careless, even cavalier experimental work by this group of physicists and physiologists is not hard to find, given our general understanding of the University of Nancy at the time. Flushed with the euphoria of their recent achievements and impatient to establish their region and their university in the industrial, scientific, and cultural avant-garde, Nancy scientists threw caution to the winds. They reinforced one another's confidence in the validity and importance of their results. Recent favorable publicity for the university seemed at first to protect and sanction the group's claims, just as Blondlot's reputation initially shielded him from criticism. N-rays, like "polywater" in the 1960s, was a big mistake.[196] But, unlike Boris Deryagin, Blondlot never admitted in print that his discovery was an error.

As for the long-term loyalty of Blondlot's French colleagues both within and without Nancy, we may note Suleiman's observation that the governing elites of France, highly selective and demanding in their criteria for membership, in Paris and in province, close ranks under duress. Once competence "has been demonstrated at an early age, it is never again called into question."[197] The episode demonstrates especially well the cohesion of the Paris-province relationship and the overlap of Parisian school loyalties with provincial family and faculty ties, forming a tightly interlocking network of influences within the French scientific community.

# 3

# GRENOBLE: FROM RAOULT'S PHYSICAL CHEMISTRY TO INTERNATIONAL SCIENTIFIC CENTER

In 1339, a papal edict founded a university at Grenoble, then a small city of some 4,000 inhabitants centered within a ring of mountains to the southeast of Lyon. The city's original site on the left bank of the Isère River, near the confluence of the Drac and the Isère, was reported in a letter of 43 B.C. to Cicero, and the settlement became an important stopover on the Roman route linking the Rhône Valley to Italy.[1] The fortified city of Gratianopolis, as it was called in the fourth century, passed under the control of the Burgundians, the Franks, and then Provence; and in the thirteenth century its counts took the title "Dauphins of the Viennois." The last of the independent dauphins, Humbert II, obtained from Pope Benedict XII a *studium générale* at Grenoble, establishing a university for his kingdom. For the next hundred years the university offered degrees in arts, medicine, and canon and civil law, until the new French dauphin, the future Louis XI, moved the university to Valence in 1454. Grenoble was by then part of France, since Humbert II had ceded his state to the king of France in 1349 for 200,000 gold florins.[2]

In the sixteenth and then in the eighteenth century, a university was briefly reconstituted at Grenoble, but it was not until the Napoleonic reorganization of 1808 that the university Faculties definitively existed once more.[3] The city had a Jesuit college in the eighteenth century, a school of surgery from 1771 to 1792, and an école centrale for some years after 1796. However, there was little basis for recruitment of students among the sparsely inhabited mountains surrounding the city.[4]

Local erudites met together on the eve of the Revolution in the Académie Delphinale. Reconstituted as the Société des Sciences et des Arts in 1802, the group's meetings were attended for a few years by the geologist Déodat de Dolomieu, the jurist Berriat-Saint-Prix, and the librarian Jacques Champollion-Figeac, the elder brother of the Egyptologist J. F. Champollion. The Académie retook its name in 1836, only to be abandoned by a positivist splinter group in 1838, which formed the Société de Statistiques.[5] Marie-Henri Beyle, known as Stendhal, portrayed the city in his novels with the repugnance and distaste of a young man anxious to escape his family and strike out on his own.

Stendhal's grandfather, a physician who was one of the Grenoble erudites of the late eighteenth century, persuaded the city's library to buy an Egyptian mummy. As Stendhal recalled, he "was very fond of geological theories . . . and used to talk to me with passion . . . of the geological ideas of a M[onsieur] Guettard, whom he had known."[6] Stendhal also recalled the books in the house of his father, whom he disliked, and his grandfather, whom he adored:

> My father and grandfather had Diderot and D'Alembert's folio *Encyclopedia;* it is, or rather it was, a work costing seven or eight hundred francs. It takes a tremendous influence to persuade a provincial to invest such a big capital in books, from which I now conclude that before my birth my father and grandfather must have belonged entirely to the philosophical party. . . . I had the most entire confidence in this book, because of my father's aversion from it and the marked hatred which it inspired in the priests who used to come to the house.[7]

Stendhal was ambivalent in his portrayal of the Dauphiné character: unsympathetic in the old Sorel of *Rouge et noir* ("the cold, discontented, far from civilized air which is the habitual expression of the cunning natives of Dauphiné") and occasionally praiseworthy in the autobiographical *Life of Henri Brulard* ("The character native to Dauphiné has a tenacity, a depth, an intelligence and subtlety which one would seek in vain in the civilizations of Provence or Burgundy, its neighbors").[8] Like many French boys of his generation, Stendhal set his sights on mathematics and Paris as his salvation. Like many, he was disappointed.

I had adored Paris and mathematics. Paris with no mountains inspired
me with a distaste so profound it went almost as far as home-sickness.
Mathematics came to be no more to me than the scaffolding of yester-
day's fireworks. . . . *Having no mountains* damned Paris absolutely in
my eyes. *Having clipped trees in its gardens* was the last straw.[9]

As a garrison town, Grenoble had a reputation for loose morals,
which was commemorated in the novel *Liaisons dangereuses* written by
the artillery officer Choderlos de Laclos. It was a banking town, with
around eight banks, in addition to a branch of the Banque de France, at
midcentury.[10] The traditionally important industry was the making of
gloves, and, in 1866, 45 percent of the city's 32,000 employees were
working in the glove industry, many of them in their homes. Light
industries included hemp and straw hats; heavier industries at midcentury were the manufacture of lime and cements. Traffic on the Isère
River, which reached a high point in the early 1840s, included wood,
iron, salt, and the wines of the Midi.[11] Railroads changed this pattern of
commerce, first by developing a Lyon-Marseille line and then by integrating Grenoble into the railroad network. At midcentury, travel by
rail still was expensive and slow, with a journey from Paris to Grenoble
taking some eight days.[12] Grenoble remained a fairly isolated town. The
Dutch painter Jongkind, who stayed in the Grenoble region between
1878 and 1892, portrayed Grenoble's mountains in impressionistic
watercolors suffused by his own intense emotions.[13]

At the end of the century the city was completely transformed by the
development of hydroelectric power, generated by mountain streams
and waterfalls. Grenoble became a major supplier of industrial motors
and electrical equipment, and metallurgy began to develop rapidly in
the 1880s into a national and international industry. This industrial development marched hand in hand with activity and growth at the university Faculties, especially in Letters and Sciences. During the years of
the Third Republic, while the city's population doubled (from 42,660 in
1872 to 95,806 in 1936),[14] enrollments at the university increased ten-
fold, from 340 in 1868 to approximately 3,000 regular students in 1930.
Of these, 500 were foreign students enrolled in science and engineering
courses associated with the Sciences Faculty. In addition, special
courses offered by the university for foreign students attracted another
2,068 students in 1928–29.[15] The isolated mountain town became a center for students from all over the world.

This outcome was unforeseen in the original organization of the

Napoleonic University Faculties following the Revolution, nor was this development planned by the decentralizing educational ministers of the Third Republic. A forward-looking administrator might have anticipated that the Grenoble Sciences Faculty would specialize in Alpine botany, zoology, and geology, as it did. But it was not to be expected that one of the Faculty's chemistry professors, François Raoult, would develop laws for a new discipline of physical chemistry and become an internationally known scientist in this field. Equally unlooked for at Grenoble was the indigenous evolution of the two new science-related fields of electrical engineering and electrometallurgy. We will look at these developments in some detail, first concentrating on the engineering specializations and international clientele that developed within the Sciences Faculty, and then considering Raoult's career and accomplishments.

EARLY SCIENTIFIC LEADERS IN GRENOBLE

The university's revival at Grenoble, begun under Napoleon, was not immediately successful. Three Faculties were established in 1808: Law, Letters, and Sciences. The Letters Faculty was abolished by the decree of October 1815, which suppressed seventeen French Faculties. Grenoble's Letters Faculty was among them, partly because the Letters dean had welcomed Napoleon too warmly when he marched through Grenoble the previous spring. J. F. Champollion, who taught history in the Letters Faculty until 1815, was Napoleon's enthusiastic supporter, as was the local prefect and noted mathematician Jean-Baptiste Fourier, who had accompanied Napoleon to Egypt. Perhaps it is not surprising that, in 1818, the Law Faculty also was suppressed, allegedly for its liberalism.[16]

Only the Sciences Faculty survived the Restoration. It was rejoined by Law in 1824 and by Letters in 1847, supplemented in 1866 by a School of Medicine and Pharmacy, which became a Faculty in 1894.[17] During the first three-quarters of the century, the only real higher education at Grenoble was in the Law Faculty. Indeed, from 1851 until 1870 the Letters Faculty awarded only 64 licence degrees and two doctorates, and the Sciences Faculty 29 licence degrees and no doctorates. In any given year, only a handful of students were engaged in university-level course work.[18]

The Lycée Impérial, which replaced Grenoble's école centrale, was considered by Fourcroy one of the best lycées in the Empire.[19] When the Faculties were first established in 1808, most of the professors were lycée professors. The Sciences Faculty consisted of three men: the lycée professors H. Bret and André Laurent Chabert in pure and applied mathematics and F. M. H. Bilon in physics and chemistry.[20] Stendhal has left us a dismal portrait of Chabert:

> most of the pupils in the mathematics class at the central school went to M. Chabert . . . a bourgeois who was rather badly dressed, but always looked as if he were in his Sunday clothes, and terrified of spoiling his coat and waistcoat and his nice cashmere breeches of gosling green.[21]

Chabert was "like an apothecary who knows a few good recipes; but there was nothing to show how these recipes arose one out of the other."[22] Still, classes with Chabert and the école centrale were superior to those he had with the priests:

> One day my grandfather said to the Abbé Raillane: "But Monsieur, why teach this child the Ptolemaic system of the universe, which you know to be false?" "But it explains everything; and besides, it is approved by the Church."[23]

Whereas the Law Faculty was installed sumptuously in the Palais de Justice on the left bank of the Isère, the Letters and Sciences Faculties were placed across the street from the lycée in the "Halle aux Grains" of a former Dominican convent. This facility was said to be an

> unsightly, dirty and misshapen shed; it is badly lighted, its surroundings on the Place de la Halle are narrow and dark, its façade on the lycée street produces the most ridiculous effect; the whole is uncomfortable and disgraceful . . . a baroque thing shocking at once both for its appearance and its smell.[24]

Despite its manifest drawbacks, the Place de la Halle remained the Faculties' home until the inauguration of a new university palais at the Place de Verdun in 1876.

Those Sciences Faculty members who wanted to do research worked together in a single laboratory, described later by Raoult:

> Each professor possessed his own little table; this room which was a laboratory during the daytime served simultaneously as the living quarters for the concierge. Here the physics professor arranged his instruments; the zoology professor dissected his rabbits and fed his

pigeons; the geology professor broke his pebbles and laid out his fossils; and the chemistry professor carried out all his operations.[25]

The Grenoble professors had few students to teach and few examinations to grade. Even baccalaureate examinations were scarce. From 1811 to 1852, the Grenoble Sciences Faculty pronounced only 111 admissions to the Sciences baccalaureate, or less than three per year.[26]

In 1848, the Grenoble professors inaugurated a series of evening public lectures for audiences of several hundred. Scientific topics were in vogue.[27] The listeners were lycée and Faculty students, workers, soldiers, and retired pensioners. They were crowded together into an old lecture hall,

> where tiers of seats, angled in a square, overlook a reserved space. At the left, before a large black table, some seats for the ladies. At the back, an immense fireplace, which displays a mass of retorts and twisted glassware under a protective hood. Before the fireplace, the table and the armchair of the professor.[28]

The first advertisement for popular courses (fig. 4)[29] was signed by Emile Gueymard, at that time the Faculty's dean. Gueymard was strongly committed to engineering science and civic duty, as well as to fundamental science. A graduate of the Ecole Polytechnique and the Ecole des Mines, he was a state engineer who came to Grenoble in 1824 in the Faculty's first chair of natural history. He established the Laboratory of Testing and Analysis, the first in the provinces, annexed to the Sciences Faculty;[30] took charge of the potable water supply for the city; helped direct construction of a suspension bridge over the Drac; and published geological studies of the Grenoble region.[31] For Gueymard, geology was "the first of the philosophical sciences. It means in our century what that universal philosophy which includes all human knowledge meant in the epoch of Plato and Aristotle."[32] With this view guiding him, Gueymard promoted the popularization of science.

When the Fortoul ministry inaugurated its applied science "certificate of capacity" program in 1854, the Grenoble Sciences Faculty decided to charge fees for the new program.[33] However, the minister insisted that both the public lecture and the smaller conférences must be free, although fees might be charged for laboratory work. Given the extra demands made on Faculty members, the municipal council decided to turn over 2,600 francs to the academy rector, to be distributed among professors teaching in Fortoul's new program.[34] The courses stopped in

1870, began again in 1874, halted once more, and recommenced in 1881, with some protest from Faculty members who thought they were being overworked and underpaid. Raoult complained that time was being taken away from his research,[35] and he received only 300 francs for teaching an evening municipal course in applied chemistry which met once a week for six months of the year.[36]

Course of physics, mechanics and chemistry applied to hygiene, agriculture and industrial arts. Encouraged and recommended by the Mayor and the Municipal Commission; in the amphithéâtre of the Place de la Halle. The courses will be made especially for workers and will have for their goal initiating the most practical kinds of knowledge to use for their work and for the conservation of their health. The courses to begin 16 March and continue Tuesdays and Thursdays, each week. Also M. Lory will make a course in applied chemistry once a week on Saturdays at 8 p.m.; and M. Grange, chargé de cours in physics, will make a course in applied physics Tuesdays and Thursdays at 8 p.m. He will discuss atmospheric variations, modern industrial discoveries, steam engines, railroads, electric telegraphy, etc.

Fig. 4. 1848 advertisement for popular lectures

Some rewards were soon to come to the professors, however. The municipality of Grenoble became the first city in France to construct a new palais for its Faculties. The city contributed 1.2 million francs to this project, the state 250,000 francs, and the département 50,000 francs. Planning for the new buildings began in the late 1860s, when agitation for revitalizing French science was just beginning nationwide. The formal inauguration of the new Faculties took place ten years later, in 1879.[37] The Sciences Faculty became increasingly vigorous and active, enthusiastic about the institution of state scholarships in the 1870s, and ready to publicize their current research in the annual list of publications discussed by Jules Ferry in 1881. That the ministry was newly stressing research as a Faculty duty was underlined at Grenoble by a doubling in the 1881 budget for "collections": "these collections are not to be museums; they are meant for teaching and personal work."[38]

When inaugurated in 1879, the Sciences Faculty had to meet the needs of only eleven fully matriculated licence candidates, taught by seven chaired professors—an enviable student/teacher ratio. Raoult was one of the seven professors; he had been teaching since 1867 and held the chair of chemistry since 1870. Other chair holders were Jean Collet in pure mathematics, Joseph Carlet in zoology, Charles Lory in natural history, and Alphonse Legoux in mechanics. The chairs of physics and botany were temporarily filled by chargés de cours.[39]

At this time, and for almost twenty years, from 1871 to 1889, Charles Lory was the dean and best-known member of the Grenoble Faculty. A native of Britanny and a normalien, he was highly respected by his colleagues. Like his predecessor, Gueymard, he directed the départemental testing laboratory attached to the Faculty, and he advised the municipality on Grenoble's water supply.[40] Lory's geological interests were more specialized than Gueymard's—he wrote monographs on the strata around Grenoble and the Dôle. In a study of the Cretaceous strata of the Jura he showed that a great freshwater lake had occupied the region after the retreat of the Jurassic sea, to be replaced by a vast marine invasion. His *Description géologique de Dauphiné* employed methods of both microscopic mineralogy and geology, and he developed a theory of longitudinal zones in the Alps, each zone with a different history, structure, and composition. In 1877, he presided over a Grenoble meeting of the Société Geologique de France, in which eighty geologists from different countries participated, and directed excursions for them into the mountains and valleys around Grenoble.[41]

When he replaced Lory as dean in 1889, Raoult was almost sixty years old and just acquiring an international reputation for his experimental work on the theory of solutions. In 1883 he had received the 10,000 franc La Caze Prize from the Académie des Sciences and was named corresponding member; in 1892 he was to be awarded the Davy Medal of the Royal Society in London. During his deanship, the vigor of the Sciences Faculty was channeled more and more into applied sciences, especially electrical technology, industrial physics, and metallurgy. The clientele of the entire university was beginning an exponential expansion, due principally to Grenoble's attracting engineering students and foreign students.

However, Raoult himself was not much responsible for the new direction the Sciences Faculty was taking. He was interested in pure research and not in applied science or in engineering teaching.[42] The rector, E. Boirac, described him in 1899 as holding

> back a little outside of the efforts underway for orienting the teaching
> of the Sciences Faculty in a practical direction, either because, as he is
> nearing retirement, he has less interest in the creations whose full

development he will not see, or because the nature of his personal work inclines him to prefer the researches of pure science.[43]

Still, Raoult's colleagues wanted him to continue as dean past the formal retirement age. And no doubt some wanted his continued support for the minority view that the role and highest mission of higher education is scientific activity and personal research, independent from demands for immediate application or the requirements of a teaching program.[44]

## THE GROWTH OF ELECTRICAL SCIENCE AND TECHNOLOGY

The movement of the Grenoble Faculty toward applied and engineering sciences was dramatic. Shortly after Raoult took over the deanship, the départemental general council initiated an effort to add a second chair of chemistry to the Faculty, in chemistry applied to agriculture and industry. The council voted 1,500 francs to this end. This proposal came up just as the Faculty learned that the ministry was considering creating chairs of industrial electricity in some Sciences Faculties. Janet proposed that his public evening lectures in physics, sponsored by the municipality, be transformed into a course in industrial electricity.[45] The chemistry chair proposal now took a back seat to industrial electricity among the Faculty, local interest groups, and the ministry. Although the request for a chemistry chair was renewed in 1900 by Raoult, who had long wanted an Institute of Chemistry at Grenoble, it was not until 1912 that a second chair of chemistry was established, and then with the emphasis in electrochemistry and electrometallurgy.[46]

Perhaps the development toward electrical physics and electrochemistry was inevitable, given the technological and economic transformation under way in Grenoble at the turn of the century. This development began to take dramatic, visible form in the large shafts and pipes that local industrialists linked to the high plateau of nearby Belledonne and other mountain areas after 1860. One shaft, with a drop of two hundred meters, was installed by a paper manufacturer in 1869, just as the adaptation to water power of a turbine dynamo was accomplished. In 1883, Marcel Desprez transmitted electrical current from Vizille to Grenoble, a distance of seventeen kilometers, and by 1895, hydroelectric power in one Grenoble area installation exceeded 1,000 kilowatts. (In 1906, Grenoble was exporting electric power to Lyon.) Hand in hand came the establishment by Paul Héroult of a factory for the electrochemical manufacture of aluminum at Froges in 1886, H. Gall's factory for chlorates manufacture at Prémont in 1894, and the production of carbides and sodium using the electric furnace developed by Charles A. Keller at Grenoble and Henri Moissan at Paris.[47]

While Grenoble had a couple of technical schools at which the "self-made men" of industry sometimes studied, including the Ecole Vaucanson and the Ecole de Voiron, the municipality now chose to support a more sophisticated program of industrial electricity directly affiliated with traditional university education. In 1892, funds from the city (2,000 francs), the départemental council (500 francs), and a group of industrialists (2,265 francs) helped underwrite the Sciences Faculty course in industrial electricity.[48] The list of private subscribers who committed themselves to three years of funding included individuals and firms in Lyon, Nancy, Grenoble, and smaller towns.[49]

In a speech to the municipal council in 1892, the mayor emphasized that only in Paris was there high-level teaching of electricity, with one course at the Conservatoire des Arts et Métiers and a second at the Ecole Centrale des Arts et Manufactures. Only the latter had a laboratory for electrical studies. Special institutes were devoted exclusively to electricity in Zurich, Liège, Ghent, London, and Berlin. By creating a course in industrial electricity at Grenoble, the mayor said, we will have begun the establishment of an institute in keeping with "the modern conception that we ought to have of higher education and its role."[50]

When Janet left Grenoble for Paris, his teaching was taken over by Joseph Eugène Pionchon, who lectured to twenty listeners in 1895–96 and sent reprints of his lectures to 155 subscribers.[51] He encouraged the Sciences Faculty council to propose the creation of a university diploma in industrial electricity, stressing that the level of study required for the diploma should be fully comparable to that for state certificates of higher studies; only the admission criteria would be different. It was expected that students would have a mathematical knowledge comparable to that of higher mathematics taught in lycées.[52] The curriculum included electrochemistry and industrial design as well as visits to factories.[53]

Support from the university and regional communities was strong. The Law Faculty signed a recommendation in 1898 that "the greatest share of the resources of the University be devoted to the development of the teaching of industrial electricity."[54] Correspondingly, the university borrowed 65,000 francs to refurbish a wing of the Lycée des Jeunes Filles for the Sciences Faculty. Here in 1900 the newly named Institut Electrotechnique was housed, along with laboratories for general physics and for the PCN program. This new installation was assisted by two educational ministry grants for 10,000 francs each for laboratory equipment and by the commitment of a 1,500-franc yearly subsidy from a new regional society of industrialists.[55] The budget for the institute became separate from the budget for the Sciences Faculty in 1904.[56]

The Institut Electrotechnique was not to stay long in these quarters. Casimir Brenier, the president of the chamber of commerce, donated 5,000 square meters of land on the western side of the city center, near the Isère, for the construction of a new set of buildings. The city's old hydroelectric factory, with dynamos, steam engines, and turbines in place, became a laboratory for the institute, which was renamed the Institut Polytechnique in 1912.[57] In addition to industrial electricity, teaching programs were worked out in hydraulics, electrochemistry, and metallurgy. In 1905, a chair of industrial physics was created for Louis Barbillon, who had taken over Pionchon's course when he moved to Dijon in 1904.[58]

When the Institut Electrotechnique was formally inaugurated in 1901, there were eighteen students enrolled in its course work for 1900–01. By 1907–08 there were 143; in 1912–13, 368; in 1916, only 172 because of the war; and in 1920, 1,427.[59] In the early 1900s, the Sciences Faculty prided itself on admissions examinations that were comparable in difficulty to those of the Ecole Centrale in Paris. In 1908, graduates of the Ecole Polytechnique and the Ecole Centrale in Paris were enrolled for further study in Grenoble.[60]

A continued strength of the school was that its graduates never lacked jobs. In 1906, the graduates of the Institut Electrotechnique were all hired within a few days after their examinations; the following year, they had positions even before taking examinations.[61] In the 1920s, 30–40 percent of the graduates took jobs in the Alpine region and another 30–35 percent took jobs in Paris; others, we may assume, returned to their native countries, since many were foreign students.[62]

By the late 1920s, when jobs were becoming scarcer, a decision was made to decrease the number of admissions and diplomas in electrical engineering by upgrading admissions standards and the quality of training. The number of students was reduced, so that by 1925 there were only 600 in the electrical engineering program.[63] This elitist strategy was similar to admissions procedures at the grandes écoles, where, as we have already noted, there was a 20 percent increase from 1900 to 1932 in the number of admissions candidates and a 15 percent decrease in the number of admissions.[64]

While the earliest development of engineering education at Grenoble was in electrical engineering, associated with physics, other disciplines became strong as new components of "polytechnical" offerings around 1906–07. Among these was hydraulics, first taught as a course in 1906, and developed into a laboratory program in 1908. An institute specializing in paper research, especially cellulose chemistry, was annexed to the Institut Electrotechnique in 1907, at the initiative and expense of the Union des Fabricants de Papiers de France. At this time, only two such schools existed in Europe—one at Vienna and another at Manchester.[65]

The third disciplinary area of applied teaching and research was electrochemistry and electrometallurgy, established as a course in 1906, with a laboratory in 1908, and a chair (for Georges Flusin) in 1912.[66] Raoult, who did his doctoral work on a topic in electrochemistry, provided a distinguished antecedent for this specialty, if entirely on the side of fundamental science. It is interesting to compare the establishment of the chair of electrochemistry in 1912 with developments in Paris. In 1900, Sorbonne scientists discussed changing the title of Louis Troost's chair from "mineral chemistry" to "electrochemistry" after his retirement. The Sorbonne's professor of organic chemistry, Albin Haller, thought the change should be made, commenting that Nancy already was superior to Paris in this field and that many foreign universities now had institutes of electrochemistry. (Walther Nernst's Institut für Physikalische Chemie und Elektrochemie had been established at Göttingen in 1894.)[67] Alfred Ditte, the Sorbonne's other professor of mineral chemistry, agreed with Haller.

The Paris Sciences Faculty already was considering Henri Moissan, whose research was in electrochemistry, for its vacant chemistry chair. But the dean was not persuaded by Haller's and Ditte's recommendations, arguing that the three present chemistry professors could share the teaching of electrochemistry and that it was a branch of physical chemistry, already represented at the Faculty in a course taught by Jean Perrin. Emile Picard offered his view that "there is a place at the Paris Faculty for two chairs of pure chemistry" (by which he meant mineral chemistry).[68]

At Grenoble, there was no such dissension. The program in electrochemistry and electrometallurgy was closely tied to the Institut Poly-technique and supported by local industrialists. During the war, Flu-sin's laboratory provided one of the few electric furnaces available in France, in addition to Moissan's in Paris, for the synthesis of nitrogen products, liquid chlorine, and aluminum alloys. Corporations began providing scholarships and subsidies to the laboratory during the war, when government and university subsidies were lacking.[69] At the war's close, the Ministry of Public Works and Transports began making large grants to the university: 240,000 francs for research in hydraulics and 250,000 francs for the establishment of the Institut de Electrochimie et Electrometallurgie (IEE). By 1921, the newly established IEE had re-ceived a total of 1.6 million francs from the Ministry of Public Works and the Ministry of Public Instruction.[70]

Construction of the IEE took almost ten years. The buildings were inaugurated in 1931 on a site across the city from the Institut Polytech-nique, on a former field of maneuvers for the army's Alpine artillery division.[71] Like the Institut Polytechnique, the IEE was separated into two divisions, one giving a more elementary degree than the other. In 1930, the number of annual admissions to the higher division was fixed at thirty students, with priority given to graduates of the grandes écoles and to students with an engineering diploma and some course work in the licence curriculum of the Sciences Faculties. If thirty places were not filled by candidates with these qualifications, a concours was held for the remaining places.[72] IEE "research" was to include "both researches of a purely scientific order and researches directly interesting industry, remunerated by the interested party."[73]

## FROM MOUNTAIN VILLAGE TO INTERNATIONAL STUDENT CENTER

The interest in industrial electricity and electrochemistry dovetailed with a universitywide movement to attract foreign students. This began in the 1890s, at a time when French educators in general began to wor-ry that too many foreigners preferred German universities to French ones. In 1896, two of Grenoble's Letters professors organized a Comité de Patronage des Etudiants Etrangers. They mailed out five thousand circulars across Europe to announce special courses for foreign stu-dents. Initial enrollments were meager—five students in 1897—but in

1898 Grenoble launched a veritable educational revolution in establishing *cours de vacances* just for foreigners. These vacation courses were to take place from July through October, in four series of one-month sessions. They included practical courses in the French language and small lectures on contemporary French intellectual and social history and current French politics. In the Sciences Faculty, the botanist Paul Lachmann and the geologist Wilfrid Kilian participated in the summer-course plans by organizing Monday scientific excursions around Grenoble. Students completing a summer examination received a "certificate of French studies" under the seal of the University of Grenoble, and students who did not want to present themselves for examinations received a document attesting their residence at the Faculties.[74]

One hundred ten foreign students enrolled at Grenoble for summer 1899. By 1913, there were 1,200 foreign students in the cours de vacances, and by 1925–26, 2,371 foreign students enrolled, before the unsettling political events and economic crises of the late 1920s began to take their toll.[75] In 1900, Grenoble became the Letters Faculty most popular with foreigners, after Paris.[76]

Many foreign students enrolled, too, during the regular semesters in the Faculties. For example, during the academic year 1910–11, 1,420 foreign students enrolled at Grenoble, but of these students, only 768 enrolled in the summer courses. In 1912–13, the Sciences Faculty included 110 foreigners among its 502 students, with most of the foreign students among the Institut Polytechnique's 368 students. Half of the 110 foreigners were Russians.[77]

After the war, as might be expected, the numbers of Russians and Germans at Grenoble declined. But in the late 1920s, the Russians once more could be found in appreciable numbers in the Sciences Faculty. In 1928–29, for example, of 479 foreign students there, 155 were Polish, 50 Russian, 50 Rumanian, and 27 Bulgarian. Grenoble also had 28 Chinese students.[78] Only Paris, Toulouse, and Nancy enrolled more foreign Sciences students than Grenoble, and the university as a whole was second in foreign enrollments only to Paris.[79]

In attracting foreign students, the University of Grenoble defined an "original style"[80] in French university education, with significant influence both in spreading the values and methods of French culture and science and in opening up contacts with intellectuals, scientists, engineers, and industrialists abroad. In 1912, the year the Institut Polytechnique was inaugurated, the Grenoble university rector Charles Petit-Dutaillis remarked that "there is almost no part of the world today which does not make appeal to Grenoble's students, especially those of the [Grenoble] Ecole Polytechnique."[81] France's export of science and technology was undergoing a Saint-Simonian revival.

It must be noted, too, that Grenoble's success was a surprise to those who made educational policy in Paris. When the Institut Electrotechnique was inaugurated in 1901, Louis Liard was present. "Ten years ago," he said, "if anyone had told me that Grenoble was going to be an international foyer of instruction, I would have smiled and remained skeptical. However, you have brought about this miracle."[82] Citing the founding of technical schools abroad, Liard praised Grenoble for its innovations in industrial electricity and expounded his views on the general tasks ahead for French science and engineering.

> To my mind, the Faculties have too many doctors and lawyers. . . . it is necessary that youth turn toward science. . . . Science has remained purely theoretical for too long a time. It should be adapted to industry, and French industrialists are beginning to understand this. . . . Last year, when I visited the installation of technical courses at Liège, the rector, who accompanied me, made the following remark to me à propos of the Franco-Russian alliance, a remark full of truth—and moreover of irony: "The French supply money to the Russians, but it is the Belgians who supply them machines."[83]

Liard wanted to renew the Saint-Simonian tradition of France's exporting scientific and technological expertise. His vision for Grenoble was amply fulfilled in the course of the next decades. The number of patents in the region constituted the great majority of all départemental patents during the period 1895–1940, many of them filed, to be sure, by people who did not have any higher education.[84] The scientific and engineering specialties of the university's institutes and Sciences Faculty closely meshed with the region's leading sectors of industrial expansion: hydraulics and electricity, electronics and chemistry, and "the atom." By 1962, 42 percent of Grenoble's "workers" were foremen and skilled workmen.[85] The university has had the strongest expansion rate of all French universities, growing from 4,200 students in 1950 to 27,000 students in 1976. Of these, less than one-third were originally from the Grenoble region; 3,400 were foreigners in 1976. Under the leadership of Louis Néel, Nobel Laureate in physics in 1970, Grenoble is a center of nuclear studies and the site of a powerful neutron accelerator, where French and German engineers have been joined by British and other foreign scientists since 1974.[86] In short, the mountain town has become a crucial scientific, technical, and commercial center of national and international stature.

APPLIED SCIENCE VERSUS FUNDAMENTAL SCIENCE

We may ask whether this achievement, at least before the Second World War, was realized at the expense of fundamental science. If we compare the kinds of prelicence preparatory certificates offered at Grenoble in 1900 and 1928, we find very little changed. The only program that disappeared after 1900 was in fundamental science—astronomy. New programs were general mathematics, electrotechnology, and metallurgy.[87]

In the natural sciences of geology, zoology, and botany, the Grenoble Sciences Faculty continued to train some students interested in fundamental research. In the period 1900–1924, five doctoral degrees in geology and two in zoology were awarded to students who became professors in higher education. But during this period, no state doctoral degrees were awarded in mathematics, physics, or chemistry to students who went on to teach in higher education in France.[88]

In geology, Kilian was a distinguished successor to Lory. He was appointed to Lory's chair as soon as he was thirty years old and held it from 1892 to 1925.[89] Kilian agreed with the general technical orientation of the Faculty and suggested in 1901 that the teaching of mineralogy be slanted toward applied mineralogy, to the benefit of engineers, agriculturalists, and industrialists.[90]

The thrust of Kilian's personal interest, however, was fundamental research, specifically, tectonics and the formation of the Alps, and he was elected to the Académie in 1919 as a nonresident member.[91] He accepted the Grenoble deanship in 1918 but resigned two years later on the grounds that its duties interfered with his research.[92] In 1921, he was awarded the Prix Gaudry, and his laboratory became attached to the Ecole Pratique des Hautes Etudes four years before his death. One of his last research students was an American, Gayle Scott, from the University of Texas.[93]

Zoology and botany did not fare as well as geology. The botanist

Charles Musset had obtained some notoriety in the 1880s as a result of his part in the controversy with Pasteur over spontaneous generation. His successor, Lachmann, through the support of the Société des Touristes du Dauphiné, created in 1893 the first Alpine garden in France, in the high plateau of Belledonne, at an altitude of 1,850 meters. Lachmann, who held the botany chair from 1893 to 1907, fought hard to obtain Faculty funds for this garden and for two others,[94] arguing that industrial electricity was not the only discipline requiring a "technical" laboratory: there were also the Alpine gardens and pisciculture. But he managed to squeeze out only 600 francs for his gardens in a Faculty meeting of 1900, when a 10,000 franc subsidy was discussed for technical education.[95]

Lachmann's successor, Marcel Mirande, organized a new botanical laboratory, abandoned the first two Alpine gardens, and relocated and improved the Jardin du Lautaret with aid from the Touring Club of France and other organizations.[96] The zoologist Louis Léger managed to found the Institute for Zoology and Pisciculture,[97] with one successful project being the restocking of Alpine rivers.[98]

As for physics and chemistry at Grenoble, the emphasis on applied physics and chemistry was overpowering after 1900. The teaching of physics was in a peculiar situation because its chaired professor, Pionchon, for reasons of "health and family," worked out an arrangement to exchange teaching at Dijon with Henri Bagard, a lecturer and adjoint professor there. When this happened, Grenoble's lecturer in physics, Barbillon, took over Pionchon's course in industrial electricity at the Institut Electrotechnique. In 1905 this course became a chair of industrial physics, as we have seen.[99] But Pierre Vaillant, who became associate (associé) professor in general physics, did not have the authority to organize a decent laboratory.[100] The situation was not resolved until 1919, when Pionchon formally received a now vacant chair in general physics at Dijon and Vaillant was named to Grenoble's chair.[101]

In contrast, Grenoble had two chemists with long tenures: Raoult, from 1867 to 1901, and André Recoura, from 1901 to 1932. Raoult's distinguished work and career, to be discussed in detail, brought great prestige to Grenoble. His successor, Recoura, was a very good chemist, more interested than Raoult in applied chemistry. Recoura studied complex salt compounds, which are anomalous in their behavior by not responding to tests for their constituents. His particular interest was the chromium complex salts (an example of which is $CrCl_3(NH_3)_5$), which neither show a qualitative test for chromium nor precipitate their chlorine atoms with silver nitrate. Recoura is not as well known in this field as his Zurich colleague, Alfred Werner, a Nobel Prize winner in 1913, but in the same year that Werner was awarded the Nobel Prize, Recoura also was nominated.[102]

This does not mean that research was not going on in fundamental physics and chemistry, but the doctorates awarded were university degrees, not state doctorates. In 1907, for example, eleven university doctorates were awarded in chemistry.[103] In addition, doctoral research might be carried out in Grenoble laboratories, but defended in Paris. This was the case, for example, for the chargé de conférences in electrochemistry, M. Andrieux, who won the Prix Ancel of the Société Chimique de France in 1929 for his Paris doctoral thesis on the electrolysis of metallic oxides.[104]

This result was not simply the effect of enthusiasm for industrial and applied science. There was little real choice about the direction expansion and specializations could take, given the interest of private donors in funding applied or engineering education and research and the state's unwillingness to increase, or indeed to continue, strong funding for basic science. While the number of candidates at Grenoble for the licence certificates quintupled from 1910 to 1920 (from 106 to 584), the number of personnel increased only by one professor (the second chair of chemistry), one lecturer (geology), and one aide.[105] In addition, and most telling, there were only small numbers of students in postlicence mathematics and physical sciences because of decisions by the educational ministry which insured that Paris would remain the center of study for the agrégation.

The Grenoble Sciences Faculty was unhappy with the 1903 reform attaching the Ecole Normale Supérieure to the University of Paris:

> the transformation of the Ecole Normale Supérieure and the modifications introduced in the agrégation has had as its effect the centralization almost exclusively at Paris of the preparation of these concours, depriving us thus almost absolutely of our very precious clientele of scholarship students. In other times this would have resulted in incurable anemia for us; at this moment, it is to the Institut Polytechnique that we owe our ability to maintain scientific high culture here in a state of prosperity.[106]

When the ministry encouraged provincial Faculties to specialize in one or more disciplines for the agrégation,[107] Faculty interest often shifted toward expanding the bread-and-butter curriculum. For example, when Jean Collet retired from the chair of infinitesimal analysis in 1919, the Grenoble Faculty favored a transformation of that chair into one of "general mathematics," a lower-level preparation needed by students who were not aiming toward the mathematics agrégation. Appell, in the ministry now, warned Kilian, Grenoble's new dean, that if Grenoble asked for an additional mathematics chair in general mathematics, the Faculty would lose altogether the chair in infinitesimal analysis.

Once Emile Gau was formally named to the chair of infinitesimal analysis, he informally worked out an arrangement with Janet's son, now lecturer in mathematics, that Janet would teach infinitesimal analysis and Gau would teach general mathematics.[108] Gau claimed that there was little hope for teaching higher mathematics in Grenoble, given the problem with the agrégation; that Grenoble had worked out a path in the direction of applied science because of its situation and the kind of students who were drawn to the Faculty; that higher mathematics should be kept in the Faculty for the few who want it, but that it was necessary to develop the teaching of general mathematics, which interested a larger number of students.[109]

Inevitably, confusion and dissension arose at Grenoble about the intellectual relationship between fundamental science and engineering science and about the interconvertibility of Faculty and institute academic programs. As we have seen, the Ministry of Education began an effort after the war to impose order on the nationwide mosaic of programs that had developed more or less independently at Faculties around the country. In 1924, the ministry drew up a master list of institutes, chairs of applied sciences, names of personnel, and certificates and degrees granted under the names of the different French universities.[110] A principal aim was to establish "equivalencies" among the diverse university and state programs. Grenoble initially resisted this aim, recommending instead a repeal of a 1922 decree that granted a certified *ingénieur des arts et manufactures* the equivalence of a licence-preparatory certificate in physical mechanics, applied chemistry, or general electrotechnology. Since these engineering students, when first admitted to the Paris Ecole Centrale, automatically were granted the equivalence of a certificate of general mathematics by a decree of 1917, this meant that the Ecole Centrale engineers need actually prepare only one certificate of higher studies in order to become *licencié ès sciences*. Grenoble protested that equivalencies should only be used to enable students to dispense with certain work while studying for a higher certificate, not to get the certificate itself. In addition to questioning the intellectual exchange value of traditional scientific education and engineering education, the Faculty perceived the new equivalencies among institutions as measures undermining university autonomy.[111]

But by the end of the decade, in 1930, the Sciences dean, René Gosse, had become accustomed to the equivalency notion. Perhaps because jobs were disappearing in industry, some of the university's engineering graduates were interested in enrolling in the Faculty to study for the licence. The issue was whether these electrical engineers and hydraulic engineers should be admitted to the Faculty without the baccalaureate diploma. Reminding his Faculty council that for the past two years the engineers were tested on French composition in their final examinations, Gosse recommended that those students with a high score on this section be deemed of sufficiently "general culture" to enter into the state licence program.[112]

Dean Gosse's recommendation proved unpopular with members of the Grenoble Sciences Faculty. Parisians and provincials alike were imbued with the tradition of "general culture," associated with the baccalaureate, as a prerequisite for higher education and for professional careers. While Greek became less important for the baccalaureate preparation, in the mid-twentieth century the Latin-Sciences baccalaureate still remained the favored background for a serious scientific career.[113] Engineers, as in the nineteenth century, were accorded greatest social and intellectual prestige if they had gone the route of the lycées. And while the Grenoble Faculty continued to welcome foreign students in increasing numbers, its professors were just as unwilling as their Parisian and ministerial colleagues to integrate foreign scientists into the French teaching establishment.

Where the Grenoble Sciences Faculty differed most clearly from its Parisian counterpart was in its willingness to pursue practical applications of science as part of scientific education and research. The influx of foreign students interested in these kinds of science programs contributed to a new cosmopolitanism and a rapid scientific and technical development that transformed Grenoble from an isolated mountain town to an international science center in modern times.

FRANÇOIS RAOULT AND THE RISE OF PHYSICAL
CHEMISTRY AS A DISCIPLINE

Of those scientists who taught at Grenoble in the late nineteenth century, the most remarkable was François Raoult, who rose from obscu-
rity to scientific fame in the 1880s and 1890s. For many years, he was
better known and respected in Germany, the Netherlands, Sweden,
and England than in France. He was one of three French scientists and
the only provincial French scientist to serve on the first editorial board
of the *Zeitschrift für physikalische Chemie*. Had he lived past 1900, it is
likely that he would have been considered for a Nobel Prize in chemistry, because his research played a crucial role in the development of the
new "ionist" physical chemistry identified with the school of van't
Hoff, Arrhenius, and Ostwald. This new discipline won four of the first
ten Nobel Prizes in chemistry.[114]

J. H. van't Hoff, the first Nobel Laureate in chemistry, described the
international renown that came to Raoult:

> [He] lived in that somewhat out of the way town, Grenoble. . . . the
> romance of his life was that almost sudden rise to fame, spreading
> from this nearly unknown corner, first over the frontier of his country,
> and then back to France, which made him one of the most prominent
> men of science of his age.[115]

As is true today, most chemists knew of Raoult's work through a textbook. Editions of Wilhelm Ostwald's *Lehrbuch der allgemeinen Chemie*
(1885) made ample use of Raoult's experimental results regarding
changes in the freezing points, vapor pressures, and boiling points of
solutions. Most nineteenth-century physicists knew of Raoult's work
through van't Hoff's thermodynamic publications on osmotic pressure,
in which van't Hoff drew a powerful analogy between the behavior of
gases and solutions.

In a 1902 textbook, Harry Clary Jones, an American physical chemist
who taught at Johns Hopkins University, wrote:

> It is now generally known that within the last fifteen years a new
> branch of science has come into existence. This branch, occupying a
> position between physics and chemistry, is known as physical chemis-
> try. . . . The new physical chemistry really begins with the chapter on
> solutions.[116]

Looking in some detail at Raoult's work on solutions is important for
better understanding the development of physical chemistry. In addi-

tion, his career stands out as one of the most important among French provincial scientists of the nineteenth century. Because Raoult's career was in many ways typical, it sheds light on the careers of provincial scientists, in general, who moved up the ranks from collèges and lycées into Faculties and, because his career is also one of unusual achievement, it helps illuminate the process of conceptual innovation and the emergence of a new discipline in the French scientific community.

Raoult came to the Sciences Faculty at Grenoble in 1867 after teaching mathematics and physics in lycées for fourteen years. When he was appointed at Grenoble, his career was undistinguished, although his doctoral work in electricity was later praised for an originality that went unrecognized initially. As a student, he had none of the advantages of scholarships or patrons and all the disadvantages of having to work for a living while moving from lycée to lycée in the years he was doing a doctoral thesis. The thesis was defended at the Sorbonne under the directorship of Jules Jamin, who had little contact with Raoult during the period.

Raoult was born May 10, 1830, in Fournes, in the Département du Nord. His father was an employee in the local customs service of Villers Cotterêts and intended that his son enter the Bureau de l'Enregistrement. However, young Raoult persuaded his parents to let him study in Paris. There he gained access to laboratories, and in 1853 his paper on the transport of electrolytes and electrical endosmosis was presented at the Paris Academy of Sciences by H. Becquerel and Pouillet. Raoult later told William Ramsay that he experimented in this period with the passage of electricity through solutions, before learning what already had been done. Friends told him that his results were well known as Faraday's and Ohm's "laws." Raoult recalled to Ramsay that he "took comfort in the thought that if he were able to make such discoveries, with an importance already universally recognized, he must also be able to advance science in other directions." He soon had to leave his work behind, however. His discouragement surfaced in the concluding lines of his 1853 paper: "I leave to others more fortunate than I the care of leading science ahead in the new path that I have begun to open up."[117]

In need of money to support himself, Raoult took a position at the Lycée de Rheims in 1853, teaching two lessons of mathematics and one lecture each week. His salary was only 700 francs, for he did not have the licence in physical sciences until 1855.[118] During 1855, he taught physics and chemistry at the Collège Saint Dié; he then returned to Rheims, where he remained until 1860, moving up through the ranks in the lycée from substitute for the adjoint professor of physics, to adjoint professor, to chargé de cours. In winter 1861, he was transferred to the lycée at Bar-le-Duc. He took a leave of absence for reasons of health in 1861–62 and requested a transfer to a lycée closer to Paris, where he soon would be defending his thesis. Transferred to Sens in 1862–63, he completed his thesis and expressed the desire for a position in a university Faculty.[119]

His teaching at the Sens lycée was not entirely successful in the view of his administrative superiors. In 1864, the ministry's inspector reported that Raoult lacked "method" in teaching, that his lectures often were vague and indecisive. The inspector claimed that Raoult preferred laboratory research to his teaching duties and concluded with a recommendation for kicking Raoult upstairs: he should be advanced to a Faculty chair.[120] The minister wrote Raoult in July 1864, not mentioning the alleged penchant for research but cautioning Raoult against sacrificing his students' performance on examinations to his interest in public lectures:

> The inspectors-general have told me that the courses confided to you at the lycée have not presented entirely satisfactory results this year, and that your students have not appeared sufficiently prepared on the difficult parts of science. I fear that you have spent too much time on the preparation of your public lectures. . . . You must not lose sight of the fact that your first duty is to assure the progress of your students.[121]

The principal obstacle to moving Raoult to a Faculty was the fact that he did not have the proper credentials. The rector of the Dijon academy wrote the ministry that

> M. Raoult does not have the licence in mathematical sciences, and it seems to me that, even for work in chemistry and especially in physics, he will feel the need one day or other for the kind of knowledge that the licence presupposes. Also M. Raoult is not agrégé, and while the agrégation is not obligatory for higher education, I like to see the Faculty professors equipped with the agrégation diploma, which seems to me more serious than that of the doctorate. I leave it to you to decide, but I think it would be in Raoult's interest to be pushed first into the mathematical licence and the agrégation.[122]

Equipped with a research degree but not with the agrégation, Raoult's chances for success in higher education seemed limited.

While Raoult never took the mathematical licence, by 1867 he passed an agrégation examination associated with Duruy's program in special secondary education. He asked to be considered for a Faculty position at the Ecole Normale Spéciale at Cluny, or at a university Faculty.[123] At Sens, he continued to give public lectures, no doubt to supplement his income. His courses proved very popular, attracting more than a thousand listeners in physics and agricultural chemistry. This seems, in the end, to have contributed to the positive decision to transfer him from a lycée to a Faculty. As Hervé Faye, the astronomer and general inspector, recommended, "I believe that his true direction is not secondary education. He will be much more successful as professor at a Faculty. The astonishing success which he has obtained at Sens, where his lectures have sometimes brought together 1,800 listeners, is a palpable proof."[124]

In 1867, Raoult was appointed to teach LeRoy's courses in chemistry at Grenoble, after failing to get positions in physics at Rennes in 1865 and Clermont-Ferrand in 1866. He married Pauline Truet, and they had a child. In 1868, a second infant, a daughter, was born.[125] Raoult continued to work on problems of electrolysis, heat, and chemical action, trying to separate the effects in batteries of chemical reactions and physical processes. The appointment at Grenoble brought him a laboratory, as well as a salary increase from 1,200 francs to 3,000 francs. He continued to supplement his salary by teaching popular lecture courses and routinely carried out industrial and agricultural sample analyses at the départemental laboratory founded by Gueymard, for which he received no remuneration.[126] In 1870, the Grenoble Sciences Faculty supported Raoult's candidacy as the permanent replacement for LeRoy in chemistry on the grounds that his research was "as interesting for chemistry as physics."[127] A. J. Balard (who held three chairs of chemistry simultaneously in Paris) recommended that Raoult be transferred to a physics chair at Poitiers so that a "proper chemist" could be named at Grenoble, but the minister followed the Grenoble Faculty's warm recommendation, which was supported by Jules Jamin.[128]

Like so many French professors, both Parisian and provincial, Raoult continued to take on extra teaching duties to supplement his income and fulfill general responsibilities agreed on by the Faculty. He taught one of the Faculty's municipal evening courses (in applied chemistry) once a week for six months of the year, for a stipend of 300 francs.[129] In 1873, he also began teaching a chemistry and toxicology course in Grenoble's School of Medicine and Pharmacy, for which he received 2,500 francs, in addition to his Faculty salary of 8,000 francs, in the mid-1880s.[130] He usually had seven to ten students enrolled for three hours of conférences each week, about thirty students in a public lecture at the Sciences Faculty, and about twenty-five students in his Medical School course, which met twice a week.[131]

Raoult's teaching position in the School of Medicine caused some resentment when he took a leave of absence from his teaching duties there in 1887, while drawing 1,000 francs in sick leave pay and continuing to teach in the Sciences Faculty. The reason given for the sick leave was laryngitis, but his "elocution" at the Sciences Faculty was reported to be "very good" during this period. In 1892, he resigned from the School of Medicine after its director and the academy rector consulted with the ministry on the matter. The rector said he appealed directly to the ministry because he did not want to oppose Raoult, given his "high administrative and scientific position."[132]

In 1892, Raoult was dean of the Sciences Faculty, a position to which he was elected unanimously by his colleagues in 1889. As we have seen, he presided over a Faculty which launched an intensive effort in the applied sciences, especially electrical sciences and engineering, in the 1890s. His scientific reputation was blossoming. In 1888 he received the La Caze Prize in chemistry from the Paris Academy of Sciences, an award of 10,000 francs, which was roughly equivalent to his yearly income. And in 1895 he received the annual prize of the Institut de France, which carried an award of 20,000 francs. With his promotion in 1889 to "first class," he received a raise in Sciences Faculty salary to 11,000 francs, supplemented by 1,000 francs for the deanship and 1,000 francs from the School of Medicine (which he lost in 1892).[133] Given these figures, one gains some insight into the considerable prestige of the prizes he received for his research.

Raoult's concern with his salary and the general practice of both Parisian and provincial professors to double up on teaching duties to supplement their primary income were not unusual. As we have noted earlier, there were two salary scales in France, one for Parisian professors and one for provincial professors. A decree of 1881 established four classes of provincial professors, and advancement from one class to another was based on present salary, seniority, and ministerial discretion. In 1881 there were 110 professors or chargés de cours in the provincial Faculties: ten in first class at 11,000 francs; eleven in second class at 10,000 francs; fifty in third class at 8,000 francs; and thirty in fourth class at 6,000 francs. Advancement took place when a vacancy occurred in one of the classes, and salaries did not vary from the specified amounts. Placed in first class in 1889, Raoult remained at the same salary until his death in 1900. The only ways to increase one's income were to do extra teaching or move to Paris; moving from one provincial Faculty to another made no difference in salary or class.[134]

Raoult's colleagues Boirac and Kilian later commented that Raoult's "place était marquée à Paris" but he preferred to stay in the calm and "hardworking" environment of Grenoble. However, Raoult's possibilities for advancement to Paris were lessened by his lack of the regular agrégation diploma, a disadvantage that, as we will see, was later to plague his young provincial colleague at Lyon, Victor Grignard. When the establishment of a permanent course in physical chemistry was discussed at the Sorbonne in 1898, Raoult's name was not brought up. The reason for this might have been that he was so near the age of retirement, or it might have been that his chemical work was too closely identified with the German school of physical chemistry.

Raoult's most important work was carried out at Grenoble when he was in his forties and fifties. A ministerial report in 1875 placed little faith in him, saying that while his publications had been noticed by both physicists and chemists, his research lacked "powerful initiative" or importance.[135] But his long-standing interest in the relationship in electrolytes between chemical and physical forces was to take an important new turn about 1877 and 1878. This new direction can be linked, I think, to applied research on solutions of ammonium gas which he carried out at Grenoble, to the interests and research of some of his Grenoble colleagues, and to contemporary debates at the Academy of Sciences.

On his arrival in Grenoble, Raoult had continued with the problems of his thesis, publishing a summary of his views on the difference between voltaic and chemical heat in the *Bulletin* of the Société Statistique de l'Isère in 1870. In this and earlier publications, he distinguished three heat-related processes occurring in a solution, claiming that only one of these produces an electrical effect. He described these processes as (1) *désagrégation*, by which he meant vaporization, melting, or dissolving; (2) diffusion of the salt molecules in the solvent; and (3) combination of the salt with water, chemically. The first two processes are similar, consisting of "separation of similar molecules," while the third process occurs between unlike molecules and produces measurable electricity or heat.

Raoult found the unexpected result that values for heat due to chemical reaction are generally different from heat due to electrical energy. This contradicted the generally held assumption that in the simplest electrical cell the chemical energy freed by chemical reaction passes over exactly into electrical energy. It was more than a decade after Raoult's experimental findings that J. Willard Gibbs at Yale University independently showed from a theoretical standpoint that an element may either take up heat from the surrounding medium or give out heat, so that the electrical energy produced in the cell is, then, equal to the chemical energy that has disappeared, plus a term that is proportional to the electromotive force and to the absolute temperature. Helmholtz arrived at this result in 1882, about the time Ostwald and other European scientists discovered Gibbs's results. It was only later that Raoult's work could be seen to fit in with these results, and van't Hoff was to comment that Raoult's work on electromotive force appeared at least ten years too soon.[136]

Raoult's research was soon influenced by his colleagues and duties at Grenoble, in a direction that initially seemed to be at odds with his earlier work and then to nicely dovetail with it. Like some of his Faculty colleagues, he belonged to the Société Statistique de l'Isère. He may have heard there a lecture demonstration by Claude Alphons Valson in the late 1860s. Valson taught mathematics at Grenoble from 1858 to 1877, before taking a position at the Catholic Institute in Lyon. He was interested in capillary action as a manifestation of molecular processes and, according to Ostwald, was the first scientist to describe any properties of salt solutions as additive properties. He and his collaborator, A. P. Favre, did this for results they had obtained for the surface tensions of solutions in capillary tubes. Raoult later quoted their 1872 conclusion that the process of "solution results in giving to the elements of the dissolved bodies a reciprocal independence."[137]

Another colleague at Grenoble whose research may have interested Raoult was Jules Violle, who taught physics there in the 1870s, before moving to Lyon in 1877 and to the Ecole Normale Supérieure in 1884. Violle's major project at Grenoble was the design and use of an actinometer to measure the "solar constant," the sun's radiation in calories per square centimeter per minute. The temperature of the sun may be calculated indirectly from this value. To arrive at the relationship between temperature and energy radiation, Violle made numerous studies of the specific heats of metals at high temperatures, comparing experimentally found increases in energy to theoretical extrapolations from the law of Dulong and Petit. Violle was doing this work precisely at the time when the validity of Dulong and Petit's law relating specific heat and atomic weight was under discussion at the Academy of Sciences, in the matter of dissociation of elementary vapors at high temperatures.[138] This became of interest to Raoult.

In his duties as professor of chemistry, Raoult had certain obligations for keeping up with recent developments in chemical theory and for doing research that tied in directly with the discipline. His research papers in the late 1860s and early 1870s focused on some clearly chemical topics, some with physiological significance which tied in with his duties in teaching toxicology at the School of Medicine. Some of this research was performed as a practical service to the community—for example, the examination of gas given off by hot springs near Grenoble. As a judicial expert, he proved the ordinary presence of zinc and copper in the human liver.[139]

Studies of gases dissolved in liquids, the kinds of chemical work frequently significant for medical physiology, included experiments on the absorption of ammonia by saline solutions. Raoult studied the variations in absorption of ammonia water, calcium nitrate solution, ammonium nitrate solution, and potash solution; the dependence of solubility on concentration; and the near independence of heat produced during absorption from the nature of the absorbing liquid.[140] According to Ramsay, who later talked with Raoult about his work, it was this series of experiments in 1873–74 which led Raoult to study the freezing points of the salt solutions of ammonia. From then on, he busied himself with the freezing points, boiling points, and vapor pressures of salt and organic compounds in water and other solvents.[141]

In July 1878, Raoult published a paper reporting the results of measurements of the freezing points and vapor pressures of salts in water. He referred to two earlier, separate studies by Adolf Wüllner and Friedrich Rüdorff. Wüllner had studied the lowering of the vapor pressure of water by a salt dissolved in it; Rüdorff had studied the lowering of the freezing point of water by a salt.[142] In each case, the experimenter had concluded that the difference was proportional to the weight of salt dissolved in the water. As Raoult was soon to learn, Charles Blagden had arrived at this result for freezing points in the eighteenth century, concluding that, with mixtures of salts, the effects were additive and that the depression of freezing point was greater than expected at higher dilutions. These results were revived in 1871 by Louis de Coppet, a Swiss chemist trained at Heidelberg.[143]

Raoult explained both freezing point and vapor pressure depressions by the same cause: the affinity of the salt for the water. Here Raoult seems to have accepted the chemical theory of solution current in France, that is, that solution is a chemical process of hydration. This theory, taught at the Collège de France by Berthelot, attributed the release of heat of solution to a chemical combination of the dissolved substance with water; this hydrate theory received detailed and influential treatment in Berthelot's classic textbook published in 1879, *Essai de mécanique chimique fondée sur la thermochimie*. Berthelot's view of solutions, shared by many chemists on the Continent and in England, was that more than one kind of hydrate might be formed in a solution, and the presence of definite hydrates—with two equivalents, six equivalents, and so on, of water combined with the salt—might be verified by the shapes of heat-of-dilution curves.[144]

Raoult concluded his 1878 paper with a table of values for freezing point depressions and vapor pressure depressions for a variety of salts, noting that anhydrous salts follow the same order of arrangement when classified either by change in freezing point or change in vapor pressure and that the depressions are greater the smaller the "atomic weight" of the salt.

Switching from salts to alcohols as solutes in solution, Raoult next studied the freezing points of liqueurs, temporarily abandoning the study of vapor pressures (perhaps because of the difficulty of measuring boiling points and vapor pressures for volatile alcohols).[145] In an 1882 paper, he extended his analysis from alcohol to other organic compounds in water, then from water as a solvent to five other liquids: benzene (which freezes at 4.96°), nitrobenzene (5.28°), ethylene dibromide (7.92°), formic acid (8.52°), and acetic acid (16.75°). He now argued that the method of freezing point analysis was far more important than simply the prediction of freezing point or the analysis of the purity of a substance by its deviation from normal freezing point (or melting point). "Its most important application will be the determination of molecular weight in the numerous cases where the measurement of vapor densities is impossible."[146]

It was the discovery of de Coppet's work that helped Raoult develop this new point of view. De Coppet calculated for the depression of the freezing point of 100 grams of water, in which 1 gram of solute was dissolved, a value he termed the "atomic depression." This value was the product of the depression in freezing point by the "atomic" (really, the molecular) weight of the salt solute. A precedent for this kind of calculation existed in Dulong and Petit's 1819 law for specific heats and atomic weights: the product for a number of solid elements is a constant, the "atomic heat." De Coppet pointed out that the "atomic depressions" within five similar classes of salts in water are equal. Thus, for example, potassium chloride, sodium chloride, potash (potassium hydroxide), and sodium hydroxide all have atomic depressions of approximately 34, whereas barium chloride and strontium chloride have atomic depressions of approximately 45; those for zinc sulfate, magnesium sulfate, and copper sulfate are lowest at approximately 17.[147]

To studies of salt solutes, Raoult added his investigations of twenty-nine organic substances, listing lowering of freezing point and calculated "molecular depression" for each. Where $C$ is the freezing point depression for a solution containing 1 gram of solute in 100 grams of water, $P$ is the number of grams of solute dissolved in 100 grams of water, and $M$ is the molecular weight of the solute, then

$$\frac{C}{P} \cdot M = K,$$

where $K$ is the "molecular depression." For all Raoult's twenty-nine organic substances dissolved in water, the molecular depressions $K$ were between 17 and 20, for an average of 18.5. The $K$ value for alkaline salts in water was, on the average, nearly twice as great, that is, 37.[148]

In addition to offering a new method for calculating molecular weights of an organic solute in water by using the average $K$ value of 18.5 for a known solution, Raoult suggested an explanation for the behavior of solutions of organic solutes which now leaned more heavily on physical considerations than on chemical considerations, and he clearly distinguished chemical and physical "molecules" from "atoms." "This tends to show," he wrote, "that in most cases the molecules of organic compounds are simply separated by the act of solution and are brought to one same state, in which they exercise the same influence on the physical properties of water."[149] He reiterated this point in a paper the following month: "In a multitude of cases, the depression of the freezing point of a solvent depends only on the relation between the numbers of molecules of dissolved bodies and the solvent; it is independent of the nature, number, and arrangement of atoms which compose the dissolved molecules."[150] This was precisely Avogadro's distinction between molecules and atoms, which had been emphasized by Stanislao Cannizzaro at the Karlsruhe international chemical congress in 1860 and was still under debate in France around 1880.

However, as Ostwald later commented, it was Raoult's use of solvents other than water (benzene, nitrobenzene, etc.) that was crucial in throwing light on the pattern of regularities and anomalies in solution behavior, and in making clear that the process of solution for salts, acids, and bases (electrolytes) in water was really different from that in nonaqueous solutions. As Raoult put it, "The results in water are less harmonious than with other solvents, at least for mineral materials."[151]

What Raoult found was that values of $K$ (molecular depression) for most substances in acetic acid lay between 36 and 40, most often near 39, but that for some mineral substances like sulfuric acid, hydrochloric acid and magnesium acetate in acetic acid, $K$ was half the usual value, that is, 19. The pattern of a dominant value for $K$, and a secondary value for $K$ which was about half the dominant value, followed for formic acid, benzene, nitrobenzene, and ethylene dibromide as solvents.[152] In water, the values for $K$ were not as well clustered around a few $K$ values; the average values at which Raoult arrived were 37 for most mineral materials in water; 18.5 for magnesium sulfate, metaphosphoric acid, hydrogen sulfide and all organic materials; and 50 for barium and strontium chlorides.[153] The existence of the secondary $K$ values in nonaqueous solvents was taken to suggest for the solute in question that either the values commonly accepted for their molecular weights were half the true value or that the solute molecules were polymerized in solution. Raoult (erroneously, it turned out) inclined to the latter view, for example, for organic materials in water.[154]

It is important to realize that the calculation of molecular weights for mineral substances, especially the traditional calculation methods employing vapor densities for volatile substances and specific heats for solid substances, was under considerable discussion at this time in France. The distinction between elementary atoms and molecules still was not always made clearly and accurately, and French chemists remained divided over the choice between a system of equivalent-combining weights, based on H = 1, C = 6, O = 8, and a system of atomic weights, based on H = 1, C = 12, O = 16. Raoult's work bore directly on the points under discussion as well as on the ongoing debate in France and elsewhere concerning the value of physical assumptions in solving chemical problems.[155]

Raoult's statement about the *physical* action of solute molecules on solvent molecules identifies him with the French disciples of Adolphe Wurtz who accepted the general validity of postulates identified with Avogadro's theory and the extension of this physical theory from gases to liquids. By using the value H = 1 and O = 16 in his November 1882 paper on freezing points, Raoult adopted the recommendations of Cannizzaro and the Wurtz school of chemistry.[156] In addition, Raoult went on to develop a notion of salt "decomposition" into electropositive and electronegative "radicals," which Svante Arrhenius was to incorporate in the new and controversial idea of "ionization."

Raoult's reasoning was the following: the molecular-lowering constant $K$ is a maximum value when there is the least association of solute molecules with each other (polymerization) or of solute molecules with solvent (as in hydration). It was the presence of separate, distinct particles of matter as *physical molecules* or particles that affected the solvent's freezing point.

> The simplest manner of explaining the observed facts consists in allowing that in a constant weight of a determined solvent, all physical molecules produce the same molecular lowering of the freezing point. According to this hypothesis, if the chemical molecules of the dissolved body are completely separated from each other, the molecular lowering is a maximum and the same for all. If, in contrast, the chemical molecules are joined with each other in more or less substantial numbers, molecular lowering is more or less inferior to the maximum. It is half when the chemical molecules are joined in pairs, and, in general, the abnormal lowerings correspond to this condition [what I have called "secondary"].[157]

For in aqueous solutions, he theorized, molecules in solution act as if their properties are those of their electropositive and electronegative radicals: "Contrary to what I have believed up to now, the general law of freezing does not apply to salts in water. . . . on the contrary, it applies to the radicals constituting the salts, almost as though these radicals were simply mixed in the solution."[158] This notion of chemical radicals was based in the traditional chemistry of Faraday's cations and anions, Berzelius's electrochemical dualism, and Laurent's substitution theory. In concluding that "molecular depression" values of "radicals" were independent and additive, Raoult also drew on recent studies by Favre and Valson and by Hugo de Vries.[159]

Raoult sorted out two different types of anomalies in molecular depressions, one type occurring in fairly concentrated solutions and the other occurring in dilute solutions. De Coppet had also recognized the first type and referred it to "alteration of the dissolved bodies" as concentration increases, without being more precise about the nature of the alteration or the anomalies. Raoult attributed this type of anomaly (secondary $K$ values) to hydration and polymerization.[160] Using a graphing procedure, Raoult drew different curves for different aqueous solutions, where the x-axis denotes molecular freezing point depression and the y-axis, molecular weight. Points along each curve correspond to different concentrations of solution. In most cases, each curve has a fairly flat part, more or less parallel to the x-axis, and a rapidly rising part, corresponding to greater dilution.[161]

> Now if we look at curves for freezing point lowering where solutions are dilute enough to freeze at 0° to −1°, we see something that no other observer has remarked until now; that the coefficient of depression of an ordinary compound is increased by dilution and takes values higher and higher than the normal. Such an increase reveals an increase in the number of molecules and, consequently, a partial decomposition of the dissolved matter. One is then obliged to conclude, in the presence of the fact to which I am calling attention, that all bodies are decomposed more or less in very dilute solutions.[162]

What does Raoult mean when he says "all bodies are decomposed more or less in very dilute solutions"? Since he refers to a resulting increase in the number of "molecules," he probably does not mean charged particles, or charged "radicals." Further, he refers here, as in an 1884 paper, to Berthelot's finding that "this is so for a good number of salts," that some salts "are partially decomposed in water." Since Berthelot's view is one of chemical associations of solute and solvent, we may conclude that Raoult means that at great dilutions these hydrate-type compounds or polymerized molecules in solution decompose into their constituent molecules, and, based on freezing point depressions for nonsalts, this is true as well (he says) for alcohol, tartaric acids, and sugar, to a lesser degree.[163]

It is at this time that Raoult began explicitly associating his work with the "new chemistry," arguing that his results indicated that the equivalentists' values for biatomic (divalent, in modern terms) metals should be doubled to conform with values for atomic weights accepted by Wurtz and most non-French chemists. He arrived at this conclusion by the additive principle of electropositive and electronegative radicals, that is, by the result already presented in 1885 that metallic salts in water have molecular depressions that are the sums of the partial lowerings produced by their radicals. Van't Hoff later described this result as "general and unexpected."[164] Thus, from the relations

| | | |
|---|---|---|
| − monoatomic radicals | Cl, Br, . . . OH, $NO_3$ | K = 19 |
| − biatomic radicals | $CrO_4$, $SO_4$, $CO_3$ . . . | K = 9 |
| + monoatomic radicals | H, K, . . . $NH_4$, Ag . . . | K = 16 |
| + biatomic radicals | Ba, Sr, . . . Mg . . . Al | K = 8 |

we can calculate for KOH, for example, 16 + 19 = K = 35, a value comparing favorably with the experimental value of 35.3.[165]

Raoult found that the $K$ values for the radicals fall into two groups, (19/16 and 9/8), with the values in the ratio of 2:1. But,

> if I were to use equivalent weights for the biatomic metals, rather than molecular weights, . . . the natural analogy would be altogether obscured, without simplifying anything. In these two cases, the ratio is very far from unity, and this circumstance gives a new basis to the division of metals into monoatomic and diatomic [monovalent and divalent] kinds, a division still in controversy. This new basis is arrived at altogether independently from previous considerations. . . . Comparison [of these numbers] allows recognizing the atomicity [valency] of constitutive radicals of a given salt, and consequently, the salt's atomic formula and its molecular weight. The already numerous results which I have gained in this way almost all conform to what we would expect according to the new chemical theories . . . uranyl, whose atomicity [valence] has been very uncertain until now, is demonstrated in my experiments, to be definitely biatomic [divalent]; it follows that the molecular weight of uranium is quadruple its equivalent combining weight, conforming to the predictions of the periodic law of M. Mendeleef.[166]

Further, in an invited lecture at the Chemical Society of Paris in May 1886, Raoult used the freezing point depression method to calculate the molecular weight of *water* as a solute in acetic acid. Claiming that there was an advantage in using data from a chemist disinterested in the controversies plaguing French chemists, Raoult applied data from Rüdorff to calculate $M$ from known values of $K$ for water in acetic acid, arriving at the value $M = 18.3$. He concluded, "In this case, as in all parallel cases, it is always to the atomic formula that cryoscopy *donne raison.*"[167]

While still invoking polymerization, hydration, and other chemical reactions as processes occurring in fairly concentrated solutions, Raoult reiterated the anomalies in dilute aqueous solutions, a view he first expressed in August 1884:

> We come to see that, upon the point of freezing of the water, salts in aqueous solution act as if their electropositive and electronegative radicals were not combined, but simply mixed in the liquid. . . . this very singular mode of action belongs only to salts dissolved in water. One does not observe it with organic matter, nor with ethers. . . . *True salts, dissolved in water, have then an altogether special constitution.*[168]

While Raoult's principal concern at this point was to offer experimental information that would settle the atomic debates among chemists in France, his experimental results were noticed by an altogether separate community of scientists, indeed by scientists of three different nationalities whose common interests and methods were to lead them to found the new school of "ionist" physical chemistry, institutionalized in the *Zeitschrift für physikalische Chemie.* And it is not surprising that after Arrhenius published his theory of the "ionization" of salts in 1887, others saw "ions" confirmed by Raoult's work. Raoult's wording about "radicals" is the same after 1887 as in 1884–1886, but seems now to have new meaning: "for neutral salts of mono- and dibasic acids, everything happens as if the electropositive and electronegative radicals of salts in aqueous solution were not combined but simply mixed."[169] Raoult's integration into the new conceptual framework of physical chemistry came about in the following way. In 1884, van't Hoff published his *Etudes de dynamique chimique*, studying, from the thermodynamic point of view, chemical reaction velocities and the theory of affinity.[170] Arrhenius, who had just completed his dissertation on electrolysis in Sweden and was soon to take turns working with Ostwald, Kohlrausch, and Boltzmann, reviewed the *Etudes* and made special criticism of van't Hoff's comment that the vapor pressure depression of solutions might be a starting point for chemical affinity studies. Rather, Arrhenius claimed, he agreed with the French chemist Raoult that a purely physical process was occurring.[171]

A chance discussion with his colleague Hugo de Vries led van't Hoff in 1885 to a new aspect of the study of chemical affinity: that osmotic pressure might provide a measure of the affinity of a salt solution for water, by measuring the pressure required to stop the flow of water into the salt solution. The consideration of osmotic pressure resulted, however, in a purely physical analogy between solutions and gases, whereby the osmotic pressure in solution is due to impacts of the solute molecules on the semipermeable membrane. Using thermodynamic analysis, van't Hoff deduced for dilute solutions the applicability of the gas law

$$PV = RT,$$

and he offered three experimental confirmations of this law, one drawn from Hans Pfeffer's studies of osmotic pressure and the other two drawn from the research of Raoult on freezing point depressions and vapor pressures.[172]

Van't Hoff deduced Raoult's experimental laws from the osmotic pressure equation, and he showed most of Raoult's data to confirm values calculated from his equations. But the experimental effects for salt in water were larger than predicted by van't Hoff's theory, and he had to introduce a coefficient $i$, varying it to correspond with Raoult's experimental results:

$$PV = iRT.$$

Van't Hoff offered as explanation for the varying $i$ values (e.g., $i = 1.98$ for HCl, $i = 1.82$ for $NaNO_3$, $i = 1.78$ for $KClO_3$) the hypothesis that the abnormalities resemble deviations from Avogadro's law in dissociating gases. In a private letter, Arrhenius suggested that the values of $i$ indicate that NaCl is partially dissociated, just as we say that at high temperature $I_2$ vapor is dissociated.

> Now this assumption might be deemed very rash were it not that on other grounds we are led to look upon electrolytes as partially dissociated, for we assume that they decompose into their ions. But as

these ions are charged with very great quantities of electricity of oppo-
site sign, conditions are such that we cannot in all cases treat a solution of NaCl as if it simply consisted of Na and Cl. . . . Since according
to the above assumptions electrolytes decompose into their ions, the
coefficient $i$ must lie between unity and the number of ions.[173]

Arrhenius published this theory in 1887. The new idea evolving out of
his doctoral dissertation on electrical conductivity was that at infinite
dilution all molecules of electrolytes in water break up into charged
ions and exert equal action as conductors. His concept of ionization in
1887 was decisively different from the view expressed in his 1884 dissertation—that salts exist in solution as complex molecules, some of
which (the "active" molecules) decompose upon dilution. Like Clausius, he had then supposed only a small proportion of free cations and
anions to be present during electrolysis, and these were of varying
complexity.[174]

In later years, Arrhenius was at some pains to make clear that Raoult
was not responsible for any idea comparable to ionization. Giving the
Faraday Lecture of May 25, 1914, Arrhenius said:

How far chemists were from assuming a dissociation of salts into their
radicles [sic] is seen from the chief memoir of Raoult [in 1885], in which
he tries to calculate the molecular lowering of the freezing point for
salts in aqueous solution. There he says that every radical has its own
"molecular depression." . . . If Raoult really had supposed a dissociation, he would have attributed the same molecular lowering to all the
radicals.[175]

The chemists and physicists who adopted Arrhenius's novel idea of
1887 were dubbed the "wild army of the Ionians" and, as one account
put it later, it is not surprising that there was opposition from chemists
to the supposition "that mere dissolution caused a stable salt to tumble
apart in complete defiance of the laws of chemical affinity and chemical
attraction."[176] The ionists approached Raoult in 1887 with an invitation
to collaborate on the new journal Ostwald was founding with van't
Hoff, the *Zeitschrift für physikalische Chemie*. Raoult replied to Ostwald in
February 1887, accepting with pleasure the honor of being enlisted

among your collaborators in the *Journal of Physico-Chemistry*. The moment has come, in effect, of linking together Physics and Chemistry,
properly called, out of this shared territory in which they are more and
more entangled, and to establish from this a special branch of science.[177]

With Berthelot and Le Chatelier, Raoult was one of only three French scientists listed among twenty-one collaborators on the title page of the first issue of the journal.[178]

The collaboration did not mean Raoult gave himself entirely over to the ionists. He continued to acknowledge the importance of some chemical action in solution, especially at higher concentrations. In a 1900 lecture, for example, he said that we have to be careful not to go too far with the extended law of Avogadro in applying it to solutions, a word of caution to extravagant claims about purely physical processes acting in solution to the exclusion of intermolecular forces.[179]

At the Universal Exposition in Paris in 1900, Raoult's instruments for measuring freezing points, boiling points, and vapor pressures were placed on display under the auspices of the International Congress of Chemistry. So accurate was his cryoscope through continued improvement that the values of 1900 scarcely differ from those in the *International Critical Tables* decades later.[180] In 1900, Raoult claimed that he had improved instrument precision from 0.01° to 1 millionth of a degree for aqueous solutions. This increase in precision enabled him to determine experimentally the change in freezing point produced in water by oxygen, resulting in a value of 4 millionths of a degree, from which he calculated a molecular weight for oxygen of 35. Thus, at the end of his career, he once more addressed the central controversy of his mature years in chemistry: the values for atomic and molecular weights, and choosing between the system $H = 1$, $O = 8$ and the system $H = 1$, $O = 16$, on the assumption that oxygen gas molecules are diatomic.[181]

Like van't Hoff and Arrhenius, Raoult was a physicist and a chemist; he shared with them and with Ostwald a keen interest and familiarity with electrolysis, and his insights into chemical affinity were shaped by the tradition that originates in chemistry with Berzelius and in physics with Faraday. Lacking the licence or agrégation in mathematical sciences, he did not have the formal mathematical training that could result in mathematical creativity via thermodynamics. But as an experimentalist, he showed insights that led him to move from one research topic to another without losing a synthetic unity in his work and his results. As is often the case in scientific innovation, the breakthrough that led to a unification of solution theory with gas theory via ionization came from anomalies revealed by laborious and precise experimental labors.[182]

Innovation came, too, from the synthesis of the disparate methods of physics and chemistry, a synthesis that was all the easier at Grenoble, because there were not strong, established chairs in pure physics and chemistry. Had the advice of Balard been followed in 1870, Raoult would have been denied a chair in chemistry to the benefit of a "proper" chemist. Then the relevance of his interests to chemical problems might not have been so apparent either to others or to himself.

With the discovery of the electron in 1897 and the conceptualization of the atom as an electrically composed particle, ionic chemistry came to have an undisputed place in the new chemistry.[183] In 1907, at a celebration of the fiftieth anniversary of the founding of the Paris Chemical Society, Armand Gautier recalled the past division of French chemical teaching into the three opposing camps of Wurtz, Berthelot, and Deville. The war was over, he said, and future progress in chemistry would come from two directions, biological chemistry and physical chemistry, from

> the possibility of predicting and measuring a large number of reactions by substituting, in part, mathematical calculations and mechanical considerations for tentative experimental ideas and predictions based chiefly on reasoning by analogy.[184]

Raoult's contribution had been fundamental to the experimental foundations of this new field, but he left no research school behind him. There is no indication that he trained any students in the new physical chemistry or that he introduced these ideas into his teaching. This state of affairs was a product of his relative intellectual and physical isolation in a somewhat out-of-the-way place, an isolation that, in turn, may well have encouraged the direction of his thinking and the painstaking effort of his achievement. The applications of physical chemistry in industry were neither obvious nor direct in 1900,[185] but the path of application lay through electrochemistry and electrometallurgy, which were strongly supported in the Grenoble Sciences Faculty. Raoult now looms large in the hagiography of the University of Grenoble. Recognition of his name and "Raoult's law" conferred prestige and distinction on the university with which he had been associated, even though he left no research school behind. But what was to put the city and the university indisputably on the map of international scientific centers was that combination of fundamental research and engineering research which was only beginning to flourish at Raoult's death and which was renewed after the Second World War.

1. Albin Haller, Professor of Chemistry at Nancy, 1879–1899. (*Courtesy of Jacques Aubry, University of Nancy.*)

3. René Blondlot in the robes of the Academy of Sciences. (*Courtesy of Jacques Aubry, University of Nancy.*)

2. Ernest Bichat, Professor of Physics at Nancy, 1876–1906. (*Courtesy of Jacques Aubry, University of Nancy.*)

4. The Institute of Mathematics and Physics at Nancy, with a bust of Ernest Bichat to the right. (*Courtesy of Jacques Aubry, University of Nancy.*)

5. Art nouveau architecture of the Musée de l'Ecole de Nancy.

6. François Raoult in his laboratory at Grenoble. (*Courtesy of the Center for the History of Chemistry, University of Pennsylvania.*).

7. The poster "Hydroelectric Power and Tourism" for the 1925 International Exhibition at Grenoble. (*Courtesy of the Musée Dauphinois, Grenoble.*)

8. Professor Louis Léger (left) and Chef des Travaux Léon Perrier (right) in the Ichthyology Gallery of the Laboratory of Zoology and Biology, Grenoble. (*Courtesy of the Archives Départementales d'Isère.*)

9. Seat of the Old University Faculties at Grenoble.

12. Victor Grignard in his lecture hall at Lyon. The tetravalence of carbon has not been observed on this occasion. (*Courtesy of Roger Grignard.*)

13. Pierre Duhem around 1900 in the robes of the Academy of Sciences. (*Courtesy of Donald G. Miller, who received this photograph from Mlle. Hélène Pierre-Duhem.*)

14. Presentation of Paul
Saurel's thesis at Bordeaux.
Right to left: Pierre Duhem,
Lucien Marchis, Dean Georges
Brunel, Paul Saurel, C. H.
Chevallier, Charles Malus, and
M. Lenoble. (*Courtesy of
Donald G. Miller, who received this
photograph from Mlle. Hélène
Pierre-Duhem.*)

15. 1838 Faculty of Sciences
laboratories in Municipal Buildings near the Hôtel de Ville,
Bordeaux.

16. 1885 Faculty of Sciences
Building at Bordeaux.

# 4

# TOULOUSE: POLITICS, ENTREPRENEURSHIP, AND SABATIER'S CHEMISTRY PROGRAM

The general site that became Toulouse was known to the Romans as Tolosa. Lying on a natural route connecting the Mediterranean Sea and the Atlantic Ocean, it was a regular entry in Roman itineraries of the fourth century. The region is one of broad river flats, rolling country, and gentle hills. By the twelfth century the counts of Toulouse were the greatest lords of southern France. Their Château Narbonnais became the seat of the Parlement of Toulouse, which was established as a permanent court in 1443. The University of Toulouse was the first founded in any country by papal charter, and its establishment in 1229 was one of the conditions of peace imposed by Louis IX on Raymond VII, count of Toulouse from 1222 to 1249. The university was to aid Rome in suppressing the Albigensian heresy.

The charter issued by Pope Gregory IX bound Count Raymond to maintain four masters to teach theology and eight masters to teach canon law, grammar, and the liberal arts. Within a few years, civil law and medicine also were taught at the university, which became a seat for Dominican teaching, associated with the red-brick Eglise des Jacobins built in the thirteenth century. The Eglise des Jacobins was among several important religious centers in the city. A large Benedictine abbey existed from the ninth century, and the powerful monastery of Saint-

Sernin had much influence and property in Spain as well as in France. The choir of the Eglise de Saint-Sernin, now the largest Romanesque basilica in existence, was consecrated in 1096 by Urban II.[1]

The University of Toulouse never attained a general reputation to compare with that of Paris or Oxford or Padua. In southern France it vied with the University of Montpellier, 240 kilometers to the east and renowed for its Medical Faculty. In the late eighteenth century, approximately one thousand students attended lectures at Toulouse, which was regarded as a good university for the study of law. In the Napoleonic university scheme, Toulouse was given four Faculties: Theology, Law, Sciences, and Letters. The Sciences Faculty occupied a building on rue Lakanal which had served as an école centrale, and the chemist Chaptal lectured there in 1797. The inscription over the entrance still reads "Ecoles publiques de chimie et de physique expérimentale."[2]

The Sciences Faculty initially included chairs in pure mathematics, applied mathematics, physics, and natural history, to which a chemistry chair was added in 1813. The Faculty slowly expanded in the next decades, with the establishment of a chair of zoology in 1838, geology and mineralogy in 1839, and astronomy in 1848. During the period 1810–1879, 193 licences were awarded (an average of 2.75 per year) and 16 science doctorates.[3] The Toulouse academy during the nineteenth century was the second in rank in France, with a jurisdiction extending over eight départements. Matthew Arnold described the lycée of Toulouse in detail after he visited France in 1859 to investigate secondary and higher education.[4]

Not only did Toulouse have the longest history of university studies in France outside Paris, but the level of instruction in the academy was one of the highest in the country.[5] The Faculties enjoyed increasing enrollments in the 1880s, keeping pace with the doubling of the number of university students throughout France from 1875 to 1891. The teaching personnel at Toulouse also doubled from 1878 to 1888, during a period when the number of professors increased nationally only by 30 percent.[6]

Despite its long and distinguished history, however, Toulouse was not in the running as a university seat when Louis Liard proposed in 1890 the establishment of five or six national universities.[7] While the city had Faculties of Letters, Sciences, and Law, and a Faculty of Theology at Montauban, its Ecole de Médecine was not yet a Faculty despite long-standing efforts by the city to bring this about.[8] Further, the ministry feared that the south of France could not support universities at Montpellier, Bordeaux, *and* Toulouse, although Toulousains not surprisingly argued that Bordeaux was in no way part of the Midi and that the educational interests of Montpellier would conflict with Marseille, not with Toulouse.[9] How was the city to remain a university center under the new regime?

## THE POLITICS OF BUILDING REPUBLICAN SCIENCE

As was true at other university centers in 1890, Toulouse had new facilities under construction, including buildings for the older Faculties and for the proposed Medical Faculty. Support of education by the city, the third largest in France (population, 147,000), was in keeping with Toulouse's reputation of advanced republicanism, a reputation dating from Armand Duportal and the Commune. In the 1870s, the municipal council had been split between those who wanted to spend monies on completing the Second Empire boulevards and those republicans who preferred to devote city funds to education. In the early 1880s, the council was sometimes chaired by university professors sympathetic to the second aim, and big spending for higher education flourished under the Radical party leaders (left-center), who controlled the city from the mid-1880s until 1906. Mayor Camille Ournac was a close friend of Jean Jaurès, the philosophy professor at Toulouse who became national leader of the Socialist party.[10]

The republican educator Claude-Marie Perroud, appointed to the Midi in 1881, was one of the nation's rectors most applauded by the Parisian republican ministry. His first years at Toulouse caused some dismay among Faculty members troubled by fears of "rapid innovation." Further, it was said he was inclined toward authority, rather than persuasion, in dealing with his colleagues. Like so many republicans, Perroud spoke highly of "la science exacte, les vraies méthodes," but university divisiveness was revealed when he announced that he would attend a meeting of the Sciences Faculty in 1881, and only the new dean showed up.[11]

Perroud defended Jaurès against criticism from those Faculty members who objected to Jaurès's mixing of professional and political activity. In 1886, Jaurès began writing for the daily newspaper, *La Dépêche de Toulouse*, whose writers included Camille Pelletan, Georges Clemenceau, and Maurice Sarraut. Anticlerical and Dreyfusard, *La Dépêche* espoused the eight-hour day in the mines, demanded tax reform, and opposed colonial expansion on the grounds that imperialism worked only to the profit of Catholic missions, diverting France from the "devoir sacré de la Revanche."[12] Perroud gave permission for Jaurès to continue writing in La Dépêche following his return to teaching duties after a four-year stint in the Chamber of Deputies.[13] Elected municipal councillor in July 1890, Jaurès devoted himself to university affairs, including the transformation of the Medical School into a Faculty and the construction of nine buildings for the Sciences Faculty.[14] It was in early summer 1890 that the chamber began debating Liard's project of regional universities, and Jaurès informed Perroud that it was time to begin agitation for Toulouse.[15]

In July, a meeting of professors in the amphitheater of the Letters Faculty unanimously passed a resolution calling for a university at Toulouse. The municipal council, chamber of commerce, Cercle Républicain Radical-Socialist, Union des Syndicats Ouvriers de Toulouse, and other groups rallied with similar resolutions.[16] Léon Bourgeois, who was minister of public instruction, visited Toulouse in midsummer and talked with Perroud, Ournac, and Cyrille Caubet, who was director of the School of Medicine. Ournac and Jaurès later journeyed to Paris for further discussions.[17]

In conversations, as well as in a series of articles in La Dépêche, Jaurès argued that it would be a material and intellectual disaster for Toulouse and the Midi if the state did not designate Toulouse as a university.[18] Education, he said, "leads to political liberty and the social emancipation of the working classes," and the decentralized provincial university would animate the region it served. Voicing the French republican identification of science with knowledge and progress, Jaurès wrote that the noblest mission of higher education is the teaching of scientific method and scientific knowledge to students and the public. The sciences are not only useful in application, he wrote, but add to national glory. "Par là, elles font vraiment oeuvre civilisatrice."[19] Jaurès's rhetoric was one of the most elegant pleas emanating from provincial centers throughout France, leading the government to give up its plan for five or six universities and create seventeen.[20] The Toulouse case illustrates particularly well the pressures exerted by provincial university cities on the educational ministry.

When the new Toulouse Sciences Faculty buildings were dedicated in 1891, Jaurès presided over the ceremonies, along with Rector Perroud, the president of the Republic, Sadi Carnot, and the Faculty's influential dean, Benjamin Baillaud.[21] Baillaud was an astronomer who had studied with Charles Hermite and Victor Puiseux. After teaching at lycées in Paris and serving as temporary replacement for Le Verrier at the Paris Observatory, he became professor of astronomy at Toulouse, dirctor of its newly created observatory, and dean of the Faculty of Sciences. First appointed dean from 1879 to 1888, he was elected by his colleagues from 1890 to 1893, and then, in 1908, left Toulouse for a Paris chair. In 1927, after presiding over the permanent international committee drawing up a new star catalog, he retired from Paris to Toulouse.[22]

When Baillaud first arrived at the Sciences Faculty in 1879, he found nine existing chairs and forty registered students. As we have seen, the Faculty had awarded sixteen doctorates since 1810. The few illustrious names connected with the Faculty included Armand de Quatrefages in zoology and anthropology, Félix Dujardin in geology, Alfred A. Moquin-Tandon in botany, François Tisserand in astronomy, and Nicolas Joly in anatomy and zoology.[23] On his arrival, Baillaud, who was thirty years old, doubtless felt more than the usual discomfort of a new man when he found that his youngest colleague was fifty-seven. The level of teaching at the Faculty, he later confided to friends, scarcely surpassed that of the baccalaureate.[24]

But Baillaud was an ambitious man, and his Parisian colleagues kept him informed about good young scientists. The new dean's first nomination to Toulouse was twenty-three-year-old Emile Picard, already lecturer in mathematics at the Ecole Normale. Picard stayed at Toulouse two years (1879–1881) and was replaced by Edouard Goursat (1881–1885), Gabriel Koenigs (1885–1886), and Thomas Johannes Stieltjes (1886–1894), as Baillaud resolutely brought the very best to Toulouse. All of these young men, with the exception of Stieltjes, became distinguished members of the Academy of Sciences. Stieltjes died in 1894, while still in his thirties, already praised by Henri Poincaré as one of France's most remarkable mathematicians. Other future Academy members who were brought to Toulouse in the next two decades included Marcel Brillouin (at Toulouse, 1883–1887), Henri Andoyer (1888–1892), and Aimé Cotton (1895–1900). Among those who remained at Toulouse throughout their careers and became nonresident Academy members were Paul Sabatier (1882–1929), Charles Camichel (1906–1941), and Eugène Cosserat (1908–1929). Baillaud became a member of the Academy in 1908 when he moved to Paris.[25]

The loss to Paris of good scientists was one of young Baillaud's major frustrations as he fought to upgrade the quality of the Faculty at Toulouse. In letters to the rector and the ministry in 1882, he wrote that it was absolutely necessary for a new Faculty member in physics to be au courant in modern physics, spend time in the laboratory, and teach seriously. "Ever since I have been dean at Toulouse," he wrote, "I have had the ambition of transforming an inactive faculty into as brilliant a faculty as possible, both by their works and by their teaching. This faculty is deserted. . . . you can't find anyone in the laboratories except during the lecture hours."[26] He attempted to unite the Faculty behind his movements for reform, but while they liked the young dean, one professor told Perroud that Baillaud was making them vote for things they really did not want ("il excelle à tirer la couverture à nous").[27]

Baillaud was alarmed in the 1890s at the news that two law professors at Lyon and Bordeaux had agreed to go to the Paris Faculty at a lower rank. The ministry had divided Faculties into two classes, he charged, those at Paris and the others. "If we are to keep illustrious teachers in provincial universities and not have their students think them inferior, it is necessary not to have inferiority written into the decrees."[28] In 1900, when Cotton was recommended for promotion at Toulouse from physics lecturer to adjoint professor, he accepted a lectureship at the Ecole Normale Supérieure, where he did not become adjoint professor until 1910.[29]

Baillaud found himself defending the abilities and rights of his students as well as his professors. When his colleague E. A. Legoux commented at a general Faculty meeting that Toulousain students were inferior to those at the Ecole Normale Supérieure, Baillaud responded, "all our students are not inferior to the pupils of the Ecole."[30]

The quality of professors and students did improve and enrollments were on the upswing, increasing more rapidly than the general rate for all of France (from 40 at Baillaud's arrival to 96 in 1889–90, 137 in 1894–95, and 340 in 1904–06).[31] The number of faculty chairs increased (from nine at his arrival to fifteen in 1903, with the increase in total number of personnel doubling, as occurred throughout France).[32] As noted earlier, the Sciences enrollments changed dramatically in the 1890s with the arrival of the so-called PCN students. At Toulouse, the 1892–93 enrollment included 95 regular students, 58 pharmacy students, and 174 PCN.[33]

Expansion in student and faculty personnel was paralleled by increasing monies for salaries, scholarships, and facilities. This did not mean that state support increased. In fact, as Sabatier pointed out in the University of Toulouse's annual report of 1896, national funds were considerably less than in 1885.[34] The overall increase in funds for the Toulouse Sciences Faculty resulted from an 1896 law giving most student fees directly to the universities and from the increase in number of students, many of them PCN.[35] Other new funds were available from the Ministry of Agriculture and the Département of the Haute-Garonne, regional general councils, and the city of Toulouse. In 1901, Sabatier and Mathieu Leclerc du Sablon, who was dean for eleven years after Baillaud, helped draw up a budget of one million francs for projects at all the Toulouse Faculties, of which one-fourth was pledged by the central educational ministry, one-fourth by the city, and one-half by the university.[36]

The question of monies was important not only for salaries but also for research, as the latter became a paramount issue under Baillaud's deanship.[37] This was no isolated phenomenon, as we have seen, but one occurring throughout France. In the dean's report of 1883, it still could be said that "the teaching of the Faculty has for its essential goal the preparation of students for the examinations of licences and agrégations."[38] But Rector Perroud's emphasis was on the university as a teaching institution, not an examining institution,[39] and Baillaud called for a Faculty which would do more than participate in building secondary education. He envisioned a Faculty which "by their researches [would play a role] in the remarkable movement which . . . will restore to our country the rank in science which twenty years of abandon and feebleness have made it lose."[40]

By the late 1880s, the Toulouse Faculty was said to have a triple role: research, teaching, and public service. "Perhaps in devoting ourselves exclusively to the preparation of examinations we have lost sight of the large public. Shouldn't we fear that they will one day ask if the services we render are fairly proportional to the great sacrifices we demand of the state?"[41]

Attention was given to revamping the student curriculum. The reformist Faculty found that students entering the university curriculum were often ill prepared for their courses. They had glaring weaknesses in mathematics and all too vague notions about science "learned by heart in manuals."[42] To combat these weaknesses resulting from the encyclopedic character of the bac, Sabatier headed a committee in 1894 which recommended that the licence place a larger emphasis on personal work and specialization, with the last semester of the third year requiring an individual project.[43] More pains were taken with personal direction to students, for which the *assesseur* (who from 1886 until the early 1900s was often Paul Sabatier) took particular responsibility. Although formal seminars were nonexistent in France at this time, Sabatier began meeting regularly with his chemistry students for what he called *entrétiens* (conversations).[44]

Sabatier, like Baillaud, emphasized that courses must be up-to-date in modern research. As happened elsewhere, monies for teaching were used to meet research needs both for faculty members and advanced students,[45] and distribution of the scant research funds led to some frustrations and jealousies. These erupted in 1892 into a major confrontation between Faculty members at Toulouse. Perhaps the confrontation was inevitable considering the growth that had taken place in the previous ten years. Priorities had been reordered and pressures multiplied, directly under Perroud and Baillaud and indirectly through changes in higher education throughout France. The divisions revealed by the 1892 dispute were to have important implications for the future.

The dispute began innocuously in early February 1892, when a commission was appointed to study the financial regime of the expanding Faculty. Sabatier was ill the day of the commission's meeting. The committee's report, presented by Stieltjes to the Faculty council, recommended that each member of the Faculty be given a minimum of 500 francs for personal research, with other monies to be requested as needed. It also outlined expenditures for costs of teaching, laboratory exercises, laboratory equipment, and so forth. At the next meeting of the council, Sabatier heatedly objected to the report both in its details and in its spirit; his objections suggested the course the Faculty would take for at least the next forty years.[46]

The spirit of the report was one of democratization (termed "anarchy" by Sabatier) as well as disgruntlement with the way in which research monies were being distributed. Stieltjes and others wanted the suppression in each Faculty research division of the directorship, a change leading—as Sabatier put it—to the "assimilation absolue des titulaires, et des non titulaires, au point de vue de la gestion des laboratoires." Under Stieltjes's plan, each member of the assembly would receive an equal sum of money, which he could employ independently of others, and the assembly would decide periodically who would direct the laboratory exercises. There was to be no mixing of funds for personal work and for laboratory teaching, something that made no sense to Sabatier since the same equipment was needed for both. "The necessity for a supervisor directing the service seems absolute to me. . . . In all foreign institutes, there is a single director, and if we judge the tree by its fruits, the tree is flourishing." When the committee's report was adopted, Edouard Lartet, Gaston Moquin-Tandon, and Sabatier left the room.[47] The following year, a new financial arrangement was passed which included the directorships and allocated 1,000 francs to each titular professor in addition to the 300 francs everyone received. Stieltjes was now dead, and the new plan was drawn up by the dean, G. Berson, and Leclerc du Sablon.[48]

The dispute, which had been rancorous, demonstrates a growing division among Faculty members. Toulouse had changed under Baillaud from an inactive research faculty to an active one, but as we shall see, some professors, particularly Sabatier and the chemists, were drawing more than others on research monies. Apparently Stieltjes and some of his colleagues wanted to ensure that their disciplinary fields would not take a secondary role in the Faculty and that younger members would have a voice in Faculty matters as the reformers of the 1880s began aging.

A marked change was taking place in the 1890s at Toulouse, paralleling what was taking place in Nancy and Grenoble, and in other countries, such as England and Germany. Baillaud had presided as dean over the teaching of a scientific discipline that dovetailed with traditional "general culture," but, in contrast, Sabatier as dean was to orient the Faculty "toward an ever more intense and fruitful participation in the life of the city and the region."[49] It was between 1893 and 1905, under the deanship of Leclerc du Sablon, that the emphasis shifted toward applied science. Leclerc du Sablon spoke of the duty of professors to reserve their so-called leisure time for scientific research, but the examples he cited were studies that had clear practical applications, such as work on black rot and on the influence of copper salts on the production of potatoes.[50]

Of course, the direction that the Toulouse Faculty was to take was not simply the result of local factors. It owed much to the tenor of the times, to the demands for further reform in French scientific education, and particularly to growing provincial specialization in applied and engineering science. The republican reformers of the 1880s had sought to extend the capacities of science and culture. Many activists of the early 1900s now worked to enlarge the potentialities of engineering science and technology. So well did they succeed that the Toulouse Sciences Faculty became a prototype for a newer style of technocratic scientific education and research in France. Robert Gilpin judges that more recent plans for reform in French science and science education have represented "a victory . . . for the scientists who favor emphasis on specialization rather than general culture in higher education."[51] This represents a victory for the Toulousains who envisioned the teaching of scientific disciplines to students who had practical aims in mind. It is a victory not altogether in consonance with the ideals of the liberal scientists and educational reformers of the early Third Republic.


THE ART OF INSTITUTE BUILDING

One of the most widely read analyses of French higher education was Ferdinand Lot's 1906 report in the *Cahiers de la Quinzine*. In physics and particularly in electrical technology, he argued, the gap separating France and Germany was large, and in chemistry, especially industrial and agricultural chemistry, the superiority of the Germans was striking. Lot concluded that these fields should be developed vigorously in France's provinces.[52]

In 1905 there existed in France some dozen chairs or lectureships in agricultural and industrial chemistry and applied physics. These were associated with provincial Faculties, and expenses were borne mainly by the cities and individual universities. At Lille, for example, there was an Institut Electrochimique, with intimate links between the Institut Industriel du Nord de la France and the Sciences Faculty. At Nancy, there now were technical schools in chemistry, electrochemistry, agriculture, and brewing. Grenoble had an Institut Electrotechnique. And there were various Ecoles de Chimie Appliquée, including ones at Lyon, Bordeaux, Besançon, Caen, and Clermont.[53] The Paris Faculty hesitated to introduce courses in the applied sciences, as we have seen. In contrast, while Paris protected and nourished science as general culture, Toulouse was to embrace almost wholeheartedly science as technology.

Sabatier was a leader in the direction Toulouse was to take with such alacrity and success. Brought to Toulouse in 1882 as *suppléant* in physics, he was named professor of chemistry as soon as he turned the legal minimum age of thirty, in 1884. Baillaud's first reports to the ministry enthusiastically noted that Sabatier was already breaking with the lethargic tradition at Toulouse and that he did more work than anyone else in the Faculty.[54] Young Sabatier juggled basement rooms at the old Faculty to set up a laboratory for teaching and research. The decrepit Faculty rooms were said to have looked, at Sabatier's arrival, like Teniers's alchemist's lair, where Bunsen burners were rare and charcoal stoves were used for heating retorts and flasks.[55] Now in the avantgarde, Sabatier brought in new apparatus and new ideas, discarding the notation of chemical equivalents, for example, and teaching Mendeleev's periodic table when it was still taboo at the Ecole Normale Supérieure.[56]

By 1900, chemistry laboratories at Toulouse included general chemistry under Sabatier, organic chemistry under Dastrem, and industrial and agricultural chemistry under Charles Fabre. In 1905, a new annex for the chemistry section had to be constructed.[57] Sabatier's research reputation had by then spread beyond the city's boundaries. He was offered, and he refused, chairs in Paris; in 1912, he shared the Nobel Prize with Victor Grignard. But in addition to his ground-breaking work in fundamental organic chemistry, he was interested from his first years at Toulouse in teaching courses in applied science. In 1886, for example, Berthelot visited a meeting of the Faculty's assembly, and Sabatier took the opportunity to propose the establishment at the Faculty of a chemist's diploma. Berthelot was uninterested, asking if this diploma was appropriate in a city so little industrialized. In the following year, Sabatier proposed in detail the creation of such a diploma, and it was adopted.[58] In 1888, Baillaud spoke to Liard about the Faculty teaching more applied science, and Liard was encouraging. When Baillaud brought up to his Faculty the notion of letting engineers teach in cours libres, there were not the strident objections that took place at Paris in 1907, although Sabatier did caution against entrusting these courses to men tired out by their profession.[59]

Agricultural and industrial chemistry courses were inaugurated in 1888 and in 1889 at the suggestion of the rector, and Sabatier and Fabre found their amphitheaters packed for the winter lectures.[60] By 1900, the city had pledged half of the necessary funds for a chair in agricultural and industrial chemistry, and when the Ministries of Public Instruction and Agriculture promised to contribute the other half, Fabre was named (at Sabatier's nomination) to the chair.[61] By this time, Henri

Bouasse was teaching a course in industrial electricity and, as was to happen at Paris in 1908, the city offered in 1906 to underwrite the principal costs of transforming the course into a permanent chair.[62]

Sabatier's interests in applied science extended to both agricultural and industrial chemistry. From 1889 through 1910, he contributed papers to the *Journal d'Agriculture Pratique* and served often as president of the agricultural society that published the journal. Most of his agricultural papers had to do with nitrogen and with fertilizers; he also published a manual of agricultural chemistry in 1889. Beginning in 1905, Sabatier published occasional papers on industrial processes of hydrogenation as well as a study of methods for preparing illuminating gas. He took out nine patents during the period 1905–1908, one for illuminating gas and the others connected with his work in catalysis (discussed later).[63]

In 1906, a new series of profound transformations was under way at Toulouse, as, in rapid succession, Sabatier founded the Institut de Chimie (1906), the Institut Electrotechnique (1907), and the Institut Agricole (1909). The Institut de Chimie resulted from a report of spring 1906 which recommended the establishment of a *diplôme d'ingénieur chimiste de l'Université* and a *diplôme d'État de chimiste-expert*. The report noted that Nancy, Lille, Bordeaux, and Rennes had institutes of industrial chemistry. Toulouse, it was argued, now had large enough staff and facilities to offer an engineering degree without new expenses.[64] Accordingly, the new Institute of Chemistry at Toulouse received temporary housing in the main building of the Faculty and welcomed eleven students its first year; five years later, its enrollment was 69. Sabatier collaborated with Henri Giran, his former pupil Alphonse Mailhe, Fabre, and a lecturer, laboratory supervisor, and three aides in the teaching program. In a tradition that was to hold through Sabatier's retirement in 1929, the institute paid its own way through fees, although once Sabatier received the Nobel Prize, new funds flowed in from the state, city, and university to finance a new building and new equipment.[65] Sabatier contributed his Nobel Prize money to help build the new Institute of Chemistry. When he was congratulated on the Nobel award by Toulousain colleagues, his allegiance to the new institute was clearly expressed. He wished his honor would be reflected on the university, he said, and especially on the Institute of Chemistry, to which he was devoting all his efforts.[66]

It was not only applied chemistry that had begun to flourish at Toulouse. In 1906, Camichel and Bouasse were teaching courses in industrial electricity that were each attracting more than thirty students.[67] Many Toulousains sensed the industrial potential in hydroelectric power from the Pyrenees. With the examples of the Dauphiné and the Savoie in mind, Sabatier, Camichel, and others spoke of possibilities for a far-reaching transformation of the Toulouse area. Lille, Grenoble, and Nancy were already sites of electrotechnological institutes. The mayor of Toulouse sent to the faculty in July 1906 a proposal for the creation of a municipal chair of industrial electricity, with costs underwritten by the city and state.[68] From this proposal sprang the new electrical institute, which counted 150 pupils within two years and was so financially successful that its receipts became important to the Faculty.[69]

With chemical engineers and electrical engineers assured for the region, the Faculty sought more funds for the teaching of agriculture, which "is the chief concern of our area and gives us the major part of our richness."[70] With the patronage of the General Council of the Haute-Garonne, the city, and diverse agricultural societies, the Institut Agricole became a reality in 1909. In 1929, only Toulouse and Nancy housed such institutes. When it opened its doors in 1909, the institute welcomed ten students, and Sabatier served as its director until Prunet took over many years later.[71]

Once the technical institutes had been established, the Faculty devoted an increasing proportion of its administrative concerns to them and to the requirements for new technical degrees. In 1914, the Faculty council gave Camichel full powers to make propositions concerning the Institut Electrotechnique directly to the university council, bypassing the Faculty. The Institut Electrotechnique nevertheless remained part of the Faculty, as did the Institut de Chimie and the Institut Agricole.[72] The Sciences Faculty resisted any suggestions to detach these institutes and create a separate Faculty of Applied Sciences.[73] By the late 1920s, the enrollments at the institutes included 265 students in chemistry, 492 in electrotechnology, and 195 in agriculture. The institutes offered twelve diplômes-brevets.[74]

The successes of the technical institutes transformed the University of Toulouse. For centuries Toulouse had been a university dominated by its law school. In 1886, for example, 64 percent of students enrolled in the university were in law, 9 percent in letters, 14 percent in medicine and pharmacy, and 9 percent in sciences. By 1905–06, law school enrollment was holding steady, but the sciences had doubled; and in

1927–28, the law school enrollment had dropped to 20 percent of the school population, while letters included 15 percent, medicine and pharmacy 25 percent, and sciences 40 percent.[75]

Toulouse was not merely duplicating a national surge in science enrollment; rather, the Toulouse Faculty and its associated institutes began to take a larger share of students studying the sciences in France. Before the decentralization policy of Liard and the ministry got under way, Paris dominated higher education in science, enrolling 52 percent of all students in higher education in the 1880s. In 1894, Paris' share in the sciences was 34 percent, and the leaders in the provinces were Lyon (9 percent), Bordeaux (9 percent), Nancy (7 percent), Toulouse (6 percent), Lille (5 percent), Montpellier (4 percent), and Grenoble (3 percent). By 1910–11, the gross enrollment at Toulouse was six times the 1895 figure. This rate of change was exceeded only by Grenoble, which increased by a factor of almost seven. Toulouse's share of the sciences enrollment in France was now 10-½ percent and Grenoble's 7 percent, proportions that held until 1929. Nancy's share increased to 12 percent in 1910–11, but fell after the war to 8 percent. It could be that Toulouse increased its share of students not so much at the expense of Paris, but of Bordeaux, whose enrollments fell in the sciences from 9 percent of the French total in 1894 to 4 percent in 1929. Nevertheless, the increase at Toulouse must indicate some success in attracting potential students away from Paris, as Paris fell from 34 percent in 1894 to 29 percent in 1929.[76]

The growth of the Sciences Faculty was in large measure the result of the success of the institutes. Of the 1,479 students enrolled in 1928–29, a total of 952 were in the institutes, while 235 were PCN and 292 were in "other" programs, including the licence and agrégation.[77] Still, records show that 79 licences and 221 certificates of higher studies (of which three or four make up a licence) were conferred in 1929. (A student might take two certificates in one year.) The number of licences awarded at Toulouse in the late 1890s was less than ten in any given year, so that the 79 awarded in 1929 demonstrates substantial growth, even though this growth was outstripped by that of the institutes. The growth in science licences at Toulouse was more rapid than that of the "average" science faculty, as the number of licences conferred by all French faculties in the period between the 1890s and the 1920s only tripled.[78] But, again emphasizing the dominance of the institutes in Faculty enrollments and degrees, we see that the 79 licences at Tou-

louse represent only 5 percent of the students receiving numerous certificates and diplomas, whereas the proportion at Paris in 1929 is 25
percent.[79]

Also significant are the figures bearing on Toulouse's share of foreign
students enrolled in French university Science Faculties. In 1894–95,
Paris housed 80 percent of foreign science students enrolled in Faculties. By 1929, the Paris Faculty taught only 26 percent of this group,
while Toulouse's share had risen from 4 percent in 1894–95 to 16-½
percent in 1910–11 and 21-½ percent in 1929. Together, Paris and Toulouse shared almost 50 percent of the foreigners enrolled in Science
Faculties, with Grenoble their closest competitor at 13 percent.[80] In
1909–10 the university's annual report remarked that the new technical
institutes had so inundated Toulouse with foreigners that the city had
contributed 1,000 francs to aid in teaching many of them the French
language. In 1895, there had been only 47 foreigners in the entire university, most of them in law; in contrast, in 1909–10, there were 223
foreign students in Toulouse, with 174 of them at the Sciences Faculty
(of whom 149 were Russians). All this, as mentioned in the annual
report, was unforeseen.[81]

By 1929, the proportion of foreign students enrolled within the Toulouse Faculty of Sciences had increased to roughly 55 percent of those
inscribed and was even higher in the individual institutes; for example,
in 1927–28, 208 of 302 students in the Institute of Chemistry were foreign. Again, Eastern Europeans dominated, with the greatest numbers
being Polish, Rumanian, Bulgarian, and Russian.[82] After the First
World War, Sabatier and others became concerned with organizing
courses and exercises in favor of American students, as it had earlier
been announced that scholarships would be given to French students
to study at Harvard University and to Americans for at least one year of
study at the Universities of Toulouse and Montpellier.[83]

Clearly, then, Toulouse had begun to compete with Paris, especially
through its institutes, and Liard's goal of decentralization had been
dramatically furthered through the work of Baillaud's second successor, Sabatier. In later years, it became a cliché that Sabatier had been
principally responsible for the successes in enrollment at Toulouse. Camichel said that Sabatier had been struck with the possibilities for Toulouse since his youth, when Liard initiated the first reforms. He recalled Sabatier saying, "It is necessary to destroy the Napoleonic
centralization of the Ministry of Public Instruction, to organize the pro-

vincial University which suits its teaching to the soil and the characteristics of a region at the same time that it puts to profit, for its scientific
work, the resources of the region."[84]

If a major goal of decentralization was to strengthen regional economies, did this occur at Toulouse? Indeed, Toulouse, like France as a
whole, had been in a state of stability, if not stagnation, in the late
nineteenth century. As in most of Western Europe, the early years of
the twentieth century had been characterized by general inflation and
disaster in agriculture. But by the mid-twentieth century, Toulouse's
leading industries (with the exception of the airplane industry) were
precisely those in which the Sciences Faculty specialized through their
technical institutes. These include the manufacture of fertilizer and nitrogen products, which were topics under study by Sabatier, Prunet,
and other chemists of the Institute of Chemistry. Hydroelectric power
became of major importance as did electronics—both areas studied and
taught by Camichel and Bouasse. Viticulture emerged as another thriv-
ing area of the economy, again reinforced by the work of Sabatier's
institutes. Sabatier argued that agricultural science must be local: "nous
ne devons pas l'oublier, l'agriculture doit nécessairement être ré-
gionale."[85]

Because of the tradition of scientific education and technical institutions at Toulouse and the successful interaction of the regional economy with university curricula, the city has become a major focus of
French planning in recent years. Under the Fifth Plan of 1966–1970, for
example, Toulouse was designated along with Paris and Grenoble as a
center for computer and automation research. Along with Grenoble,
Strasbourg, Orléans, and Nancy, it is a focus for provincial regional
science.[86] Together with Lyon, Toulouse houses one of France's two
National Institutes of Applied Sciences. No longer do the institutes fit
into cubbyholes among traditional university buildings. They now
dominate a modern, 300-acre campus called the Université Paul Sabatier which specializes in sciences, languages, medicine, and technology.[87]

However, the long-term, impressive success of the Toulouse Sciences
Faculty does not mean there were not shortcomings and failures in the
meantime. Paradoxically, the very success of Toulouse in attracting
young people and training them as scientists and technicians led to
some difficulties when France was still underdeveloped industrially
and technologically. After the First World War, more students entered

higher education, but there were fewer and fewer jobs for graduating students, especially in secondary education. In 1918, the Faculty found it necessary to post a ministerial circular warning of the impossibility of assuring employment in secondary education to graduating students.[88]

Placing students in industry soon became a problem as well, since French industry was not developing rapidly. M. Nadal recounts a conversation with Sabatier after the war when Faculty members were becoming disturbed at the increasing number of diplomas: "We thought that it was about time to limit the number of technicians put on the job market. Sabatier invariably answered us by saying, 'We will always lack good chemists.' "[89] But this was hardly an acceptable solution for young French men and women who found themselves qualified to work but unemployed. This kind of frustration contributed to the malaise engulfing France's stalemated society.

The economic problems that resulted in unemployment for the Faculty's graduates also led to financial disasters for the Faculty itself. For twenty years or more, under Baillaud and Leclerc du Sablon, the budgets for the Faculty had steadily increased through combinations of state, city, regional, and private monies. With the founding of the new institutes, it became a rule of thumb that their students would pay expenses, and, year by year, fees throughout the Faculty and its institutes escalated. While some Faculty members, such as Leclerc du Sablon, were insistent that each laboratory receive all the receipts from its students, others, including the dean, argued that some laboratories had special needs and must to some extent be supported by fees levied in other laboratories. In practice, there was a good bit of juggling. Expenses in the institutes' laboratories were determined by fees, whereas fees from one pure science laboratory might be applied to equipment in another. As the years passed, Sabatier also switched funds from the Institut de Chimie to pay expenses elsewhere.[90] But with the war, the university enrollment dropped considerably and Sabatier felt hard pressed to keep the curriculum going "at the very moment when French public opinion is interested in the institutes of the Faculties and when it is possible to hope for serious help from the state as soon as hostilities are over." In 1917, Sabatier was asking the university to help balance the budget of the Institut Electrotechnique and the Faculties of Science and Medicine, and it was important that metal and steel mill industries and the railroad donated significant monies in 1917 and 1918.[91]

However, the monetary inflation affecting all of Europe and the rapidly increasing prices of coal, gas, and electricity began to take their toll
in the 1920s. This was a problem in science laboratories throughout
France. In Paris, for example, scientists repeatedly threatened to shut
down laboratories because of escalating costs for heating.[92] Sabatier
seems not to have made known the magnitude of the Faculty's debts,[93]
which state of affairs resulted in an ugly scene in late 1926 when, with
Sabatier at home ill, a budget for 1927 was presented to the Faculty
council reading 215,249.42 francs in receipts and 215,249.42 francs in
expenses. The geologist Charles Jacob asked the figures for the *real*
budget and Albert Lécaillon moved not to accept the budget.

To meet Jacob's objections, Nicolas agreed to append a statement
detailing the Faculty's true situation, but this was by no means the end
of the matter. In January, seven professors including Jacob sent to the
dean, the rector, and other council members a letter expressing astonishment that the financial conditions of the Faculty were not understood by all members of the council. They charged that the laboratory
budgets had not increased, and that it was the theoretical science of the
Faculty and not the applied science of the institutes which was suffering. "There is here a hardly healthy situation, which is the reason that
our Faculty's laboratories in pure science are more and more gravely
inferior in comparison to Sciences Faculties of other universities."

At the next meeting, in March 1927, Sabatier announced that he had
received from the director of higher education an extraordinary subsidy
of 200,000 francs, of which 100,000 francs would be applied to past
debts for gas (76,000 francs), electricity (8,000 francs), and chemical
products and other things (7,600 francs). The rest, he said, would go to
current expenses, including 65,000 francs for gas. To help make adjust-
ments for the future, Sabatier suggested raising students' fees 50 percent.

At this, Jacob exploded and demanded to know the origin of the
enormous debts and why the council had been misled. He had scoured
the minutes of the previous meetings for some indication of what had
been happening, he said, and had found no answer there to any of his
questions. Sabatier responded, with support from Adolphe Buhl, Nicolas, and Robert Deltheil, that all the council members knew there had
been debts at least since 1922 and that he had been able to pay some of
them off in 1925 from various credits at his disposition, particularly
those from his laboratory. Jacob countered: "It is this form of adminis-

tration, highly irregular in my view, which has accounted for opposition against the dean; the situation brought forward today to the council is its justification."

There followed a vote on the budget, and it was approved. An amendment was proposed "to relieve the council entirely from responsibility for the previous administration which has resulted in this distribution." It was defeated by a deadlock, with Jean Clarens (agricultural chemistry), Deltheil (general mathematics), Alexis Duffour (mineralogy), Jacob (geology), Charles Lamotte (physics), Lécaillon (zoology), and Louis Roy (rational mechanics) voting for the measure; and Sabatier (chemistry), Cosserat (astronomy), Buhl (calculus), Léon Jammes (applied zoology), Giran (chemistry), Nicolas (agricultural botany), and J. H. Sourisseau (agricultural mechanics) against. The next fall the budget for 1928 showed receipts of 384,250.40 and expenses of 384,250.40.[94] In 1928, Jacob left Toulouse for a chair at the Sorbonne.[95]

Jacob undoubtedly found the atmosphere at the Sorbonne more to his liking; indeed, his objections to the favoritism paid at Toulouse to applied science could not be refuted. There had for some time been complaints about funds paid to personnel at the institutes, many of whom were auxiliary in the sense that they did not belong, properly speaking, to the university.[96] Unlike what was happening in general in France, the Faculty staffing at the university and its institutes had not stagnated between 1910 and 1930,[97] although it did stabilize for at least ten years afterward. The gains made were primarily in the technical fields, although botany did pick up one chair. Faculty positions were added in particular by the Institut Electrotechnique, the Institut de Chimie, and by agricultural chemistry and physical chemistry.[98] If we judge budgetary expenses by comparing buildings, chairs, and total personnel, as well as by student enrollments, then clearly applied science was doing better at Toulouse than theoretical science.

The success of the Faculty institutes was a striking phenomenon. But if we note the departure of Mailhe for Paris in 1926 and of Jacob in 1928, we cannot help but ask if Toulouse was still failing to keep outstanding research talent from leaving for Paris once they were given the opportunity. Of those scientists who remained at Toulouse throughout their careers, there were some very good men. Three became nonresident members of the Academy of Sciences and three became correspondents: Sabatier in chemistry, Camichel in mechanics, Cosserat in astronomy, and Mathias in physics, Leclerc du Sablon in botany, and Roy in mechanics. Two well-known chemists collaborated with Sabatier at Toulouse—Mailhe, who later went to Paris, and J.-B. Senderens, who taught at the Institut Catholique in Toulouse. Most of the prominent permanent Toulouse Faculty members had been born in the south, including Sabatier, Camichel, Leclerc du Sablon, and Roy (Cosserat was born at Amiens). Of those prominent Faculty scientists native to the Midi, only Mailhe left for Paris. He and Senderens were among the few who had not studied at Paris originally.[99]

The most well-known scientists who stayed at the University of Toulouse were Cosserat, Sabatier, Camichel, and Bouasse. Cosserat no doubt found the Pyrenees most agreeable for astronomical observation, and he seems to have been a close friend of Sabatier. Camichel, a native of the Midi, was likewise intimate with Sabatier and was devoted to the Institut Electrotechnique. Sabatier, as we have seen, remained at Toulouse because of his commitment to the institutes he had founded, because monies and facilities were amply available for his own research, and because he was very fond of the region.

Bouasse taught at Toulouse from 1897 to 1929. Although not a member of the Academy of Sciences, he was well known in French science for a series of textbooks that were the standard preparation for the baccalaureate. He was an intensely conservative Catholic, and he was considered a troublesome eccentric by Parisian scientists. His vituperative attacks against quantum and relativity physics became infamous in France and were matched in tone by his attacks on individual French scientists and on the Paris Academy ("the receptacle of a crowd of mediocrities and ignoramuses whose places have been made as collège professors, herb collectors, village veterinarians and assistant engineers of bridges and roads"). He published statements insulting to Baillaud and Sabatier, and he and Sabatier were said to have fought "like cat and dog." In a 1911 meeting at Toulouse, Bouasse remarked to his colleagues that "with the exception of Monsieur Camichel, you are completely incompetent in physics," going on to speak of their "invincible ignorance." Not surprisingly, Jules Drach called Bouasse's remarks "insolent."[100]

This study of Toulouse suggests that those prominent scientists who elected to remain there did so for idiosyncratic reasons, having to do with a love for the region or circumstances unique to their own lives and work. Paris remained a goal for most science professors of talent and theoretical bent who had no special interest in applied science. In addition, once Bouasse and Sabatier stayed at Toulouse, difficulties must have arisen for new young men coming into the faculty, particularly those of the generation of André Job or Charles Jacob, when the science of Paris, where they had been trained, was largely dominated by men of the political left-center.[101]

Both Sabatier and Bouasse, in contrast to most professors of the Ecole Normale Supérieure and the Sorbonne, were practicing Catholics,[102] and as such, their politics were identified with the right or right-center. Sabatier had earlier in his career been regarded as unfit for the deanship for this very reason, with Rector Perroud observing that while Sabatier was a serious researcher and a good professor, "his opinions and beliefs have pledged him a little too far outside the liberal camp."[103] This assessment was not simple anticlericalism on Perroud's part; he defended Baillaud on one occasion by saying, "Everyone knows that he is a Catholic and a practicing Catholic. But from there to 'militant clerical' is a long way!"[104]

University and ministry records show that there was some factionalism and dissension among Toulouse Faculty members after 1900. This state of affairs was in part a matter of politics and in part a matter of personalities. Before the First World War, the rectors Jeanmarie and Lapie blamed Sabatier for Faculty discord. Lapie expressed the fond hope that Toulouse's Nobel Prize winner might give satisfaction as dean "to all the aspirations and legitimate propositions of the Faculty members if he would broaden his manner."[105] It is quite possible that in using the administrative power at his disposal to further the fortunes of the institutes Sabatier may have alienated some young Faculty scientists who felt their own research work was playing second fiddle and already felt uncomfortable with his and other colleagues' politics. Let us turn now to the career of this influential chemist and administrator.

## PAUL SABATIER AND MODERN CATALYTIC CHEMISTRY

Paul Sabatier's roots in the Midi can be traced back at least to the mid-sixteenth century when Thomas Sabatier became a town councillor of Toulouse. Paul Sabatier was born in 1854 in one of France's oldest cities, Carcassonne. Both Paul and his older brother Théodore followed mathematics and scientific subjects under the tutelage of their uncle, who was a professor at Carcassonne's lycée. The boys pursued their studies with academic careers in mind, a path not unusual for sons and nephews of men already teaching in French secondary and higher education.[106] When their uncle quit Carcassonne to teach at the Toulouse lycée, Paul Sabatier went along, adding public lectures at the Sciences Faculty to his education. The physics courses of Daguin and the chemistry lectures of Filhol inspired Sabatier to write his parents that he wanted to become a Faculty professor. After three years as a boarding student at the Collège Sainte-Marie, he placed high in exam competitions for the Ecole Normale Supérieure and the Ecole Polytechnique. Paul Sabatier chose the former, where his brother Théodore had preceded him. In 1877, he left the Ecole Normale first in the physics agrégation to teach at the lycée of Nîmes.[107] When Berthelot asked the Ecole Normale's director to recommend a normalien for his laboratory, the director proposed Sabatier. Berthelot wrote Sabatier that he was obtaining for him a leave of absence from Nîmes with the understanding that he should return to secondary teaching if a scientific career did not offer chances of success.[108]

Difficulties other than the usual academic ones beset Sabatier during those years in Paris when he was part of the elite group associated with the Ecole Normale Supérieure, the Collège de France, and the Ecole Pratique des Hautes Etudes. Although the normaliens of the 1870s and 1880s were not involved in political activities, unlike many of their counterparts in the 1890s, a number of students and professors were outspokenly freethinking and anticlerical, as well as republican.[109] The milieu of anticlericalism and materialistic positivism was trying for youths like Sabatier who were pious Catholics, and Sabatier's experience appears particularly uneasy since Berthelot was an extreme and arrogant advocate of positivistic scientism. Indeed, Berthelot regarded the struggle between clericalism and the new Republic as one deciding the destiny of France.[110]

Sabatier worked assiduously in Berthelot's laboratory and attended some of the Berthelot salons frequented by Ernest Renan and Georges Clemenceau, but he often found his habits of thought foreign to Berthelot's. Sabatier preferred Pasteur as a man of science to emulate, praising him as a "vehement fighter, intolerant like all those who have a profound faith."[111]

Differences of opinion between Berthelot and Sabatier extended beyond politics and religion to scientific ideas. This disagreement evolved over the older man's insistence on using the traditional notation of chemical equivalents rather than atomic notation. I shall analyze later Sabatier's writings on atomic notation and on the periodic classification of the elements, but here we must note that the immediate result of the two men's disagreement was one Sabatier considered devastating. He later confided to a friend that he had been "disgraced" because of his "independence of mind" and that Berthelot would not support him for a position in Paris. Given the choice of teaching at Algiers, Lyon, or Bordeaux, he decided on Bordeaux.[112]

Certainly Sabatier's doctoral thesis must have been influential in forming his teachers' opinion of him and in determining his first post. In this we find nothing Berthelot would consider radical or unorthodox. Indeed, the thesis was dedicated to Berthelot and was written firmly within the framework of his thermochemical studies. Entitled "Recherches thermiques sur les sulfures," it sought to complete the study of sulfides, and included accounts of the preparation of various metal and nonmetal, especially alkali and alkaline-earth, sulfides and polysulfides, with determinations of their heats of reaction and solution. Tables in the thesis are in terms of equivalents, and at various points Sabatier mentions that his experiments confirm Berthelot's principles or predictions.[113]

In the thesis committee's report, Henri Sainte-Claire Deville wrote that Sabatier's thesis was overly monographic, but he expressed confidence that once placed in a Faculty, with time and research apparatus at his disposal, Sabatier would demonstrate much originality of mind.[114] Berthelot recommended Sabatier to the post at Bordeaux. In 1882, when Sabatier learned that Toulouse's Sciences Faculty dean would support his candidacy for a suppléance in physics, he telegrammed Berthelot asking for a recommendation to the ministry. Berthelot must have spoken well of Sabatier, since Dean Baillaud had insisted that the ministry's appointment to Toulouse be a man who was a serious teacher and researcher.[115]

Sabatier's political and religious difficulties continued when he arrived at Toulouse. As we have seen, the university's rector, Perroud, was of the same republican and anticlerical stamp as Berthelot. Although Perroud initially expressed enthusiasm for Sabatier's performance as a Faculty member, he balked when Baillaud recommended Sabatier in 1885 for the deanship. Perroud apparently submitted a letter to the ministry detailing "formal reservations." In a report of the following year, Perroud observed that Sabatier was a serious worker and a good professor who showed concern for the students, but as far as the deanship was concerned, he was too far outside the liberal camp.[116] Sabatier's later friends spoke of his "independence of character," which brought him trouble first with Berthelot and years later with the rector.[117] Clearly, Sabatier's conservative political and religious persuasions presented obstacles to his early professional career.

In 1905, the year the Academy awarded Sabatier and his collaborator, Senderens, the Prix Jecker for their work in catalysis, the Toulouse Faculty of Sciences elected Sabatier to the deanship. Rector Perroud retired shortly afterward, and Baillaud accepted a position at the Paris Observatory. From that time on, Sabatier seems to have exerted firm control over the Sciences Faculty at Toulouse.[118] He remained a nonconformist, at least by Parisian standards, in his administration of the faculty, and his practical success in guiding it from 1905 until his retirement in 1929 was remarkable.[119] His personal international reputation in chemistry undoubtedly contributed to the success of the institutes in attracting students.

Innovative in administration, Sabatier might even be called radical in his teaching and research. In 1884, he succeeded Edouard Filhol in the chair of chemistry at Toulouse and, according to Baillaud's reports to the ministry, broke with outmoded traditions at Toulouse and worked harder than any other Faculty member. His courses were forcefully avant-garde, breaking down older demarcations between physical and chemical phenomena, discarding the notation of chemical equivalents, and employing Mendeleev's periodic table.

As we have noted, French chemists, in contrast to their English and German colleagues, generally had not been receptive to Mendeleev's work. Mendeleev's first periodic tables appeared in 1869, in the journal of the Russian Physical-Chemical Society. They were published quickly in German journals, and H. E. Armstrong published the first English-language account in 1876 in the *Encyclopaedia Britannica*. But many leading French chemists were unenthusiastic and skeptical.[120]

Adolphe Wurtz, who was not one of the skeptics, expressed admiration for Mendeleev's "powerful" principle of classification in his *La Théorie atomique* in 1879. Wurtz included in an Appendix a foldout graph of Lothar Meyer's table charting atomic volumes and atomic weights. He made explicit the linkage between atomic and periodic concepts by commenting that the periodic relations would have remained hidden if Mendeleev had attempted to deduce them from equivalents.[121]

It was this atomic connection which aroused Berthelot's reaction against the classifications of John Newlands, A. E. Béguyer de Chancourtois, Meyer, and Mendeleev. Berthelot did not deny the periodic system's usefulness, but he feared that it was too imaginative and would undermine positive science by throwing chemistry "into a mystic enthusiasm parallel to that of the alchemists."[122] Berthelot was not alone: in the next years, Paul Schutzenberger, Armand Gautier, and C. I. Istrati published classifications of the elements based not on their atomic weights but on their valences; almost twenty years later, Moissan was still citing Berthelot's 1885 criticism against Mendeleev's classification.[123]

In contrast to many of his older French colleagues, Sabatier began using the classification as a teaching device at least in the early 1890s. He published a memoir on the periodic classification in the Toulouse Faculty's *Annales* in 1890, saying that he customarily used a number of graphs in his teaching for demonstrating the law of periodicity, the most useful of which was the graph of atomic volumes (the volumes occupied in the solid state by the atomic weights of successive elements). Like Meyer, he linked coloration of chemical substances with the size of atomic weights.[124] He felt that the discoveries of gallium in 1875, scandium in 1879, and germanium in 1885 had demonstrated the capacity of the periodic principle for correct prediction, although convincing proof of its "reality" awaited the discovery of a new alkaline or alkaline-earth metal or even a halogen. In the meantime, the periodic classification was pedagogically justified since "the students . . . accept the presentation of the elements a lot better when thus grouped in a rational and somewhat unexpected manner."[125]

Sabatier's enthusiasm for Mendeleev's and Meyer's work may have been conditioned by his habit of reading in foreign journals and by his contacts with English scientists. Sabatier thus broke out of the French scientific community in a way uncommon to many French scientists. We may even speculate that it was Sabatier's relative estrangement from the mainstream of French chemistry which encouraged this internationalism. The English had awarded Mendeleev the Davy Medal in 1882 and the Faraday Medal in 1889, and in visiting London to accept them, Mendeleev on both occasions dined with William Ramsay, who later became Sabatier's good friend. Ramsay, who was fluent in French, met Sabatier in 1894 and probably reinforced his admiration for Mendeleev's work.[126] Doubtless, Sabatier found some comfort in knowing that his nonconformity in France was passé in England. Sabatier was likewise knowledgeable about developments in German science. Indeed, in conversations with journalists in 1912, he ascribed to the German Nobel Prize winners Emil Fischer (1902), Adolf von Baeyer (1905), and Otto Wallach (1910) support for his and Grignard's election to the 1912 Nobel Prize.[127] As we will see, he was wrong in this view.

By the early 1900s, Sabatier was less isolated in his approach to atomism. In 1902, young Paul Langevin gave his first course in Paris on ions and electrons, and Jean Perrin's lectures presented the atom as "jiggling around like an old friend."[128] In his use of the atomic theory and the periodic classification, however, Sabatier had stood apart from the majority of his French colleagues, particularly those of his generation. By 1909, W. A. Tilden could refer to the periodic law as a "commonplace of modern theoretical chemistry,"[129] but this was only becoming true in France for both the periodic law and atomic theory at the end of the decade. Sabatier's early espousal of atomic ideas provided a strong voice for the atomic conceptions whose late acceptance in France contributed to a reputation of backwardness for French science at the end of the nineteenth century.

Sabatier was to be vindicated in his conviction of the heuristic value of the atomic hypothesis, as questions about the mechanism of chemical reaction led to work on the hydrogenation of hydrocarbons by catalysis with finely divided metals. Sabatier's breakthrough into organic catalysis emerged most unexpectedly from two decades of work in inorganic chemistry. Before he even began the experiments that led to his Nobel Prize-winning work, Sabatier had published over ninety articles or essays in inorganic chemistry, for which he was given the Academy of Science's 1897 Prix La Caze.[130] Much of this research was on chemical equilibria, reaction velocities, and absorption spectra, employing the tools of thermodynamics.

He began moving away from thermodynamics in the mid-1880s in his work on the sharing of a base between two acids, an area of research where thermal information was inadequate.[131] Research on the sharing of a base between two acids had led Meyer to conclude in the late 1880s that there was no relation between chemical affinity and the release of heat, and that Berthelot's and Ostwald's thermochemical doctrines were becoming "as outmoded as Berzelius's electrochemical ones." Sabatier found Meyer's views attractive, but his own interest was in developing general laws of affinity where Berthelot's principle of maximum work (that the evolution of heat is a criterion of spontaneity) would be a particular case.[132] It was Sabatier's interest in relations of affinity between chemical atoms and molecules which led to the Nobel work.

In 1890, Ludwig Mond had synthesized an unusual metal carbonyl compound, $Ni(CO)_4$, and he and Berthelot independently succeeded in making iron carbonyl [$Fe(CO)_5$]. Sabatier and Senderens wondered if unsaturated gaseous molecules other than carbon monoxide would give analogous results. At this time Sabatier and Senderens had just begun what was to turn into a very fruitful collaboration.

Senderens, an abbot who had taught chemistry at the Institut Catholique since 1882, had been working in chemical research with Sabatier's predecessor, Filhol, when Filhol died in 1892. Senderens completed research for his doctoral thesis in Sabatier's laboratory that same year. Simultaneously, Senderens was directing laboratories for chemical research and industrial chemistry at the Institut Catholique, and he was active on scientific committees concerned with local, practical problems.[133]

Following up on Mond's and Berthelot's experiments with carbonyl compounds, Sabatier and Senderens succeeded in fixing nitrogen dioxide on metals such as copper, cobalt, nickel, and iron. The technique involved reducing with a current of hydrogen a tube full of the finely divided metal oxide and then passing an unsaturated gas through the tube. With nitrogen dioxide, an unstable black solid ($Cu_2NO_2$) was obtained at temperatures of 25–30°C.[134]

After failing to obtain an analogous synthesis with other nitrogen oxides (NO and $N_2O$), Sabatier and Senderens decided to try experiments on finely divided metals with ethylene and acetylene gases. Then they learned that Moissan and Charles Moureu had recently attempted this with acetylene and failed. What they had obtained was incandescence and a blockage of the tube with carbon, accompanied by the production of condensed hydrocarbons (such as benzene and styrene) and pure hydrogen gas. Moissan and Moureu concluded that a physical phenomenon caused the results: that the metal absorbed the acetylene gas, releasing heat, leading to a polymerization and decomposition of the acetylene. They got the same results with platinum black.[135]

Sabatier made some discreet inquiries into whether Moissan and Moureu would be continuing these experiments. After learning that they would not, he and Senderens set to work. The platinum black that the two Parisians had used was a well-known inorganic catalyst, and apparently its use sparked Sabatier into thinking about catalytic mechanisms. "I had very different personal ideas on the mechanism of the phenomena of catalysis—ideas which I doubtless owe to the influence of my teacher Berthelot," Sabatier later explained. "I attributed the decomposition of Moissan's acetylene to an affinity of the metal either for acetylene itself or for the constituents of the latter—carbon or hydrogen—which it could pull away from the endothermic molecule of this hydrocarbon."[136] Sabatier was convinced that what was occurring was a phenomenon of chemical affinity, not merely physical absorption.

In first repeating the experiment, Sabatier and Senderens used ethylene rather than acetylene, hoping that it would prove less energetic. At around 300°C, they too observed a blockage of the tube containing finely divided nickel and the production of free hydrogen. But they also found methane, and concluded that there formed between nickel and ethylene an unstable combination, analogous to nickel carbonyl, which doubled itself into carbon, methane, and nickel—which can then repeat an identical process.[137]

Pursuing this idea, the two tried the reduction of the finely divided nickel oxide at temperatures below 300°C, cooling the reduced nickel in a current of hydrogen and then of ethylene. After washing the gases, leaving the tube with bromine to absorb any traces of ethylene, they discovered the product gases to be a mixture of ethane, *formène* (a mixture in equal volumes of ethane and hydrogen), and hydrogen, with traces of hydrocarbons similar to Pennsylvania petroleums. Above 325°C, they found that the ethane decomposed into methane and carbon and the formène into carbon and free hydrogen. Directing equal volumes of ethylene and hydrogen on the reduced nickel with heating to only 30–45°C resulted in a gas mixture that was practically all ethane. This result, they believed, "ought certainly to be attributed to the temporary formation of a direct and specific combination of nickel and ethylene."[138] They now found that cold acetylene with nickel likewise produced ethane, and cobalt, iron, copper, and platinum black gave similar, but less intense, results. It was not only the easy hydrogenation of ethylene and acetylene that was extraordinary in their results but also the use of metals outside the platinum family for catalysts. This was completely unexpected.[139]

Following these initial successes, Sabatier and Senderens turned to the challenge of saturating the stable aromatic benzene ring with hydrogen. Berthelot had attempted earlier to hydrogenate benzene with a concentrated solution of hydrogen iodide in a sealed tube heated to 250°C, but he had only obtained the isomer of hexahydrobenzene, methylcyclopentane ($C_6H_{12}$). The two younger men made the attempt with reduced nickel and hydrogen at 180°C in a vertical U-tube cooled by melting ice. As Sabatier later reminisced, they noticed the tube becoming clogged by colorless crystals that they assumed must be benzene solidifying at 5°C, since cyclohexane had been reported in the literature to crystallize at −11°C. But when they opened the tube, they detected the odor of roses: "It was from cyclohexane obtained practically pure at the first attempt, the fusion point of which is in reality 6.5° . . . that hour was one of the greatest joys of my life."[140]

After this breakthrough, successes followed rapidly in the next three years, including the transformation of unsaturated ethylenic or acetylenic carbides into saturated carbides, the transformations of nitrate derivatives and of nitriles into amines, the change of aldehydes and ketones into the corresponding alcohols, the reduction of carbon monoxide and carbon dioxide into methane, and the change of phenol into cyclohexanol and of aniline into cyclohexamine. They also found that they could produce the major types of natural petroleum by modifying conditions of the hydrogenation of acetylene.[141]

In 1905, Senderens left Sabatier's laboratory and began working exclusively at the Institut Catholique de Toulouse. With Jean Aboulenc, he concentrated on the study of the catalytic action of metalloids on metals and of metallic oxides, salts, and mineral acids on alcohols and organic acids. He later said that, with the principal procedures for catalytic hydrogenation and dehydrogenation now worked out, he preferred to turn to a new topic after Sabatier and he were awarded the Prix Jecker. Senderens was to make many technical contributions in the field of refined chemicals in association with the large firm of Poulenc Frères.[142]

Mailhe had begun assisting Sabatier, and these two continued to work together for over ten years.[143] In their initial work, they studied the hydrogenation of polyhalogenated aromatic rings and concluded that reduction proceeds more easily the fewer the remaining halogens to be replaced.[144] Sabatier and Senderens earlier had discovered that they could dehydrogenate primary alcohols and obtain the converse aldehyde simply by varying conditions. Now Sabatier and Mailhe, in studying metallic oxides, found that some were catalysts not for hydrogenation and dehydrogenation but for hydration and dehydration. While ordinary vapors of alcohol directed at 250°C on reduced copper split into hydrogen and aldehyde, the same vapor directed on finely divided alumina or thoria split into water and ethylene.[145] Sabatier and his co-workers also learned to manipulate experimental conditions so that they could carry out both oxidation and reduction simultaneously, removing hydrogen in one part of a molecule and fixing it in another.[146]

Of reactions studied by Sabatier and his co-workers, several had commercial implications. These included the transformation of nitrobenzene into aniline, of acetone into isopropyl alcohol, and of carbon monoxide into methane; the preparation of cyclohexanes, especially cyclohexanol and paramethylcyclohexanol from phenol and paracresol; and, most outstanding from the industrial viewpoint, the transformation of liquid fatty acids (oleic acid) into solid acids (stearic acid).[147]

As we have seen, Sabatier's accidental entry into catalysis was occasioned by his doubting the thermal, physical explanation of Moissan's and Moureu's 1896 results. Sabatier preferred a chemical explanation involving molecules and their affinities. Perhaps he was influenced in this by experiences harking back to his student days, when Berthelot had insisted, in opposition to Pasteur, on the role of a chemical agent in the yeast-catalyzed transformation of glucose into alcohol.

Catalysis had acquired a history by the 1890s which dated from Kirchhoff's 1811 discovery that mineral acids on heating provoke the change of starch into dextrine and sugar, without themselves being modified by the reaction. Berzelius had defined catalytic phenomena as early as 1836, deriving the name from καταλγω ("I destroy"). Berzelius argued that the catalytic force acted on the polarity of atoms through some phenomenon of temperature elevation, the most popular view throughout the nineteenth century.[148]

Those who supported the physical theory of catalysis agreed that porous materials, which have surfaces that are very large compared with their masses, absorb gases with enough energy that the resultant compression and local heating of the absorbed gases causes a reaction that otherwise would require a much higher temperature. This view was supported in France by the work of Jacques Duclaux and Moissan on the absorption of gases by finely divided metals.[149] That a catalyst did not induce a reaction but rather accelerated it was also the view of Ostwald, whose Nobel Prize in 1909 was awarded for his work on catalysis and for investigations into chemical equilibria and rates of reaction. Ostwald argued against the formation of intermediate compounds, citing insufficiently exact measurements. It was necessary, he said, to prove that the succession of assumed reactions requires less time than the direct reaction itself.[150]

In analyzing the physical and thermodynamic theories of Moissan and Ostwald, Sabatier asked what *causes* the condensation of gases and vapors in the pores of a solid. He judged spurious the idea that the absorption could be correlated to ease of liquefaction of gases, since the easily absorbable hydrogen is a very difficult gas to liquefy. How could high local pressure and temperature be developed in cases where the catalyst is held in suspension? And, most important, how could a purely physical conception account for the specificity of catalysts and the remarkable diversity of effects produced, depending on the particular metal or oxide used?[151]

From the very first papers on hydrogenation, Sabatier was convinced that the transformations proceeded through a chemically unique, unstable intermediate, and we must assume that he and Senderens together worked out ideas on this point.[152] Berthelot had provided some precedent for a chemical explanation in work on the decomposition of hydrogen peroxide by alkalies and by silver oxide.[153] Sabatier assumed that in hydrogenation there were various nickel hydrides involved, whose composition depended on the activity of the nickel. Carefully prepared nickel resulted in the very active $NiH_2$, which would work on benzene, while impure nickel or nickel prepared at too high a temperature gave an impoverished hydride,

$$
\begin{array}{c}
Ni\text{---}H \\
| \\
Ni\text{---}H
\end{array}
$$

which is inactive with benzene, but works with the ethylenic carbides or nitrate derivatives.[154]

In proposing unstable intermediaries, Sabatier had ready answers to Ostwald's methodological and epistemological critiques. Regarding the time taken by a series of reactions, Sabatier cited well-known reactions such as the use of platinum in the oxidation of hydrogen, and responded that the formation and decomposition of intermediate compounds usually corresponds to a diminution of the free energy of the system. This diminution is accomplished by steps, he explained, and this process is frequently much easier than an immediate and direct decrease of free energy, just as the use of a staircase facilitates a descent.[155]

He argued that while the presence of catalysts might indeed lower the temperature required by a reaction, the catalyst's greatest asset was in reacting with a molecular gas in order to provide a free ion for a reaction that simply would not occur otherwise. While Ostwald might complain that there was no observational proof for these intermediaries, Sabatier could cite instances outside his own research where they had been observed. One that he noted was the intense blue coloration that appears when hydrogen peroxide and chromic acid solutions are mixed; the unstable combination $3H_2O_2:2H_2CrO_4$ can be isolated by shaking with ether and then evaporating the ether.[156]

In his early investigations, Sabatier met criticism of his chemical explanation of catalysis by an almost unerring ability to predict reactions using the supposition of a particular intermediate.[157] Zinc oxide and titanic oxide, for example, are physically similar, and the physical explanation of catalysis would predict similar effects with formic acid vapor at 300°C. With titanic oxide, water and carbon monoxide result, but in the presence of zinc oxide, one gets dehydrogenation into hydrogen and carbon dioxide, just as with copper metal catalyst. Also favoring the hypothesis of chemical reaction was the decomposition of alcohols into hydrogen and aldehyde on copper, and into water and ethylene on alumina or thoria.[158]

In 1912, Sabatier published his influential textbook *La Catalyse en chimie organique*, which became both exemplar and program for the science of organic catalysis.[159] In the next decades, as the Faraday Society hosted two major discussions on catalysis, Irving Langmuir's theory of "chemisorption" became a rival explanatory scheme to Sabatier's. Langmuir postulated that an adsorbed gas is held chemically by the unsaturated valences at the surface of the metal or oxide catalyst, forming a type of compound ($M_xG_y$) where $x$ may vary with the mass of the catalyst and $y$ with its surface as well as with the pressure and temperature.[160] This was contrary to Sabatier's hypothesis of distinct, individual intermediates, and accorded more importance to physical conditions. In 1927, Sabatier reiterated his original theory.[161] Twelve years later he said that throughout his work he had "jealously conserved [his] theory of temporary combination." As a criterion for this theory's validity, Sabatier cited its continued usefulness: "As Poincaré has justly said, 'A theory is good as long as it is useful.' Theories pass, and only the facts remain."[162]

Sabatier's views on scientific method associate him with the avant-garde of the late nineteenth century. Sabatier was not an original thinker in this regard, but was receptive to the ideas of others, including the physicist Brillouin, the mathematician Picard, and the chemist Job. Like them, he owed a great debt to the writings of Henri Poincaré.

Brillouin, Picard, and Job seem to have reinforced Sabatier's optimism about the heuristic value of the atomic approach. Brillouin, an old friend, taught at Toulouse from 1883 to 1887, and he was one of the few physics professors at the Ecole Normale Supérieure in the 1890s who was completely sympathetic to the kinetic theory. He regarded the atomic, corpuscular hypothesis as valuable in its testable consequences and in its ability to group a large number of superficially unrelated phenomena.[163] Like Brillouin, Picard (who taught at Toulouse shortly before Sabatier joined the Faculty) was close to Sabatier during their student days at the Ecole Normale. Picard favored more and more the molecular-kinetic approach in physics, and in doing so, he also emphasized its usefulness.[164] Job was one of Sabatier's younger colleagues who taught at Toulouse from 1903 to 1908. Job's research during these years was directed to chemical kinetics and reaction mechanisms, and Sabatier found support for his own approach in Job's.[165]

It was Poincaré's influence that is most obvious in Sabatier's methodological statements. His use of Poincaré's ideas is yet another example of the powerful role played by this mathematician and philosopher in the methodology and theory of modern science. Sabatier often quoted Poincaré when speaking of his own use of the atomic hypothesis. In accepting the Nobel Prize, Sabatier spoke of the ephemeral life of scientific theories in Poincarèsque terms: "Theories should not have the pretention of being indestructible. They are only the plow which serves the laborer in tracing his furrow and which he is allowed to replace with a more perfect one the day after harvest." Sabatier echoed Poin-

caré's dictum that science does not attain knowledge of things themselves, but only of relations between things.[166] In his idea that scientific
theory is time bound in a way that art is not, Sabatier was apparently
influenced also by the physicist Jules Tannery, Baillaud's brother-in-
law.[167]

In 1912, Sabatier shared the Nobel Prize in chemistry with Victor
Grignard. This was the first chemistry prize awarded jointly, and there
was some criticism of the Swedish Academy of Sciences. As we will see
(chap. 5), Grignard himself said he regretted that the 1912 prize was not
conferred jointly on Sabatier and Senderens, and later, jointly on himself and his teacher, Philippe Barbier. While Sabatier was not criticized
for the Nobel Prize committee's decision, he was taken to task for a
remark made in a 1911 lecture before the Berlin Chemical Society. On
that occasion Sabatier referred to Senderens as his student and did not
describe clearly Senderens's independent role in their work. In a subsequent letter to the Berlin Chemical Society, Sabatier stated that a misun-
derstanding had occurred and he had not intended to demean the active and important role of Senderens in the discovery of the method of
catalytic hydrogenation and dehydrogenation. However, it is not surprising that the relationship between the two men cooled and that
many French chemists, well disposed to Senderens or concerned about
fairness, were bitter at the Nobel Prize announcement of 1912.[168]

While Sabatier attributed his election to the influence of German
chemists with whom he had talked in Berlin, he in fact owed his nomi-
nations in 1912 to Gaston Darboux, perpetual secretary of the Academy
of Sciences (who nominated Sabatier and Georges Urbain, jointly), and
O. Widman, a member of the Swedish Academy of Sciences and the
Nobel Prize chemistry committee (who nominated Grignard and Sabatier, jointly). Nominations of Sabatier had come as early as 1907, and
Georges Lemoine nominated Sabatier and Senderens for a joint prize.
The Nobel Laureate Ernst Büchner was the only German to nominate
Sabatier (in 1911). The award to Sabatier was the single occasion in the
period 1901–1915 for a protest note to the Academy. It came, quietly,
from Senderens, who sent an account of his long collaboration with
Sabatier, noting that "M. Sabatier paraît marquer une tendance assez
prononcée à se faire le seul auteur de ces méthodes."[169]

In 1913, Sabatier became the first member elected to a newly created
section at the Academy of Sciences for members not living in Paris. "It
was necessary to modernize," Sabatier was told in a ceremony of 1913.

"This result is due in large part to the irresistible push of your works, and this will not be the least of [your] services . . . rendered to this nation, whose intellectual forces risk becoming sterile if they continue to become too concentrated."[170] After Sabatier, five more nonresident members were elected in votes at the Academy from April 1913 to December 1913: Georges Gouy at Lyon; Henri-Emile Bazin, retired at Chenove; Charles Depéret at Lyon; René Gosselet at Lille; and Pierre Duhem at Bordeaux.

In electing Sabatier to the Academy, not only were Parisian scientists recognizing a non-Parisian French scientist who had won a Nobel Prize[171] but they were also recognizing one who had chosen to remain in the provinces knowing that only Paris residents could be elected to the Academy. Five years earlier, in 1907, Sabatier had refused both Moissan's chair at the Sorbonne and Berthelot's (his former teacher at the Collège de France).[172] That he would do such a thing is almost inexplicable within the framework of the French academic scientific community, since both chairs were specifically within his field of research and teaching.

When the Parisian opportunities came in 1907, Sabatier had just taken over the deanship at Toulouse. He was intimately involved in setting up the new Faculty institutes, particularly his Institut de Chimie. The former dean, Baillaud, was leaving the Faculty, and the liberal rector, Perroud, was retiring. Camichel, who became one of Sabatier's closest friends, had joined the Faculty and was taking over the direction of the Institut Electrotechnique. Monies from various sources were flowing into both the chemical and electrical programs at the university, and Sabatier's immediately practicable research on catalytic methods of hydrogenation and dehydrogenation was attracting quite adequate funding.

Sabatier knew that provincials could not be elected to the Academy, but he doubtless assumed that he could never have research facilities at Paris comparable to those he was building at Toulouse. He probably thought that he would have at Paris neither the administrative control nor the personal autonomy of his present situation. He certainly would not have found at the Paris Sciences Faculty the vigorous interest in teaching applied science which predominated at Toulouse. Sabatier's granddaughter has explained that he remained at Toulouse because of the institutes and because Toulouse was "chez lui."[173] His decision was an enactment of his expressed commitment that "light should come not only from Paris, but also from the provinces." As he later said of the Academy of Sciences, "chance smiled on me; they changed the rules."[174]

At the turn of the century, a period in which French science was said to be in decline, particularly with respect to German achievements in physical science and technology, Sabatier stands out as a brilliant and somewhat isolated figure among French scientists. He was known worldwide in both the theory and the application of that organic chemistry in which the Germans were said to have a monopoly. He was not an integral member of the powerful French scientific community centered at Paris, and he cannot be used as an example of the characteristics of that community.

Sabatier was cited in his lifetime as an example of the achievements of the decentralizing program in French institutional reform. But an understanding of the fabric and structure of the French scientific community at the turn of the century makes evident that the teaching and administration of French science did not encourage the influence of a conservative, Catholic, and provincial scientist committed to the atomistic mechanisms of theoretical science and the engineering institutions of applied science.

Certainly, Sabatier tried for a number of years to advance his career within the structure and rules of the Paris-dominated French scientific community. But after some time in the provinces, he became outspoken in his opposition to Berthelot's antiatomism and other pedagogical and theoretical approaches typical of the Sorbonne and Parisian schools. Sabatier's adherence to a mechanistic representation of chemical atomism was a position shared with Berthelot's Parisian opponents, Wurtz and Brillouin. Sabatier also was familiar, as we have seen, with English and German scientific views. One further important influence may have been the promechanist publications of some Belgian and French Catholic scientists writing in the 1870s and 1880s for the Catholic-associated *Revue des Questions Scientifiques*.[175] To his great benefit, Sabatier was a man familiar with scientific conjectures outside the Latin Quarter.

If we assume that flexibility and diversity of opinions lead to intellectual and scientific creativity, then Sabatier indeed seems to represent an example of this correlation. Similarly, if competition of institutions and ideas aids achievement, this relationship is exemplified in Sabatier's somewhat bitter sense of rivalry toward the individuals and institutions of the Parisian establishment. Under Sabatier, the research and teaching program at Toulouse established a new field in organic chemistry, radically outstripping Paris, and the foundation of technical institutes at Toulouse offered some promise of counterweights to the traditional centrality of Paris in the French scientific community.

# 5

# LYON: APPLIED CHEMISTRY, VICTOR GRIGNARD, AND THE EXPERIENCE OF THE FIRST WORLD WAR

In his study of the French universities at the turn of the century, W. Lexis, a Göttingen professor, spoke of Lyon as the "most important and best equipped among French provincial universities."[1] In Napoleon's decree of 1808 organizing the Université de France, the academy of Lyon included the départements of the Rhône, Ain, and Loire. There were three Faculties: Sciences, Letters, and Catholic Theology. From 1808 until 1885, the Theology Faculty existed without interruption. The Faculties of Sciences and Letters were suppressed in 1815 and reestablished in 1833 and 1838, respectively. A secondary school of medicine, created in 1821, became the Ecole Préparatoire de Médecine et Pharmacie in 1841 and a regular Faculty in 1877. A Faculty of Law was created in 1875; Theology was suppressed in 1885.[2]

Until 1900, the Faculties were dispersed throughout Lyon, a city in which medieval streets and towers line the right bank of the Saône River and back up against hills on which two Roman theaters still stand. The first Sciences Faculty was housed from 1808 to 1815 in the Palais Saint-Pierre, a twelfth-century convent between the Saône and Rhône rivers in that part of municipal Lyon built mostly during the seventeenth century. When reestablished in 1833, the Sciences Faculty took up quarters in the basement of the Collège Royal on the right bank of the Rhône, and in 1865 moved to two floors of a building beside the Palais Saint-Pierre on the broad rue de l'Hôtel de Ville.[3] In the nineteenth century, the city began spilling over to the Rhône's left bank, where in 1900 all the university Faculties were assembled together on the quai Claude Bernard overlooking the river like small, splendid parliament buildings for Lyon's Republic of Science. The construction of the new university palais cost more than ten million francs, with the city of Lyon shouldering most of the financial burden. On the day of inauguration of the buildings, the university had 63 chaired professors teaching 2,456 students, of whom 73 were women and 94 were foreign students.[4]

The funds pouring into the university came from local industrialists, municipal and départemental councils, and townspeople convinced that it was to the region's benefit to support education and research. Tight links joined universitaires and local government. Paul Cazeneuve, a professor in the Medical Faculty, presided over the départemental general council and was elected to the national Chamber of Deputies around 1900. Three successive mayors of Lyon from 1884 to 1957 were university professors. The palais was built under the administration of Mayor Antoine Gailleton, who taught in the Medical Faculty. His successor, Victor Augagneur, was another professor in the Medical Faculty. Edouard Herriot, mayor for more than fifty years, succeeded Augagneur in 1905 and held onto the mayoralty until 1957 while serving as a senator (1912–1919), deputy (1919–1942), and prime minister (1924–1925 and 1926–1932).

In his early years, Herriot used his brilliant oratorical skills to attract large crowds to his course in French literature at the Faculty of Letters. As mayor and parliamentarian, he turned those same skills to regional and national politics, using his influence in Paris to enhance the prosperity of his native city. He encouraged scientific and technical education and denounced the Ecole Polytechnique as the "only theology faculty that has not been abolished."[5]

The city of Lyon had long enjoyed special status in France, both politically and economically. It was from the balcony of Lyon's Hôtel de Ville that the Third Republic was first proclaimed on September 4, 1870, almost a day earlier than in Paris, following the French defeat by the Germans. The demise of the Second Empire became an opportunity for efforts at decentralization by Lyonnais provincial leaders. After the new Republic's appointee for prefect to Lyon arrived in the city, he reported to Gambetta that the Lyon citizenry was intent on increasing local autonomy and control. A moderate municipal assembly continued to demand new federalism for France long after the upheavals of 1870–71 were past.[6]

As mayor, Herriot compared the spirit of the city to England's Manchester: "The boldness and commercial integrity of Lyon are traditional. In its farsightedness and spirit of initiative, our Chamber of Commerce resembles that of Manchester."[7] Lyon was a city in which commerce and industry had thrived since Roman times, and Lyon's fairs attracted merchants and bankers from all of Europe and the Mediterranean during the Renaissance. The new system of railroads in France, centered in Paris, began to undermine Lyon's position in trade in the mid-nineteenth century, as did the removal to Paris in 1865 of the headquarters of one of the leading banks, the Crédit Lyonnais.[8] Traditional manufactures in textiles, especially in the silk industry, were to give way to new industries by the turn of the century. Among the city's most successful entrepreneurs around 1900 were the brothers Louis and Auguste Lumière, pioneers in the development of photography and inventors of cinematography. Their commercial ventures, scientific and technical innovations, and close connections with the Sciences Faculty typify the good relations in Lyon between industry and science.[9]

Especially important to the Lyonnais economy was the development of the chemical industry of dye manufactures, a field in which Germany took a decisive lead after 1870. It was initially in Lyon that the Renard brothers sold the triphenylmethane dye magenta (fuchsine) discovered in 1859. A Renard lawsuit in 1864 to prevent other French firms from manufacturing the dye by more efficient procedures became an important factor in the technical and commercial lead soon enjoyed by German and Swiss dye, pharmaceutical, and explosives industries.[10] French emigrants to Switzerland during this period were the founders of firms that eventually became the Swiss chemical giants Geigy, CIBA, and Durand-Huguenin. Among these emigrants was the Monnet family, who came to Switzerland and then returned to France to establish the Usines du Rhône and Rhône-Poulenc in Lyon.[11]

Well before the 1860s, Lyon industrialists lobbied for the establishment of state and municipal opportunities for chemical education and practical training. Lyon's Sciences Faculty and municipally supported school of chemistry were already in the 1850s the sites of serious training in basic and applied chemistry. It is not surprising that organic chemistry was to become a specialty of new research and engineering schools, associated in the twentieth century with Victor Grignard, who took his doctoral degree at Lyon in 1901 and returned there to teach after winning the 1912 Nobel Prize in chemistry.

## A MARRIAGE OF SCIENCE AND INDUSTRY

The core of Lyon's Sciences Faculty was seven chairs established in 1834, to which four more chairs were added in the period 1870–1895.[12] In 1906, a third chair of mathematics, in differential and integral calculus, was created; and in 1922, the chair of chemistry applied to industry and agriculture was made into two chairs.[13] Among Lyon's first professors in the reestablished Sciences Faculty of 1834 were several eminent men, including the mathematician A. A. Cournot, who left Lyon in 1836 to become Grenoble's academy rector, the physicist Victor Regnault, who moved to Paris in 1836, and the chemist J. B. Boussingault, who went to Paris in 1836.[14] Lyon maintained its academic and scientific reputation throughout the nineteenth century, and when the Academy of Sciences elected its first set of nonresident members in 1913, the Lyonnais scientists Charles Depéret and Georges Gouy were among the six elected.

Gouy came to Lyon in 1883 and taught physics until his retirement in 1925.[15] As a student, Gouy had been something of a maverick in Paris. His teacher, Jules Jamin, did not take well to Gouy's objections to Jamin's explanations of soap bubbles, for example.[16] Gouy completed his thesis in photometry in 1879 and, along with Pierre Curie, who remained his close friend after Gouy left Paris, became an aide in the laboratory of Paul Désains. Reserved and taciturn, and criticized by université inspectors for lecturing in a monotone, Gouy was a first-rate scientist by the 1890s.[17]

In addition to research on light waves,[18] among Gouy's most important work was his recognition of the importance of Brownian motion as an application of the kinetic theory of gases, at a time when the kinetic theory was not popular in France. He recognized that the motion of microscopically small particles in liquids and gases poses a fundamental problem for Carnot's principle and suggested that the second law of thermodynamics might be inapplicable to the world of small dimensions.[19] Asked by Picard in 1898 to pose his candidacy for a position in physical chemistry at the Sorbonne, Gouy responded that he would not leave Lyon and that the lack of a proper laboratory at Paris was decisive.[20]

The other Lyon scientist elected to the Academy in 1913 was Depéret, dean of the Sciences Faculty from 1896 to 1929. Depéret's initial work in geology was on the Tertiary strata of the Roussillon region. His monograph with F. Delafond on the Tertiary geology of the Bresse region became a classic in French geology. Depéret also wrote a book on paleontology, *Transformations du monde animal*, that gained him an audience in Europe and America. In the long run, Depéret's paleontological studies were called into question with the criticism that his evidence was often fitted too hastily to questionable hypotheses and preconceived theories. But his reputation was excellent in his lifetime.[21] In the natural sciences, during the years 1893–1896, Depéret was joined at Lyon by the evolutionary biologist Félix Le Dantec, who left Lyon to fill the Sorbonne chair for the "evolution of organized beings." Le Dantec was succeeded at Lyon by Maurice Caullery, who became one of the best-known French biologists of the early twentieth century.

In addition to physics and geology, Lyon had a strong curriculum in mathematics. Especially notable was Ernest Vessiot, a normalien who taught at Lyon's lycée from 1887 to 1892 and returned to Lyon in 1897 after sojourns in Lille and Toulouse. His work applied notions of continuous groups to the theory of differential equations, extending Picard's studies and developing results obtained by Drach and Cartan.

For several years, Vessiot and Cartan both taught at Lyon. The son of a blacksmith in the French Alps, Cartan attended the Lyon lycée as a scholarship student before entering the Ecole Normale and teaching at Montpellier, Lyon, and Nancy. In 1912, Cartan became a professor in Paris, two years after Vessiot moved from Lyon to Paris. Vessiot is perhaps best remembered as a director of the Ecole Normale Supérieure, whereas Cartan, in his work on the so-called Lie groups and his analysis on differentiable manifolds, is regarded as one of the great mathematicians of the century.[22]

However, it was not in physics, mathematics, or the natural sciences that Lyon's Faculty was to achieve its greatest reputation, but in chemical science, a field in which there was strong interest from local industry. Lyon's first chemistry professor, appointed in 1808, was Jean-Michel Raymond, inventor of a blueing dye that made him a fortune. Quitting the Faculty, he established a chemical factory in his native town of Saint Vallier in the Drôme, probably a wiser move than even he initially realized, given the suppression of the Sciences Faculty in 1815.[23]

After Boussingault left Lyon's chair of chemistry in 1836, he was replaced by A. Bineau, a student of Dumas and Thenard, who stayed in Lyon until 1861.[24] Bineau was followed in the chair of general chemistry by Adrien J. J. Loir and then by Berthelot's student Philippe Barbier. The municipality was especially interested in the technical applications of chemical expertise to local industries. At the Sciences Faculty's founding, the Academy's council extolled "with satisfaction the definitive institution of a Faculty which promises to render eminent services to science and to industry."[25]

The ministry, however, would not allow a systematic curriculum in applied science at the Sciences Faculty. As a result, local industrialists founded the Ecole Centrale Lyonnaise in 1857, on rue Chevreul on the left bank of the Rhône. The school was to have a curriculum intermediate in level between the écoles des arts et métiers and the Ecole Centrale in Paris; in 1888, it was taken under the patronage of the chamber of commerce.[26] The city also began a program of evening courses in 1872, requesting the participation of university professors. In the first year, Loir gave lessons in chemistry, Emile Duclaux in physics, and Théodore-Désire Dieu in industrial mechanics.[27] With the exception of a brief respite in the 1890s, when the municipality encouraged the teaching of only very popular lectures, Sciences Faculty members continued to participate in the municipal program through the 1920s.[28]

One of the most successful of the public lecturers was Jules Raulin, a former pupil of Pasteur whose doctoral research on the chemical role of minerals in plant nutrition received high praise from Pasteur. Raulin came to Lyon in 1879 to fill a new chair in chemistry applied to industry and agriculture. In 1883, he organized the Ecole de Chimie Industrielle (ECI) in the attics of the Palais Saint-Pierre, with twelve students under his tutelage, and devoted most of his research time to agronomy and the production of silk.[29]

Immediately after the 1885 decree giving legal, civil status to the Faculties, the Lyon chamber of commerce offered the Sciences Faculty 9,700 francs annually to help Raulin teach industrial chemistry. This sum included 4,000 francs for matériel, 3,200 francs for four student scholarships, and 2,500 francs for the salary of a lecturer or assistant director.[30] Because the students to be trained in the ECI were "dehors de toute préparation aux grandes universitaires," the Sciences Faculty council decided that the businessmen's designated favorite for the post,

Léo Vignon, should be given the title "assistant director" rather than lecturer, and that the Faculty should request as little money as possible from the state ministry as a salary supplement to the municipality's 2,500 francs.[31] In spring 1889, however, Raulin[32] successfully supported Vignon's request for a change of his title to lecturer, and in December 1892, he supported Vignon's candidacy for the title of adjoint professor,[33] a title Vignon received in summer 1895 after some dissension among members of the Faculty council and some difficulties with the minister. This was a typical route by which an irregular position in applied science became legitimized in provincial universities. In 1896, Vignon was unanimously voted the council's first choice to succeed Raulin in the chair of applied chemistry.[34]

In 1894, shortly before Raulin's death, the Sciences Faculty and the Medical Faculty jointly requested municipal aid for an institute, to be called the Institut de Chimie, which would provide chemical services for both Faculties. The municipal council donated land, and with monies put together from different sources, the institute was completed in 1900.[35] The new building included large laboratories on two floors for medical students in the PCN program, as well as a 300-seat amphitheater for use by both general chemistry students and Vignon's ECI students.[36] The ECI was moved from its quarters in rue Pasteur to the north half of the new institute building. Barbier, who held the chair of general chemistry, taught in the ECI, as did Albert Offret, professor of mineralogy, and H. Rigollet, lecturer in industrial physics. The ECI staff was a mixture of scientists with primary interest in fundamental science and scientists committed to engineering science, all of them formally university personnel. This combination of pure and applied science in one institution was a hallmark of the organization of Lyonnais institutes in the next decades and was considered a great strength.[37]

It was not only the teaching staff of the ECI that mixed theory and practice but its administration as well. Vignon was assisted by a committee named by the minister of commerce, on the nomination of the Lyon chamber of commerce.[38] After the First World War, a group of Lyonnais industrialists joined together to establish an organization they named the Fondation Scientifique de Lyon et du Sud-Est, under the leadership of Joseph Gillet, head of the giant Progil firm, which manufactured textiles, dyes, artificial silks, pharmaceuticals, perfumes, and other products.[39]

With the agreement of the chamber of commerce and the university they established the ECI as a corporation in 1921, detached from the university and in rented university buildings. The direction of the school included a university committee composed of the rector, Sciences Faculty dean, and the professors of chemistry in the Sciences and Medical Faculties. After 1921, students were selected for the school by a competitive concours examination, at the level of the mathematics baccalaureate. The program of study was three years; the first year was devoted to mineral chemistry, the second to organic chemistry, and the third to physical chemistry, organic chemistry, and some specialized fields. The curriculum included visits to factories and an examination that led to an engineering diploma. Most of the students in the ECI also completed the requirements for the state licence degree.[40]

The administrative reorganization of 1921 was carried out after Vignon retired from the Faculty in 1920. The direction of the ECI was given to Grignard, who returned to Lyon in 1919 from Nancy to replace his retired teacher, Barbier, in chemistry, soon renamed organic chemistry. Grignard's assistant director was Louis Meunier, an old comrade and longtime friend since their school days at the Ecole Normale d'Enseignement Spécial in Cluny. Meunier was named to a chair in industrial chemistry in 1922.[41] Another Cluny comrade at the Lyon Sciences Faculty was Paul Wiernsberger, who taught applied mathematics and directed the Ecole Technique de la Martinière, a privately endowed school of industrial arts established at Lyon in 1826.[42]

While the program of the ECI was primarily in organic chemistry,[43] the general university council organized a lectureship in industrial physics in 1898. It went to Rigollet, who had been teaching at the Ecole Centrale Lyonnaise for almost twenty years, while serving on the laboratory staff and teaching as chargé de cours in the Sciences Faculty.[44] Rigollet continued to teach in the Ecole Centrale Lyonnaise, serving as its director from 1903 to 1926, in a typically tight connection between the municipal school and the university. In the 1920s, the electrical engineering students of the Ecole Centrale Lyonnaise took courses in common with licence students who were studying for a certificate in industrial physics.[45] Then, in 1929, the University of Lyon created the Institute of Higher Studies of Industrial Physics, and all students associated with the Ecole Centrale Lyonnaise automatically became enrolled at the Sciences Faculty and received "University" diplomas.[46] Since the Ecole Centrale Lyonnaise continued to have close relations with the chamber of commerce, this physical institute, like the chemical institute, exemplifies very close links between the university and local commercial interests.

It was not until 1922 that Lyon had a lectureship in physical chemistry, partly because of the program in industrial physics. Barbier approached the ministry in 1906 regarding the creation of a chair of physical chemistry by the transformation of a lectureship in general chemistry. Théodore Vautier objected to the idea, saying that the generosity of the state could be expressed more usefully by developing the teaching of industrial physics. Dubois, Flamme, Gérard, Gouy, Koehler, Offret, Rigollet, Vautier, Vessiot, and Vignon unanimously voted against Barbier's idea, in Barbier's absence. In their formal statement to the ministry, they remarked that physical chemistry was a new and difficult science that required mathematical, physical, and chemical knowledge, that few scientists possessed the training to make this kind of teaching fruitful, and that the creation of the chair would create problems for the current organization of both the chemistry and physics services. The Lyon group further pointed out that "the Sorbonne itself does not possess a magisterial chair, but only a complementary course in physical chemistry," and they said that the Lyon Sciences Faculty did not have the funds to pay extra staff and course expenses associated with a new chair.[47]

This is important because the Institut de Chimie was to become known in the next decades for its devotion to organic chemistry, with much less emphasis than Nancy on physical chemistry and physics. Indeed, it has been argued that Lyon was typical of French chemistry in this regard, and Nancy atypical, with the result that as late as 1960, the study of physical reaction mechanisms was hardly taught by French organic chemists, while it became standard outside France.[48]

By the 1920s, the applied science institutes that had flourished before the war were in some difficulty. Part of the problem was the economic inflation of the 1920s, especially the increase in the price of coal and coke. In addition, the government decided in 1922 to cut out ninety-six posts in the universities, including twelve Faculty chairs. The ministry requested that all Faculties designate some of their chairs as "fundamental" and others as "nominal," so that the latter might be transformed when a chair holder died or retired. Not surprisingly, Lyon's scientists resisted the notion that any of their chairs were not "fundamental."[49]

The government tried another approach. Arrangements for paying salaries and laboratory expenses were often joint ones between the state and local groups. For example, a lectureship in industrial chemistry was originally funded by the Syndicat des Cuirs et Peaux in 1911 as the difference in monies between Meunier's salary as laboratory supervisor and the average salary of a Faculty lectureship. The ministry agreed to this arrangement with the provision that the Sciences Faculty never request the reestablishment of the laboratory post and that the Syndicat make a long-term commitment of money for a "University" position.[50] When Vignon retired from the chair of chemistry applied to industry and agriculture in 1920, Gillet's group, the Fondation Scientifique de Lyon et du Sud-Est, provided funds for splitting Vignon's chair into two professorships, one in industrial chemistry for Meunier and the other in agricultural chemistry (for A. Couturier). In 1925, however, the ministry eliminated payments of supplementary funds for salaries of "les titulaires de emplois d'Université" designated before July 1919.[51] Some of the state's "hard" money turned "soft."

At Lyon, too, student fees were increasingly important to the Faculty for ordinary operating expenses, and in some cases they were used to help pay professors' salaries.[52] The university rector Pierre Joubin had expressed grave concern at a drop in university revenues in 1907, when there occurred what turned out to be a temporary decline in student enrollments. The 1907 university budget deficit of 7,000 francs was a far cry from the more serious matters of the late 1920s, when the deficit for the Sciences Faculty alone climbed to 36,000 francs in 1928.[53]

Like other universities, Lyon began to develop a clientele of foreign students. The Ecole Française d'Ingénieurs was organized at Beirut.[54] By 1914, Egyptians were among the most numerous groups of foreign students to come to Lyon, and in 1926, Lyon attracted sixty-seven Chinese students, thanks to an agreement negotiated between the French and Chinese governments and the Universities of Lyon and Canton.[55] By 1930, in the "state" programs, the Sciences Faculties offered a PCN certificate, twenty certificates of higher studies, diplomas of preagrégation studies, and the doctorate. In its "University" programs, it offered a doctorate and various diplomas including electrotechnical studies (the program of the Ecole Centrale Lyonnaise), chemical engineering, and the diploma of technical studies in industrial chemistry.[56] Many of these students were foreign.

The Lyon Faculty showed little interest in the 1923 reform movement led by Paul Janet, who was concerned that the program of certificates had lowered the level of the science licence.[57] Lyon did not join Grenoble in suggesting that all agrégation students be prepared at Paris, but lobbied instead for specialization at provincial centers with increased state support for scholarships, teaching personnel, and matériel.[58] Perhaps the different response by the Lyon Faculty was a result of its long-standing reputation as a leading provincial center for education. While the Lyon council requested in 1919 that its scholarship students be transferred in that year to Paris to study at the Ecole Normale (the students were sent to Strasbourg instead), this request was unusual and due to short-term postwar problems.[59] Within a few years after the war, Lyon had ten candidates preparing for the mathematics agrégation, and during the period 1919–1926, twelve students from Lyon competed successfully in Paris in the mathematics agrégation concours.[60]

In 1927, Grignard's chemistry laboratory at the Institute of Chemistry became attached to the Ecole Pratique des Hautes Etudes, as a result of the reputation of the Lyon program for fundamental research.[61] In 1930, the educational ministry solicited the Lyon Sciences Faculty's opinion about instituting a core curriculum common to all university-affiliated chemistry institutes.[62] This inquiry was one of several that eventually resulted in the 1948 establishment of a nationwide, coordinated system of Ecoles Nationales Supérieures d'Ingénieurs, which the Lyon Institute of Chemistry joined in 1951.[63] Lyon's twentieth-century reputation as a center for academic and industrial chemistry is rooted in a tradition of more than one hundred years of cooperation and mutual support between city leaders and university scientists. Herriot did not err in characterizing Lyon as a French Manchester.

## VICTOR GRIGNARD AND ORGANIC SYNTHESIS

As a case study of a distinguished French provincial scientist, Victor Grignard is an example without parallel. He was neither normalien nor polytechnicien nor agrégé. Within a couple of years of defending his doctoral thesis at the University of Lyon, he received three national awards, and in 1906 he was voted the Academy's most prestigious prize in chemistry, the Jecker Prize. At forty-one, he shared the 1912 Nobel Prize in chemistry with Sabatier. At this time, Grignard was a professor of organic chemistry at Nancy, a chair he had held for two years. Only through the speedy intervention of Nancy's rector did the young Nobel Laureate wear the red ribbon of the Legion of Honor in Stockholm. Charles Adam alerted his friend Raymond Poincaré, who was president of the French Republic, and word reached Grignard on his arrival in Stockholm that the ribbon, as well as the Nobel Prize, was his. A popular French journalist joked:

> Ah! si M. Grignard était vaudevilliste, commanditaire de théâtres, président de tripot ou amant de coeur d'une cabotine à la mode! . . . Mais voilà, M. Grignard n'est qu'un savant, un savant de province.[64]

Grignard first arrived in Lyon in 1891, accompanied by seven classmates from the Ecole Normale d'Enseignement Spécial. He was originally from Cherbourg, where his father was sailmaker and foreman at the Arsenal. Born May 6, 1871, Grignard lost his mother early, and his father remarried. Because his father was encouraged to let Victor attend secondary school, young Grignard was able to prepare for the baccalaureate at the Collège de Cherbourg, which became a national lycée in 1886. During his lycée studies from 1883 to 1887, he regularly received the "prize of excellence" and became a favored school candidate for one of the scholarships given by the city of Paris to enable provincial students to prepare at a Parisian lycée for the concours of the grandes écoles.[65]

Unfortunately, the 1889 World's Fair overtaxed Parisian revenues and the scholarships were canceled. Grignard thus found himself competing in the less demanding concours for the Ecole Normale Spéciale at Cluny, the training school for teachers of modern secondary education founded by Duruy in 1866. When the school was closed in 1891, Grignard's scholarship was still good for another year, so he and his comrades enrolled as scholarship students at the Lyon Faculties, some 75 kilometers from Cluny, taking room and board at the Lycée Ampère. With Grignard were Louis Rousset, who was to die of cancer in 1898; Meunier and Wiernsberger, both of whom later taught at the Lyon Sciences Faculty; F. Bourion, who became professor of physical chemistry at Nancy; and Pierre Vaillant, who was to teach physics at Grenoble.[66]

Grignard's intention was to take the licence in mathematical sciences, but he failed the examination on his first try. After obligatory military service in 1892–93, he returned to Lyon and passed the examination in 1894. His difficulty was a lack of facility for memorization; it seems altogether unlikely, therefore, that he would turn to chemistry at a time when chemistry education in France singularly utilized rote memorization.

Indeed, on the basis of his experience at Cherbourg and Cluny, Grignard had no interest in the discipline:

> It was in 1894 and we had not yet emerged from the period where the influence of Berthelot was being exercised despotically on secondary education. It hindered the atomic theory from replacing that of "equivalents." At Cluny, where even then teaching was on a high level, they treated mineral chemistry in equivalents and organic chemistry in atoms. I had an impression of it as an incoherence and mnemonism that frightened me.[67]

One can hardly avoid speculating that Grignard's experience was not unusual, and that many students who might have become interested in chemical science were turned away by French secondary education. It was not until 1902 that the regime for the baccalaureate introduced into the examination's chemistry section the ordering definitions of molecules and atoms already common in English and German education.[68]

Grignard's conversion to chemistry is explained by the persuasion of his good friend Rousset and by the influence of two teachers at the Lyon Faculty, Barbier and Louis Bouveault. Rousset overcame Grignard's contempt for chemistry as a science inferior to mathematics. And on the basis of his own experience as an assistant and laboratory supervisor (chef des travaux) in general chemistry, he persuaded Grignard to accept a post as aide with Bouveault, on the grounds that doing chemistry was very different from memorizing it.

Bouveault had defended his thesis at Paris in 1890. In 1894 he became a lecturer in general chemistry after teaching briefly at Lyon's Medical Faculty. Bouveault's enthusiasm for new ideas and his passion for laboratory research won over Grignard within a matter of weeks.[69] A man of synthetic bent, full of erudition, inclined to speculation, and out to make the crucial experiment that would decide for or against a hypothesis, Bouveault's personal and intellectual qualities ran counter to Grignard's previous conceptions of chemical science.

While in Lyon, Bouveault went from his thesis topic on B-ketone nitriles and their derivatives to syntheses with camphor and terpenes, a class of oils whose study is a natural offshoot of work on the coal tar derivatives used in colors and dyes in the textile industry. Terpenes are particularly fruitful compounds for testing hypotheses about organic structure since they sometimes are "bridged rings" with a carbon atom, carrying radicals, spanning two carbons in the ring.[70]

Collaborating with Barbier, Bouveault studied citral and other terpene derivatives like rhodinal and geraniol, which make up the fragile essences used in perfume manufacture. He left Lyon for Lille, Nancy, and then Paris, where he died prematurely in 1909.[71] A polytechnicien, Bouveault held Paris doctorates in medicine and physical sciences. In recommending him in 1901 as the Sorbonne's first-line candidate for a lectureship, Haller wrote an extraordinarily warm and enthusiastic report, praising Bouveault for his powerful research, his skills as an inspiring teacher, and his openness to discussion and criticism. Bouveault "often takes pleasure in the most daring conceptions, the most risky hypotheses, without being afraid to excite objections, indeed believing in the value of the most lively critiques."[72]

After working with Bouveault for a year, Grignard became aide to Barbier, a man with whom hardly anyone at the Faculty got along.[73] Barbier had worked under Berthelot at the Collège de France and completed a thesis on pyrogenated carbides in 1876. After taking a pharmacy degree in Paris in 1879, he settled in Lyon for most of his career, first as lecturer and then as professor of general chemistry from 1884 until his retirement in 1919. He won the Jecker Prize in 1894, and in 1899 he posed his candidacy for Charles Friedel's Sorbonne chair, a position that went instead to Haller, then at Nancy. Barbier's work on chemical composition and the relationship between function and structure in chemical compounds was thought in Paris to be important and difficult.[74]

Like Bouveault, Barbier was a *nivernais*. A bon vivant, he was also moody, gruff, and tyrannical. He was legendary for his toughness in examinations and usually intimidated beginning chemistry students. As a researcher, he was brimming over with new ideas, although frequently jumping from one topic to another without clarifying or exhausting the one with which he had begun. Grignard benefited from exposure to Barbier's originality and intellect, while gradually learning how to accept Barbier's habit of treating students and assistants like manual laborers rather than collaborators.[75]

Barbier was a champion of the atomic theory. Grignard suggested later that Barbier suffered professionally from expressing this view.[76] Barbier was not the only proponent of atomism at Lyon. Of influence, too, in this regard were Gouy and Offret. As we have seen, Gouy was a leader in France in molecular physics and in the application of ther-

modynamic and kinetic theory to physical chemistry. Grignard probably heard Gouy's 1894 ceremonial public lecture encouraging scientists
to use molecular hypotheses and ideas:

> Now we must quit the solid terrain of observation and experiment, in
> order to enter the uncertain domain of hypotheses on the constitution
> of matter. These theories and hypotheses have been much abused,
> much slandered; and yet we cannot do without them. Their scientific
> importance is incontestable, and now and then they throw unexpected
> light on a whole set of questions.[77]

Grignard later spoke warmly of Gouy's influence on him.[78]

Also in the Sciences Faculty, Offret specialized in the applications of
physical and chemical methods in mineralogy.[79] A normalien and
agrégé with a doctorate in physical sciences, Offret completed his thesis
in 1891 on the variation under the influence of heat of the indexes of
refraction of minerals. For the benefit of students like Grignard, he
organized a large teaching laboratory for the study of the crystalline,
physical, and chemical properties of minerals. His chair was the first
devoted entirely to mineralogy in the provincial universities.[80]

During Grignard's student days, the Lyon Sciences Faculty bustled
with a mood of optimism amid hectic activity. Enrollments were steadily increasing and local funds were supporting research and teaching
programs. The new university palais was under construction. In winter
1889, Barbier requested that a door at the new chemical institute be left
open after the ordinary Faculty closing time, so that laboratories could
be used in the evenings.[81] When he became laboratory director after
Rousset's death in 1898, Grignard busied himself with the direction of
laboratory exercises for licence and PCN students and the preparation
of materials for Barbier's lecture demonstrations.[82]

There was strong comradeship among the laboratory staff and course
aides at the Faculty. At noon they gathered for lunch at a restaurant on
the quai Perrache. The group usually included Grignard and Miquey,
an engineer at the Ecole Chimie Industrielle; Chifflot, in botany; and
Beauverie, Faucherie, and Combe in zoology. The lunches were ordinarily gay affairs, followed by a cup of coffee and some billiards at the
Brasserie Georges, where the manager let them play free of charge.[83]

Working with Barbier, in 1898, Grignard published a paper with him
on a stereochemical problem. At Barbier's suggestion, Grignard then
worked briefly by himself on two papers on énines, hydrocarbons with
both ethylene and acetylene linkages, and on a hydrocarbon with three
adjacent double bonds.[84] But Grignard was not happy with the énine

topic and did not think it would make a good subject for a doctoral thesis.

Meanwhile, as part of his work with terpenes and perfumed essences, Barbier wondered in late 1898 if he might apply the Saytzeff technique for synthesis of tertiary alcohols to the natural substance methylheptenone in order to obtain the lemon-smelling dimethylheptenol. He knew that the Saytzeff method does not work with methyl ketones, which lose water and condense in the presence of zinc. So, given new interest among researchers in magnesium, Barbier decided to try magnesium, rather than zinc, in the ordinary Saytzeff procedure.

He covered fine magnesium turnings with an ether solution of methylheptenone, and then he gradually added methyl iodide. When about 30 grams of the methyl iodide had been added, the liquid turned yellow in a vigorous reaction, which he cooled under a current of water, continuing to add methyl iodide and then leaving the materials in contact about twelve hours. Subsequent hydrolysis by dilute sulfuric acid in the presence of ice gave a liquid from which was distilled 35 grams of pure dimethylheptenol. Barbier signaled in his paper that "the substitution of magnesium for zinc in the Saytzeff reaction is new," but he did not dwell on the fact that he had operated in the presence of anhydrous ether, another modification of the Saytzeff method.[85] Although he experimented further with this kind of synthesis, the results were undependable and the yields unsatisfactory, so he did not pursue the technique further.[86]

$$CH_3 - \underset{\underset{CH_3}{|}}{C} = CH - CH_2 - CH_2 - CO + CH_3I + Mg$$

$$\underset{\underset{CH_3}{|}}{\downarrow}$$

$$CH_3 - \underset{\underset{CH_3}{|}}{C} = C - CH_2 - CH_2 - \underset{\underset{CH_3}{|}}{\overset{\overset{CH_3}{|}}{C}} - O - MgI$$

$$\downarrow (+ H_2O)$$

$$CH_3 - \underset{\underset{CH_3}{|}}{C} = CH - CH_2 - CH_2 - \underset{\underset{CH_3}{|}}{\overset{\overset{CH_3}{|}}{C}} - OH + Mg \overset{\diagup I}{\diagdown OH}$$

At this time, Grignard informed Barbier that he was not pleased with the direction of his current doctoral research. Barbier suggested that his student take up the study of the substitution of magnesium in the Saytzeff method to see how well it might work for new syntheses.[87] At first Grignard met with the same unreliable results as had Barbier. In his laboratory notebooks, Grignard speculated that the presence of the ketone or aldehyde, in contact with the magnesium filings, might inhibit the reaction of the magnesium with the alkyl halide to form what he assumed must be a reactive organometallic intermediary in the Saytzeff reaction. He wondered whether he might better proceed by preparing an organomagnesium compound first, which might then react with the double-bonded oxygen of the aldehyde or ketone.

A major problem with organomagnesium compounds was their inflammability. In reading literature on organozinc compounds, he was struck by Edward Frankland's observations in the 1850s, confirmed by James A. Wanklyn in 1861, that organozinc compounds prepared in anhydrous ether were not inflammable in air. In the search to isolate organic radicals, Frankland had discovered symmetrical organozinc compounds—low boiling compounds like diethyl zinc, which combine energetically with chlorine, bromine, or iodine. The organozincs also can be decomposed by water to produce the corresponding hydrocarbon; in the case of diethyl zinc, ethane.[88]

Frankland found that if one mixes zinc with methyl iodide or with ethyl iodide, to which anhydrous ether is added, the metal dissolves and produces a liquid with a boiling point just above that of Diethyl zinc. This liquid seems to have the same properties as dimethyl zinc, but is less flammable and more sluggish in reactions. Frankland suggested that the liquid has a composition $Zn(CH_3)^{2 \cdot}(C_2H_5)_2O$, corresponding either to a mixture or a combination of ether and dimethyl zinc. Two years later, in preparing dimethyl zinc, Wanklyn obtained a crystallized organozinc compound, for which he suggested the formula

$$Zn(CH_3)_2 \cdot I_2Zn = 2\ CH_3 - Zn - I.^{89}$$

These studies were principally of theoretical interest, undertaken to work out notions of valence.[90]

With new interest in applying the zinc compounds in organic synthesis, particularly after Butlerov's 1863 paper on obtaining tertiary alcohols from organozincs with acid chlorides and hydrolysis, Frankland

and B. F. Duppa suggested in 1865 the advantage of first mixing together the metal, alkyl iodide, and organic substance; it was in this way that Alexander Saytzeff prepared tertiary alcohols from ketones.[91] The technique is slow, sometimes taking several months, and the yields are not good. The method is limited to dimethyl zinc and diethyl zinc, and no one immediately succeeded in making an organozinc with an aromatic radical.[92]

In 1859 and in 1860, Adalbert Schafarik and Auguste Cahours attempted to produce an organomagnesium compound, using finely crushed magnesium in a sealed tube at 100–180°C. Instead of getting a volatile liquid product like the alkyl zinc, they produced a solid mass that would not distill.[93] This line of research was taken up around 1890 by Lothar Meyer's pupils Philipp Löhr, Hermann Fleck, and Fritz Waga. Löhr succeeded in 1891 in preparing dimethyl, diethyl, and dipropyl magnesium, all of which are hard to manage because they ignite in air or carbon dioxide, and are solid, nonvolatile, and almost insoluble in neutral solvents, including ether.[94] Fleck investigated the production of ethane and magnesium hydroxide by the action of water on diethyl magnesium; he prepared trimethyl carbinol by treating dimethyl magnesium with acetyl chloride in a mixture of anhydrous ether and benzene, and he prepared and investigated reactions with diphenyl magnesium.[95] Waga continued this line of work, but none of the three German students found organomagnesium compounds especially promising for synthesis.[96]

In going back to the Frankland reaction of 1859, Grignard reasoned that while zinc-alkyl halides reacted more sluggishly than zinc alkyls, a magnesium-alkyl halide might be more vigorous in synthesis because of magnesium's higher activity as an electropositive element. Magnesium was now easier to obtain in a pure state than in Frankland's time and might provide a useful avenue for general synthesis. Perhaps he could prepare a stable, noninflammable organomagnesium compound if he used anhydrous ether.[97]

Grignard's laboratory notebook suggests the astonishment and joy he felt when magnesium began to effervesce and disappear as he slowly added to it a mixture of isobutyl iodide and anhydrous ether. On cooling, there appeared a limpid, colorless liquid that was not inflammable in air and furnished an excellent yield of phenylisobutyl carbinol after he added benzaldehyde.[98] He soon found that no preliminary heating of the magnesium with the alkyl halide was necessary.

The reaction took place spontaneously in ether at room temperature under ordinary pressure, and it also worked with aromatic halides, although not as well as with alkyl halides. When the aldehyde or ketone is added to the organomagnesium ether solution, a solid material may separate as a crystalline or viscous layer at the bottom of the flask. Consequently, small amounts of acid are added as the whole mixture is poured on crushed ice, so that this $Mg(OH)_2$ redissolves.[99]

In a 1900 paper presented to the Academy for him by Moissan, Grignard announced his discovery of new organomagnesium compounds and their proven value as a tool of organic synthesis for a number of secondary and tertiary alcohols and several new carbinols, including dimethylbenzyl carbinol, which he had prepared using benzylmagnesium bromide. He explained that he had not actually isolated the new organomagnesium compounds used in the synthesis.

> If one tries to drive off the ether, there remains a grayish mass, vaguely crystalline, which very rapidly absorbs moisture upon heating and becomes deliquescent. But the great advantage of the resulting combination is that one does not need to isolate it. Indeed, if one lets a molecule of an aldehyde or a ketone fall into the preceding etherated solution, which contains very precisely one atom of magnesium, generally there is produced a lively reaction, and once the resulting combination is decomposed by the acidulated water, one isolates the corresponding secondary or tertiary alcohol with a yield of around 70%.[100]

In this paper, he suggested the formula for the new organomagnesium compounds as RMgI or RMgBr, where $R$ might be an alkyl or a phenyl radical, and he formulated the reaction for synthesis of a secondary alcohol from an aldehyde in the following manner:

$$CH_3 + Mg \rightarrow CH_3MgI$$

$$CH_3MgI + RCHO \rightarrow RCH\begin{smallmatrix} \diagup OMgI \\ \diagdown CH_3 \end{smallmatrix}$$

$$RCH\begin{smallmatrix} \diagup OMgI \\ \diagdown CH_3 \end{smallmatrix} + H_2O \rightarrow RCH(OH)CH_3 + MgIOH.[101]$$

Favorable reaction to Grignard's first paper on the organomagnesium compounds was immediate. Moissan urged him to present his doctoral thesis in Paris, rather than at Lyon, but Grignard declined, perhaps out of loyalty to his alma mater and colleagues. In July 1901, he defended the thesis in Lyon before a jury composed of Barbier, Gouy, and Vignon. He had already published six more articles on the subject and was corresponding with chemists throughout France. He suggested the crucial role in the new synthesis of an intermediary etherate formed by one atom of magnesium combining exactly with one molecule of halogenated ether.[102]

In his thesis he also listed twenty-nine new compounds prepared by his method, the beginning of a research program that was to revolutionize organic synthesis. In the next decades Grignard's reagent proved to be a powerful reducing agent of amazing versatility which could be used to test ideas about structure, as well as to carry out syntheses of primary, secondary, and tertiary alcohols, saturated and unsaturated hydrocarbons, glycols, ketones, acids, nitriles, sulfoxides, and other compounds.[103]

By early 1905, approximately 200 papers were available on the organomagnesium compounds, 80 of them in major French journals and 91 in the reports of the German Chemical Society. By 1912, more than 700 papers had appeared; by 1926, 1,800; and by 1950, 4,000.[104] Parallels in the literature in France can probably be found only in reports around 1900 on new radiations like Roentgen's radiations and Blondlot's N-rays.

The interest in Grignard's results was both practical and theoretical. This was a period in which synthesis of organic compounds was multiplying exponentially. When Friedrich Beilstein published his first *Handbuch der organischen Chemie* in 1880–1882, it contained approximately 15,000 entries; by 1910, the number of known organic compounds was about 150,000.[105] Grignard's new reagent provided a quick and safe method of synthesis that eventually was to have so many uses industrially that large quantities of organomagnesium compounds came to be prepared commercially. One American firm was recently manufacturing forty-five tons a day.[106] In addition to its practical value, Grignard's procedure provided a new means of understanding molecular structures and confirming fundamental hypotheses about atomicity or valence in organic and mineral chemistry.

Chemists interested in stereochemistry immediately saw the organomagnesium halide as a valuable tool for research. For one thing, it could be used to estimate the numbers and positions of replaceable hydrogen, since magnesium is reactive toward active hydrogen. This suggestion was made by the Russian chemist L. Tschugaev.[107] In France, Auguste Béhal, one of the last pupils of Wurtz, the French atomist, at once became interested in Grignard's work, as did Bouveault and those other French chemists in the tradition of Wurtz and Friedel who saw in organic chemistry a magnificent illustration of the atomic theory.[108]

But the practical and theoretical ramifications of Grignard's work also had its dark side. He found himself in the unhappy position of trying to finish his doctoral thesis and defend its originality while older and more distinguished scientists, with personal laboratories and research assistants at their disposal, plunged into research suggested by his first reports. Through the collaboration volunteered by L. Tissier, Barbier's lecturer, Grignard was able to speed up his own work somewhat, but before the dissertation was even completed, Grignard found himself in disputes with men whose prestige and influence would determine his professional advancement.

In early February 1901, Grignard received a disturbing letter from Emile Edmond Blaise, who had studied with Béhal and had just completed his doctoral thesis with Friedel in 1899 at the Sorbonne. Blaise was then teaching at Lille but soon was to move to Nancy as Bouveault's replacement. Blaise's interest in organomagnesium compounds was kindled by Grignard's paper of May 1900, and he published a paper on organometallic reactions in January 1901.[109] In this paper, Blaise reported that he had been doing research with zinc-organometallic derivatives, with a special interest in nitriles, nonsaturated carbides, and isocyanic ethers, as a method of synthesis of ketones and other compounds. Following Grignard's May 1900 publication, Blaise said, he had tried organomagnesium compounds and found them highly satisfactory, for example, with nitriles,

$$R\text{—}C \equiv N + Br - Mg - R' \rightarrow R - C {\overset{\displaystyle /\!/ \ MgBr}{\underset{\displaystyle \diagdown \ R'}{}}}$$

from which a ketone is produced upon acidification: R—CO—R'. Further, he stated his intention to continue studying these kinds of reactions, especially synthesis of ketones, B-ketonic ethers, and acids.[110]

Blaise wrote to Grignard on February 7, 1901, asking him to renounce further interest in the *composition* of the organomagnesium derivatives, a problem in which Blaise was now interested. Two weeks later, after Grignard's second paper on the organomagnesium compounds had appeared in the February 11 proceedings of the Academy of Sciences, Blaise wrote a second letter, accusing Grignard of ignoring his rights to the reactions he had staked out for himself and reiterating his request that Grignard refrain from investigating the question of the composition of the organomagnesium reactant.

> The last paragraph of your recent communication leaves me to think that you are ignoring a Note that I published in the *Comptes Rendus* on some new reactions of organometallic derivatives. I do not doubt that you will leave to me the points that I have reserved to myself for study. Most of this work is already terminated and will be published without delay. On the other hand, if you have a communication to make on the nature of the organometallic derivatives which result from the action of these halogenated derivatives on magnesium, in the presence of ether, I would be grateful if you would publish them as soon as possible. Certain of my researches only await, indeed, this publication in order that they might appear.[111]

Grignard had speculated from the beginning on the composition of the organometallic compounds, and the passage to which Blaise took offense merely said:

> We can presume, then, according to the results already acquired, that the use of the mixed organomagnesium compounds permits simplifying and generalizing most methods of synthesis for which up until now the organozinc compounds have been used, and further it permits establishing new syntheses predicted from theory, but practically unrealizable by means of zinc. It is in these two directions that I pursue my researches.[112]

Grignard pressed ahead. He published two papers in the *Comptes Rendus* in March, one in April, and one in May, the last three with the collaboration of Tissier. His friend Meunier also became interested in the work and independently made aminomagnesium halides in 1903.[113] In his March 4 paper, Grignard gave reasons for preferring the formulation $CH_3MgI$ for the active organomagnesium compound, rather than $Mg(CH_3)_2$, and in the March 18 paper, he and Tissier proposed as an account of the formation of the organometallic compound:

$$CH_3I + Mg + n(C_2H_5)_2O = CH_3 - Mg - I + n(C_2H_5)_2O.$$

They seem to have adopted implicitly Frankland's suggestion that the ether was attached to the active metal molecule like water of crystallization, or was closely mixed with it, a view Grignard made explicit in his July thesis, drawing an analogy with Frankland's observation of $Zn(CH_3)_2$ in the presence of ethyl oxide and methyl oxide. The molecule of ether of crystallization gives the organometallic compound its solubility in ether. "But the ether of crystallization intervenes in no way in the chemical reactions, and we do not need to concern ourselves with this subject."[114]

In the April 1 proceedings of the Academy of Sciences, four articles appeared on the organomagnesium compounds, written by Armand Valeur, Moureu, Grignard and Tissier, and Blaise.[115] Blaise's article must especially have frustrated Grignard: he laid out a hypothesis on the nature of the compound active in Grignard's method. According to Blaise, it was an organomagnesium etherate, RMgX $(C_2H_5)_2O$, and he rejected explicitly Grignard's earlier conclusion that it was RMgX. Ether is not a "common dissolvent," he said, but enters into the reaction. This is proven by the stability of the magnesium compound in ether, because the ether cannot be completely eliminated from it even with prolonged heating in hydrogen at low pressures unless the temperature is raised to about 130°C. Blaise attributed the publication of Grignard's remarks in the first March paper to the influence of his March 4 letter: "Sur ma demande, M. Grignard a bien voulu publier les résultats obtenus par lui à ce point de vue."[116]

In March 1901, Grignard had submitted a paper to the *Bulletin* of the Paris Chemical Society on the action of mixed organomagnesium compounds on methyl naphthyl ketones.[117] Along with the paper, he submitted a note to the editor of the journal, Béhal, who had published an article in the late February *Comptes Rendus* on the action of organometallic derivatives on ether salts.[118] Grignard protested Béhal's and other chemists' taking up his line of research before he could complete his thesis. Béhal agreed to insert in the printed minutes of the Chemical Society the announcement, "M. Grignard reserves to himself the use in organic chemistry of alcohol-halogenated magnesium compounds which he has discovered."[119]

In the interim between Blaise's April 1 article and the appearance of

Grignard's priority announcement in the *Bulletin*, Grignard also wrote to Moissan and Berthelot, soliciting their intervention on his behalf. Moissan had presented Grignard's and Tissier's notes, as well as Moureu's, to the Academy. (Haller presented those by Blaise and Béhal.) Grignard explained in his letter his concern that Blaise and others were appropriating his method and, further, intimidating him into leaving a clear field to Blaise. "In these circumstances, I am asking myself if such events are the exception or the rule. [And] if it is possible for a chemist, particularly one making his debut, to have his rights of priority respected."[120]

Moissan and Berthelot each responded cordially and sympathetically, reassuring Grignard that no one would forget he was the initiator of this area of research and suggesting that he continue working as fast as possible. Once published, counseled Moissan, a discovery falls into the public domain; rights to research work cannot be divided up by a committee or by the Academy, counseled Berthelot.[121]

The question of the composition of the organomagnesium compounds continued to be discussed throughout Grignard's lifetime. His rejection of the $R_2Mg/MgX_2$ formulation was questioned in 1912–13 by Pierre Jolibois, and the issue was renewed in the 1920s. Grignard was pleased to see his RMgX formulation supported in new kinetic studies by Job and Dubien. In contrast to their work, Grignard's approach was that of traditional chemical analysis; for example, he argued that since his organomagnesium compound is not inflammable in air, it cannot be the inflammable alkyl magnesium compound $R_2Mg$; since it does not release iodine or bromine, it cannot contain free $MgX_2$. By 1930, many chemists came to favor the notion that the organomagnesium reactant corresponds to an equilibrium: $2RMgX = MgR_2 + MgX_2$. The question has recently been under discussion again.[122]

A more biting controversy was the one between Grignard and Blaise over the composition of the etherates of the organomagnesium compounds. Along with the riddle of Moses Gomberg's triphenylmethyl free radical, the oxonium discussions were among the most interesting issues in organic chemistry around 1900.[123]

In 1902, the Strasbourg chemists Adolf von Baeyer and V. Villiger used Grignard's results as an instance of the oxonium theory of quadrivalent oxygen derivatives. Von Baeyer used the name "onium compounds" for ammonium, phosphonium, iodonium, oxonium, and carbonium compounds, where the latter, as in triphenylmethyl, was suggested to have an ionizable valency denoted by a wavy line: $(C_6H_5)_3C\sim OSO_3H$.[124] They suggested for the etherate the structure:

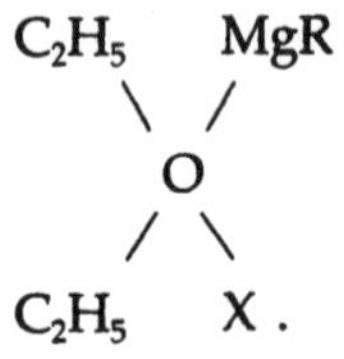

$$
\begin{array}{ccc}
C_2H_5 & & MgR \\
\diagdown & & \diagup \\
& O & \\
\diagup & & \diagdown \\
C_2H_5 & & X\ .
\end{array}
$$

The argument was taken up by Blaise in 1902 after his study of the action of the organomagnesium compounds on ethylene oxide.[125] Grignard preferred a formula where the halide was not directly attached to oxygen:

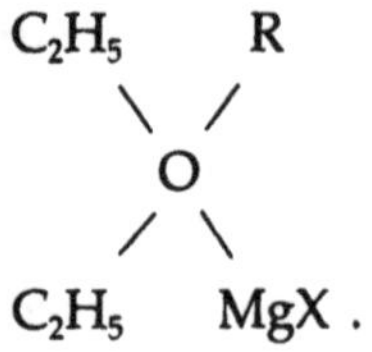

$$
\begin{array}{ccc}
C_2H_5 & & R \\
\diagdown & & \diagup \\
& O & \\
\diagup & & \diagdown \\
C_2H_5 & & MgX\ .
\end{array}
$$

He thought this better corresponded to the breakup of etherate into its reaction products and explained its resistance to sodium.[126]

In duplicating Blaise's experimental results, Grignard was able to offer an explanation of them which made use of ideas about structure, valence, and basicity. Blaise had obtained very little alcohol and mostly ethylene bromohydrine from the action of the magnesium reagent on ethylene oxide. His interpretation, using the von Baeyer-Villiger oxonium formula was:

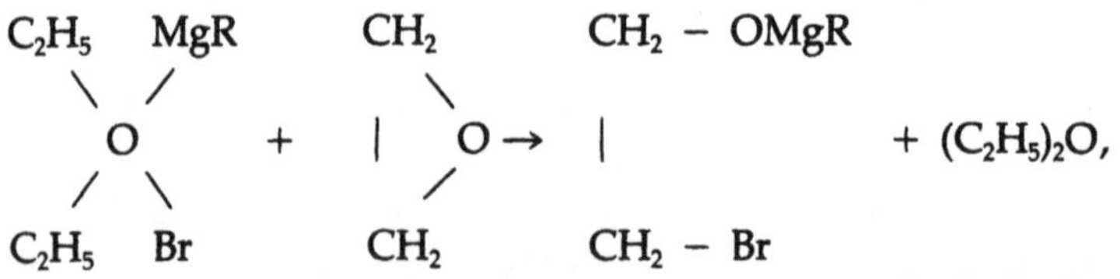

$$
\begin{array}{ccccc}
C_2H_5 \quad MgR & & CH_2 & & CH_2 - OMgR \\
\diagdown \quad \diagup & & | & & | \\
O & + & \quad O \rightarrow & & \qquad\qquad + (C_2H_5)_2O, \\
\diagup \quad \diagdown & & | & & | \\
C_2H_5 \quad Br & & CH_2 & & CH_2 - Br \\
& & \text{(ethylene oxide)} & & \text{(diethyl ether)}
\end{array}
$$

from which hydrolysis produces

$$CH_2\text{—}OH$$
$$|\qquad\qquad + RH + Mg(OH)_2 \, .$$
$$CH_2\text{—}Br$$

      ↑                           ↑

(ethylene bromohydrine)     (magnesia)

Grignard suggested that something more complicated was taking place in the reaction vessel—indeed, that the ethylene oxide's basicity causes it to replace ethyl oxide in the etherate. Thus we have, by substitution,

instead of

When the reaction vessel is cool, hydrolysis results in regeneration of the ethylene oxide, which reacts on the magnesium bromide, formed by the reaction of water and ethyl magnesium bromide, to form end products of an alkane, magnesia, and ethylene bromohydrine. But if the ethylene oxide-substituted etherate is heated, the result is

$$CH_2 - O - MgBr$$
$$|$$
$$CH_2R$$

which produces an alcohol, $R(CH_2)_2OH$. In the case of butylic alcohols, the yields were 82 percent for Grignard, without any trace of ethylene bromohydrine.[127] Working further on these kinds of reactions, and using ethylene chlorohydrine and phenyl magnesium halide, Grignard succeeded in making 2-phenyl ethyl alcohol, which has the delicate odor of roses. His method was converted to industrial use in 1905 by the Descollonges firm at Lyon, the first commercial application of his method in France.[128]

In later years, the etherate formed in Grignard's reaction was to be used as a prime example of the Lewis acid/base theory, in which acids are said to be electron acceptors and bases are electron donors. Thus, an alkyl magnesium halide, R:Mg:X, is a Lewis acid and the magnesium atom of this acid completes its octet by accepting a pair of electrons from each of two ether oxygen atoms to form a dietherate:[129]

$$R' : \overset{..}{\overset{\oplus}{O}} \quad : R'$$

$$R : \overset{..}{\underset{..}{Mg^{\ominus}}} : X$$

$$R' : \overset{\oplus}{\underset{..}{O}} \quad : R'$$

G. N. Lewis's theory was part of a broad and far-ranging system laid out in his 1923 book, *Valence and the Structure of Atoms and Molecules.* These results had developed as he, Irving Langmuir, and Walther Kossel independently followed up on the work of J. J. Thomson, H. Moseley, Niels Bohr, and Arnold Sommerfeld in an attempt to systematically relate electron arrangements to chemical reactivity.[130] While Grignard consistently employed considerations of atomic weight and molecular structure in his work, he declined to commit himself to some of the newer physical ideas as tools for solving problems of chemical synthesis. In 1935, in his introduction to the multivolume *Traité de chimie organique,* he wrote the following:

> As for the new electronic theories, they are not sufficiently developed for serving as the basis for speculations in organic chemistry, despite all the promises they offer to chemists. In this text, they will not be laid aside systematically, but for the moment will remain discreetly in the background; and it is still the very fruitful conception of Le Bel and van't Hoff which will constitute the surest guide for us.[131]

Grignard's reaction to the octet and shared electron theory of chemical valence was not unusual for an organic chemist. For the most part, it was British chemists who showed enthusiasm for electron interpretations of reaction mechanisms, and among chemists, most Germans and many Americans ignored the Lewis-Langmuir theory of valence, as did the French.[132] In Grignard's first volume of the *Traité,* he included sections on the applications of physical chemistry to organic chemistry, but the electron theory was not to have much influence on the next generation of students, who concentrated on organomagnesium research and ignored new applications of quantum mechanics to reaction mechanisms and molecular structure.

His thesis having been completed in summer 1901, Grignard continued as laboratory supervisor until fall 1902. He was then delegated to fill in for Tissier as lecturer. His salary increased from 2,000 to 4,000 francs a year, and he no longer had the responsibility of organizing and teaching laboratory sections to PCN students.[133] He was regarded as a good teacher with admirable elocution and excellent "esprit." By 1905, he had published twenty-seven research papers, twice shared the Prix Cahours (1901 and 1902), and won the Berthelot Medal (1902).[134] Long review articles and bibliographies appeared in foreign chemical journals on the "Grignard reaction" and the "Grignard reagent," listing some two hundred papers on the subject.[135] In a review article for the *Journal of the American Chemical Society*, Lauder Jones mentioned five major fields of research then of crucial interest in organic chemistry: Gomberg's work on triphenylmethyl, the work of Fischer and Curtius on polypeptides and proteins, oxonium compounds, the "method of Sabatier and Senderens" for catalytic reduction of hydrocarbons, and applications of Grignard's reaction.[136]

Haller invited Grignard to the Sorbonne in October 1903 to give a lecture on his results, an opportunity Grignard used for presenting his views on the composition of the alkyl magnesium etherates.[137] Haller also recommended to organizers of the 1905 meeting of the Association Française pour l'Avancement des Sciences that Grignard chair the section of chemistry. This suggestion was seconded by Maurice Caullery, who had recently taught at Lyon and knew that the meeting site, Cherbourg, was Grignard's hometown.[138] At the meeting Grignard had a chance to talk with Bouveault, who had been his teacher, and to meet Beilstein, who encouraged Grignard's interest in problems of chemical nomenclature.[139]

During the early 1900s, both Grignard and members of the Lyon Faculty tried to get the Ministry of Public Instruction to appoint him the permanent successor to Tissier, who was on year-to-year leaves of absence. But by summer 1905 this had not been done, and the ministry decided to appoint Grignard to Besançon for a lectureship in applied chemistry. Grignard reluctantly agreed to this arrangement, with the understanding that the ministry would allow him to return to Lyon when Tissier's post was formally declared vacant.[140] Grignard spent the academic year 1905–06 at Besançon, where the professor of general chemistry was Léon Boutroux, a student of Pasteur and the brother of the philosopher Emile Boutroux. Boutroux's studies focused on wine musts, but he was not an avid researcher.[141]

At the end of the academic year, in May 1906, Grignard received a letter from Haller which said that he alone would be awarded the Prix Jecker in the summer. The letter's main purpose, however, was to tell Grignard frankly what Haller thought might reach him by the grapevine: Haller had proposed that Grignard share the prize with Blaise.

The reason, Haller explained, was a matter of deference to Barbier, who had never received more than half the prize. This argument, he said, was rejected by prize committee members Moissan and Berthelot as sentimental and invalid. Perhaps Grignard's letter of April 1901 had wrought more effect on Moissan and Berthelot than Grignard realized.[142] Moissan wrote the 1906 Prix Jecker presentation speech, mentioning at the end that chemists were right in calling the process involving organomagnesium compounds "Grignard's reaction." Moissan mentioned Barbier's name in a short introduction, noting Barbier's substitution of zinc for magnesium in the Saytzeff reaction.[143]

In fall 1906, Grignard returned to Lyon from Besançon to a now regularized lectureship in chemistry. Grignard and Barbier returned to active collaboration in the next years, from 1907 to 1914 publishing eleven joint papers on terpene chemistry, as well as separate work.[144] But their relationship must have been strained from the outset, given Moissan's wording in the Prix Jecker presentation and the likelihood that Barbier had heard gossip about the award. Any resentment Barbier experienced may also have been fed by a lecture given by Gautier in May 1907 at the celebration of the fiftieth anniversary of the French Chemical Society. Here Gautier identified three methods of synthesis as the most powerful tools in modern organic chemistry: those of Friedel and Crafts, Sabatier and Senderens, and Grignard.[145]

That there was resentment is revealed in the fact that a few months after Grignard left Lyon for Nancy in 1909, Barbier wrote a note to be read at a meeting of the Chemical Society, with Haller sitting as president and Béhal as vice president. There Barbier criticized journals and encylopedias that attributed solely to Grignard the discovery of the use of magnesium in organic synthesis; he described his own synthesis in 1899 of dimethylheptenol, and announced that he had also produced several compounds for which he had not published his results. Barbier concluded that Grignard had the right to attach his name to the reactant in the process, but not to the principle of the reaction: "it seems to me that in all fairness it is proper, from now on, to associate together our two names in designating this reaction."[146] Taken by surprise, Grignard published a rejoinder in April, citing publications in which he or other writers had mentioned Barbier's use of magnesium, noting that some workers had recently described dimethylheptenol without awareness that Barbier had carried out this synthesis at all, distinguishing Barbier's procedure for employing magnesium from his own, and noting the importance of the work of Blaise and his students using mixed organozinc compounds.[147]

In addition to the tension with Barbier, Grignard must have experienced disappointment at two failures in 1908–09 to get positions that might have allowed him to advance in rank more quickly. He was one of six candidates for a position at Toulouse. Sabatier wrote Grignard that he personally was supporting his own collaborator Mailhe, "but I do not doubt that if you request the post, it will be given you, considering the exceptional importance of your credentials, and I think that your nomination to our Faculty would be an honor for it and an element of prosperity." However, the post went to Giran, a mineralogist from Montpellier. Sabatier wrote Grignard: "The thing is explained only by the circumstance that M. Giran is from Nîmes and so is the Minister of Public Instruction."[148]

A second candidacy was unsuccessful because Grignard was not agrégé. In March 1909, before Bouveault's death, Haller counseled him to pose his candidacy for a vacant post at Rennes. A couple of months later, Haller announced to Grignard that despite his intervention, the Rennes Faculty had presented Bouzat as their first candidate and Perriès as their second. A petition signed by Troost, Gautier, Le Chatelier, Jungfleisch, and Haller failed to change anything, because, as Haller put it, it was felt to be absolutely necessary that the new appointee be able to prepare candidates for the agrégation examination. For his part, Haller said:

> I am deeply distressed to see that researches and original minds are systematically held off from high positions in order to give them to men whose scientific works are—so to speak—worthless, and whose only qualification is being a former pupil of the Ecole Normale Supérieure.[149]

Here we see the French educational system at work, with academic opportunity and distinction created by the degrees and diplomas of the grandes écoles and the state.

In 1909, Bouveault's death left vacant a lectureship in general chemistry at the Sorbonne. Haller argued forcefully for recommending Blaise, who had been at Nancy since 1902. Bourion, Grignard's old friend from Cluny, and Georges Urbain, who was on the Sorbonne vacancy committee with Haller, both counseled Grignard to be prudent when they learned Grignard was interested in applying for the position. When

Haller suggested Grignard apply for Blaise's chargé de cours post at Nancy, Grignard followed his advice.[150]

Finally named chargé de cours in organic chemistry at Nancy for 1909–10, Grignard became a chaired professor the following year. By this time he had written forty research papers and was teaching about ninety students in two conférences of one hour each. His new position and salary of 6,000 francs were factors in his deciding to marry.[151] Grignard found Nancy extraordinarily stimulating for his work. The Institut Chimique, founded by Haller, was now directed by Antoine-Nicolas Guntz. Paul Petit, Grignard's lycée teacher, was in charge of the Ecole de Brasserie and agricultural chemistry. The physical chemist Paul-Thiébaud Müller was working on the influence of isomeric ions on the solubility of ions, Jules Minguin on tartrates, Alfred Guyot on reactions of formic acid, and André Wahl on organic coloring materials.[152]

Grignard directed three doctoral theses in the years before the war, and published papers with two of these students: E. M. Bellet and Charles Courtot. Bellet's and Grignard's work, in which Courtot also participated, dealt with the synthesis of nitriles by the action of cyanogen chloride or cyanogen on organomagnesium halides, a method that can be modified for the production of ketones. The research of Grignard and Courtot centered on the synthesis of the coloring dyes, the fulvenes, via the halogen magnesium compounds of indene, cyclopentadiene, and fluorene.[153]

The exciting atmosphere in which Grignard and his young collaborators carried out their work is described by Courtot:

> On the terrace of the Pépinière, the 150 pupils of the Institut Chimique talk chemistry as they leave the auditoriums and the laboratory. The echoes of the magnificent public garden of the city of Nancy make the words reverberate: coupling, condensation, grignardization. Moreover their clothes stay impregnated with strong and characteristic odors; we follow the initiates of Hermes by their scent. In such an environment, how is it possible not to be productive: funds are dispensed without counting them, collaborators are at the command of the teacher. Besides, what wonderful works whirl through the day![154]

On November 13, 1912, Grignard's colleagues at the Institut Chimique burst into his laboratory to tell him that the morning newspaper carried an announcement that he and Sabatier were this year's joint Nobel Prize winners in chemistry. A telegram arrived from Stockholm that afternoon. When he received a congratulatory message from Meunier, Grignard replied that he would have preferred to see Sabatier and Senderens share the prize this year, and, later, to share it himself with Barbier. He asked Meunier to find out how Barbier was taking the news.[155] Meunier entrusted this mission to Léser, who saw more of Barbier. He reported to Grignard that Barbier did not blame him (Grignard). If he were not to receive any of the prize, Barbier said, he only regretted that Grignard did not receive the whole of it.[156] After the ceremony in Stockholm, Barbier wrote Grignard to this effect, although he also wrote a letter to the editor of the Lyon newspaper *Salut Public* claiming a share of Grignard's glory.[157]

Grignard wrote Meunier that he understood his nominations came from foreign scientists. In fact, Grignard had been nominated as early as 1906 by chemists at Tokyo University and Helsingfors. Adolf von Baeyer, who was Nobel Laureate in 1905, began exerting pressure for Grignard in 1909, and Grignard might well have received the award in 1910 had Barbier not filed his 1909 protest with the Paris Chemical Society. In 1912, von Baeyer nominated Barbier and Grignard jointly, and O. Widman, a member of the Nobel chemistry committee, came up with the Grignard-Sabatier pairing.[158]

Three colleagues at Nancy told Grignard they had not supported him because they thought he had plenty of time for the award and that it should go to an older candidate. They had nominated Albin Haller.[159] Seniority considerations also prevailed against Grignard in the June 1913 election of Georges Charpy as corresponding member in the Academy of Sciences; previous commitments had been made on behalf of the elder Charpy.[160]

Election as a corresponding member to the French Academy came in November 1913, and in 1926 Grignard became a member of the Academy's section of six nonresident members, replacing Pierre Duhem. By this time Grignard had returned to Lyon, twice refusing the offer of a chair at the Collège de France, an institution that did not require its professors to prepare students for the state agrégation examination. Despite Urbain's entreaties, he refused to pose his candidacy for a position at the Sorbonne.[161] His laboratory resources were excellent and the distractions of Paris might have deterred him from his work. Perhaps, too, he felt like an outsider among the cliques of the capital.

At Lyon, Grignard took on more and more responsibility. He became the director of the reorganized Ecole de Chimie Industrielle in 1921, served on the general university council, and accepted the deanship in 1929.[162] From 1919 to 1926, he directed five state doctorates, eight university doctorates, and four engineering doctorates.[163] The research topics included some unrelated to organomagnesium reactions—for example, the determination of the constitution of unsaturated compounds by quantitative ozonization, the condensation of aldehydes and ketones, the cracking of hydrocarbons with an aluminum chloride catalyst, and the study of catalytic hydrogenation and dehydrogenation under reduced pressures.[164] However, the organomagnesium research continued to stand out as the most important program in organic chemistry in France, and Grignard's approach was to dominate in the pursuit of organic chemistry until well into the 1960s.[165]

Grignard became increasingly frustrated by 1930 in the duties that called him away from research. In addition to university administration, he accepted the burden of editing the multivolume *Treatise on Organic Chemistry* for the Masson publishing house. The work continued after his death under the editorship of his successor at Lyon, René Locquin, with the aid of G. Dupont, and eventually appeared in twenty-three volumes.[166]

Grignard's membership in the Paris Academy of Sciences made necessary frequent train trips to Paris. He sometimes undertook the round trip of fifteen or sixteen hours in one day.[167] In addition, he was still teaching four lectures a week. As he wrote a friend in 1933, "I have no time for research. The most I can do is to talk a little with my students and guide their work. It is disillusioning, isn't it?" And finally, Grignard seemed to be chagrined at the failure of the provincial universities to pull abreast of Paris and become centers of intellectual discipline and scientific research. French science as a whole was suffering from a lack of funds and a poverty of spirit. His letter continued:

> You will find that it is a lot like that in France, alas, if confessions are sincere. This is connected with the organization of scientific work in our country. As in the budget for the Faculties, there is nothing allowed for research. . . . In Paris, at the Collège de France and the Museum, professors are freed from teaching and examinations, professors at the Sorbonne have only semester courses, but in the provinces one has no need to think, much less study or even experiment. Add to that the difficulty of recruiting collaborators who love research truly for its own sake. This is the consequence of the wave of frenzied *arrivisme* unfurled since the war.[168]

The optimism of the early 1900s had vanished. It was this perception that something was very wrong with French science which motivated Jean Perrin and his Parisian colleagues to begin steps for new funding for science, under the Popular Front government of 1936–1938, and the establishment of the CNRS in 1939. As part of this movement, Albert Ranc wrote Grignard in spring 1935 to inquire about his perception of the financial situation of French science. In reply, Grignard complained that monies given to universities were designated for the construction of buildings and that nothing else could be done with the funds. There was no money for professors or subsidiary personnel, or for matériel. Heating and lighting costs were escalating, and the Lyon laboratories would soon have to close down five months out of eight because of a lack of heat.[169]

The universities had plenty of students, but no time and no conditions for training them well. The applied science institutes were turning out lots of chemists, commented Grignard, but not good ones.[170] Grignard's remarks indicate that he thought the problem was partly financial and partly cultural: there was no money, only arrivisme. To be sure, Grignard's views may in part be typical of an older generation that is often critical of younger colleagues and nostalgic for the "good old days." But his comments cannot easily be brushed aside. A state of mind threatened the scientific tradition in France and in Europe in the 1920s and 1930s; there was an ineffable pessimism fundamentally antithetical to scientific and progressive values.


THE LEGACY OF THE FIRST WORLD WAR

When the war broke out, Grignard was at a professional meeting in Le Havre near Saint-Vaast-la-Houge, where his family took summer vacations. Hearing of the general mobilization, Grignard joined his territorial regiment at Cherbourg. "Corporal" Grignard was noticed by an officer as he guarded a railroad crossing; the officer thought it odd that he was sporting the red ribbon of the Legion of Honor. Grignard soon was dispatched to the navy's chemistry laboratory at Cherbourg, then sent back in November to his own laboratory at Nancy. Recalled to Cherbourg in the spring, he was reassigned to the navy laboratory, and it was not until July 1915 that he was sent to Paris to work in the Direction du Matériel Chimique de Guerre. This occurred after the intervention of Haller, Moureu, and Urbain, who feared that Grignard would be sent to the battlefront.[171]

While at Nancy, he was commissioned to study the cracking of heavy benzol to serve as a source of toluene, which was in short supply. This involved the degradation of methyl benzols by aluminum chloride, a topic to which he returned in the 1920s with his student, R. Stratford.[172] In early December 1914, officers came to the Institut de Chimie and informed Grignard that the Germans were throwing special grenades into the trenches at the Lorraine front, making them untenable. The men had not been able, however, to obtain any samples. Grignard sent the information to Paris, suggesting the need for studying methods to respond to the German gases. He received the reply that it would be folly for the French to take up chemical warfare, because the Germans would outstrip them, as they had in the chemical industry.[173]

It turned out that the substances used in December were lachrymates, gases not held in stock by the French government except for a small amount of bromoacetic ester held by the Paris police.[174] But in April 1915, in the first instance of military employment of a deadly gas, the Germans used chlorine gas in Belgium. Assigned to Paris, Grignard was given office and laboratory space at the Sorbonne. Georges Urbain attached his brother, Edouard, and Jacques Bardet, an expert in spectral analysis, to work with Grignard. Grignard requested that one of his Nancy collaborators, Georges Rivat, join them, along with several chemists who had been sent home after being wounded at the front.[175]

Grignard's group investigated the preparation of phosgene, a nonpersistent lethal gas, by the action of oleum on carbon tetrachloride, and in October 1917, the French army used phosgene in a seven-day bombardment of the Laffaut salient northeast of Soisson.[176] A by-product in the phosgene reaction, $HClSO_3$, was converted into an ester by treatment with ethylene. This ester proved to have tear gas properties. Grignard's group developed a practical way to prepare it from sulfuric acid, carbon tetrachloride, and catalysts; they worked with other lachrymatory gases as well.[177]

On July 12, 1917, again at Ypres, the Germans introduced a new gas that disabled a soldier simply by coming into contact with his skin. Perhaps without even realizing he had been poisoned, a soldier could receive a lethal or disabling dose of the gas, which could retain these properties for several weeks.[178] Four days after the use of the gas at Ypres, everyone in Grignard's laboratory was sure it was related to dichloroethyl sulfide, or mustard gas, which had been prepared two years earlier by Moureu. The problem was to find a practical procedure for making it. Thanks to the work of Senderens, it was possible to prepare ethylene at the battlefield and to make sulfur chloride. As Grignard had embarked on a trip to the United States in June 1917, it was decided that Moureu would continue his original line of research, Job would study the synthesis of mustard gas devised by the English chemist Frederick Guthrie in 1860, and Rivat would study the chloride derivatives of methyl sulfide.[179] In winter 1918, after his return in February from the United States, Grignard developed a precise test for mustard gas which was convenient at the battlefield. It consisted of converting the compound into the more readily crystalline BB'-diiododiethyl sulfide, allowing the detection of 0.01 gram of mustard gas in one cubic meter of air:[180]

$$2NaI + (CH_2Cl \cdot CH_2)_2S = 2NaCl + (CH_2 \cdot I \cdot CH_2)_2S.$$

The test was used successfully in autumn 1918 just before the Armistice, by which time the British alone had suffered 160,000 casualties from the gas.[181] In all, there were approximately one million gas casualties.

During the war, many French scientists were called on to use their professional expertise. Pacifists and militarists, monarchists, republicans, and socialists joined together, with few exceptions, to help fight what began as perhaps the last popular war in modern history.[182] The physiologist André Mayer and the chemist Paul Lebeau improvised gas masks, which served as protection against many of the gases, while Moureu, Job, Urbain, Grignard, and their collaborators worked on making and detecting war chemicals. Jean Perrin and Marie Curie drove behind combat lines with a wagon carrying radiology equipment for the treatment of the wounded, and Perrin later joined Paul Langevin in a new office, the Direction des Inventions, to develop sonar as a tool to detect U-boats at sea.[183]

As during the Franco-Prussian War of 1870–71, the Academy of Sciences held regular meetings and organized special committees for aiding national defense. Indeed, the Franco-Prussian War had potentially marked the beginning of chemical warfare, since the idea was discussed by Berthelot's physics and chemistry committee.[184]

The horrors of the First World War left scientists and engineers with a foreboding about a war that might follow the war to end all wars. At Lyon, R. M. Gattefossé expressed the common view: "The war of tomorrow will surpass the horror that has preceded it; without any doubt, the perfection of aviation and chemical warfare will allow entire regions to be rendered uninhabitable."[185] As a consequence, the Geneva Convention was adopted by international delegates in 1925 to outlaw the use of gases in warfare, although the treaty's opponents predicted the secret production of gases by hostile states,[186] an argument still standard in weaponry debates.

Determination to prevent another war mixed equally, however, with distrust and vengefulness on the part of French scientists against the Germans and Austrians. As Schroeder-Gudehus has detailed in her book *Les scientifiques et la paix*, a scientific internationalism broke down during the war and was a long time being repaired.[187] A number of influential French scientists who were Alsatians had old debts to settle with the Germans. Among these were Schutzenberger, Lauth, and Haller, the three consecutive directors of the Ecole de Physique et de Chimie Industrielle in Paris, and Arth and Guntz, who followed Haller as directors of Nancy's Institut Chimique. Friedel, the Poincarés, and Appell were among prominent Parisian scientists from Alsace-Lorraine. Haller, as we have seen, had been apprenticed to a pharmacist in Münster when the 1870 war broke out, fought at Belfort, and then moved to Nancy. During the First World War, he was chief counsel to the French government on explosives, and he felt great satisfaction in returning to his birthplace, which had been in German hands for almost fifty years. Tragically, his revenge, like that of so many other Frenchmen, was achieved in the same year that his only son was killed in the war.[188]

Paul Appell's experience exemplifies the bitterness that contributed to the First World War. His brother Charles stayed in Strasbourg after 1871 and was arrested in 1888 for treason against Germany (for plotting with the French for the recovery of Alsace). He was sentenced at Leipzig to one year in prison and to nine more years in a fortress at Magdeburg. A letter sent to the German kaiser which was signed by eminent scientists including Pasteur, Hermite, and Poincaré was to no avail. Charles Appell fell gravely ill in 1898 and was released a year early, with the understanding that he would return to prison if he recovered his health, but he did not improve before his death in 1906. Appell's bitter view was that his brother had been the "victime de la puissance de l'organisation allemande et de l'implacabilité des dirigeants allemands."[189] It was hardly an accident that the first postwar meeting of the French Association for the Advancement of Science was held at Strasbourg.

The French mathematician Emile Picard, who lost a son in the war, was a moving force in the organization in 1918 of the Conseil Internationale des Recherches, an organization of national academies of sciences which excluded Germans and Austrians from membership. The League of Nations' Commission Internationale de Cooperation Intellectuelle (CICI), which first met in 1922, also excluded German and Austrian scientists. French nationalism was matched in 1924 by the German refusal to participate in the CICI after being invited to do so, and by the Verband der Deutschen Hochschulen, which cautioned German intellectuals against taking up old friendships with their colleagues in France and England. Political accords were signed between German and French ministers in 1925 to foster scientific interchange, which was not taking place spontaneously among most scientists.[190] At Lyon, for example, the Sciences Faculty did not begin discussing the renewal of subscriptions to German periodicals until five years after the war.[191] In 1926, Lyon and other Faculties received a circular from the minister of public instruction, addressed to universities, scientific societies, and organizations, asking them to renew the French-German exchange of theses and academic writings that had been going on before 1914.[192]

One of the champions of Franco-German reconciliation was the University of Illinois chemist William Albert Noyes, who sought to persuade chemists to renew personal correspondence and communication as well as to participate in international committees of the League of Nations and scientific organizations. In June 1922, Noyes attended a meeting in Utrecht of some forty chemists from ten different countries, to which the French and Belgians refused to send representatives because the Germans had been invited. Noyes also attended a meeting of the International Union of Pure and Applied Chemistry at Lyon, and found that Germans and Austrians were not being admitted to that organization. After travels through Europe, which included a ten-day stay in Germany, Noyes decided to initiate correspondence with some French chemists with whom he was acquainted, including Victor Grignard. He hoped to help settle issues disputed by both sides, including the initiation of gas warfare, France's indifference to Wilson's fourteen-point peace plan (Clemenceau said "The Good God has only ten"), and the French use of black African troops in the occupation of the Ruhr.[193]

Particularly among French scientists, there was a great deal of anger over the fact that the Academy of Sciences in neutral Sweden awarded the 1918 Nobel Prize in chemistry to Fritz Haber, the scientist-architect of the German gas warfare offensive. Many French scientists also continued to demand a repudiation by German scientists of the 1914 "Appeal to the Civilized World," signed by 93 German intellectuals, 22 of whom were scientists. French scientists regarded as unpardonable their German colleagues' defense of the German violation of Belgian neutrality and the German intellectuals' "protest before the whole civilized world against the calumnies and lies with which our enemies are striving to besmirch Germany's undefiled cause."[194]

In his correspondence with Noyes, Grignard expressed anger about the manifesto. The document, he said, was not only a tissue of lies but was a document of shame for the German scientific spirit.[195] Carrying over his suspicions of German territorial ambitions to the chemical community, Grignard criticized German scientists' conduct in committee work on organic nomenclature in the revision of Beilstein's handbooks. "This domineering pride that the Allies should have been able to break in the political domain . . . constitutes a serious defect in a true scientist."[196]

Like many of his compatriots, Grignard voiced suspicion of England and the United States as well. He thought the aim of peace would best be served by the guarantee of the security of France's borders; he wondered if England's industrial and imperial ambitions were not leading them to encourage hostility between France and Germany; and he was wary of the sensationalist journalism and jingoism of nationalist newspapers everywhere. Grignard also showed insight and impatience with regard to Noyes's all-too-typical double standard of political idealism:

> In your wars with Mexico and Spain, if you have given some compensatory indemnities for the territories annexed, it is because you have recognized that these had a value incomparably greater than the cost of the wars, and . . . in my opinion, that argument has no value, for it amounts to saying that one can take anything by paying for it. In strict justice there is only one thing that should be considered, and that is the wish of the people annexed.[197]

What is clear is that the issue of military use of scientific discoveries and inventions became acute before the development of nuclear armaments in the Second World War. The specter of the obliteration of the fields, forests, and inhabitants of Europe was already haunting Europeans by 1918, and the golden image of Science as human activity on a higher plane had been severely damaged by the events of the war. In 1928, Julien Benda wrote of the "trahison des clercs": the disciples of learning had betrayed the universalist and transcendental calling of their priesthood.[198] Excesses of chauvinism, public squabbling, recriminations, and mutual distrust among academic intellectuals called into serious question the traditional Baconian and Comtian claims for scientific method and scholarly research flourishing above political partisanship and improving society.

The pessimism and relativism that were troubling undertones of fin-de-siècle French culture spread throughout Europe in the 1920s and 1930s. No wonder Grignard lamented the loss of scientific esprit in the mid-1930s. The arrivisme he decried was one manifestation of waning confidence that the idealistic aims of a fundamental and improving positive science could be realized. As logical positivism replaced Comtian positivism, and pragmatism edged out idealism, so the technicians, engineers, and specialists—in increasing numbers—studied science and its applications for limited practical ends. The First World War and its aftermath made clear that the scientist and the scientist's work were not nonpartisan, objective, and value-free. There was a loss of innocence.

The decline of science in France after the First World War was not only a matter, then, of limited monies available in an inflationary economy. Nor was it simply a matter of political reassertion of state control and renewed centralization of disparate institutions. There also was a loss of confidence in science and scientists, and many young men and women who might have yearned to be savants now readjusted the breadth of their vision or sought truth and understanding elsewhere. This development was not unique to France; it also afflicted other European countries and Great Britain.

Provincial universities such as Lyon, which long had been offering vocational and specialized scientific and engineering education, were now inundated with increasing numbers of students oriented toward specialized, professional training. For those, like Grignard, whose activities in earlier years had moved back and forth between research interests and practical education, equilibrium was lost. Advanced students interested in science as a high intellectual calling now went almost exclusively to Paris, leaving one of France's most distinguished chemists to work with students whom he often found disappointing. And as we have seen, this state of affairs did not exist in the Lyon Sciences Faculty alone; by 1930, it was a dilemma faced by Faculties throughout provincial France.

Before developing some systematic conclusions about the Paris-province relationship, its evolution, and the dynamics of provincial research programs, we will turn to a Sciences Faculty that did not make great gains in the period 1860–1930, despite a long-standing distinguished reputation. While other Faculties might have been chosen by way of contrast, Bordeaux is especially interesting because of the tenure there of one of France's most influential physical theorists, Pierre Duhem.

# 6

# BORDEAUX: CATHOLICISM, CONSERVATISM, AND THE INFLUENCE OF PIERRE DUHEM

The history of Bordeaux's Sciences Faculty stands in strong contrast to that of the other four provincial Faculties discussed so far. Unlike Nancy, Grenoble, Lyon, and Toulouse, by 1930 Bordeaux attracted a smaller proportion of students studying the sciences in France than it had before 1900. In 1894, Bordeaux and Lyon led all provincial universities in enrolling science students, each inscribing 9 percent of science students in French higher education, but by 1930, Bordeaux's share had fallen to 4 percent.[1] Other statistics similarly suggest that Bordeaux's neighbor institutions eclipsed her in the crucial period of provincial growth and efforts at decentralizatiori (1875–1905).

There is a striking lack of optimism and good spirits at Bordeaux Faculty meetings in comparison to the other provincial centers in the late 1890s and early 1900s. There were fewer administrative entrepreneurs who maneuvered the establishment of lucrative engineering and applied science institutes. And Faculty research, particularly in the areas discussed here—organic chemistry, physical chemistry, and electrical physics—showed little innovation and leadership.

Nor was there much turnover among Faculty members at Bordeaux. Very few scientists who later became illustrious in Paris or elsewhere taught at Bordeaux and moved on. Among longtime chair holders, there were only a few distinguished scientists, notably the chemist Auguste Laurent (who taught at Bordeaux from 1838 to 1847), his successor A. E. Baudrimont (1847–1880), and the physicist Pierre Duhem (1894–1916). Duhem's career at Bordeaux is of particular interest both for his skeptical attitudes toward the aims and methods of physical science and for his particular influence on decision making within the Bordeaux Sciences Faculty. In addition, Duhem's career, like that of Sabatier, illustrates the role played by anticlericalism in the broader political stage of the history of French science. As we have already begun to see, republican, radical, and socialist educational ministries consistently avoided effective decentralization that might undercut the administrative power and ideological values of the current government regime and its elite. Nowhere did the dangers of decentralization seem more real to republicans and to the Left than in the person of Pierre Duhem and the kind of conservative Catholicism that was sometimes a powerful force in the city of Bordeaux.

## THE CONSERVATIVE TRADITION IN SCIENCE

Bordeaux has long been one of the great cities of France. During the nineteenth century, it was one of the most conservative. In his 1895 history of the city, Camille Jullian described Bordeaux as the "calmest of French cities since 1815."[2] Lying on the west bank of the Garonne River, the city is about sixty miles from the sea, in the plain that is the wine-growing district of the Médoc. The vineyard-lined Gironde, the estuary formed by the junction of the Garonne and Dordogne rivers, provides access northwest to the sea. Looking toward the city's center from the nineteenth-century stone-arched bridge crossing the Garonne, Jullian and his contemporaries could view a crescent of wide and busy quays, lined by warehouses, factories, and mansions, with towers and steeples rising behind. The harbor, formed by the basin of the Garonne, accommodated in 1900 approximately one thousand vessels, with steamers setting off for Great Britain, Spain, Argentina, and the United States.[3]

Wine sales accounted for one-fourth of the city's export revenues, with the maritime exportation of wines quadrupling from 1848 to 1879, and commerce in general doubling from 1866 to 1886. Industries supporting the wine trade also flourished, including cooperage and the making of bottles, corks, and wooden cases. There was little heavy industry other than shipbuilding and refitting of ironclads and torpedo boats as well as merchant vessels. Jullian notes in workers' attitudes toward their *patrons* what he deemed a praiseworthy attachment, almost medieval in character.[4]

The city had been the headquarters of the Girondists during the Revolution, and its citizenry suffered heavily during the excesses of the Reign of Terror. The return of the Bourbons was welcomed. It was to the safe haven of Bordeaux that the modern French government retreated in times of crisis, first in 1870, and then again in 1914 and 1940. At the beginning of the nineteenth century, the city's population was 90,000. In 1891, it was about 250,000, and by 1921, it had not increased very much: it was 267,409. Bordelaise banks favored investment in traditional colonial enterprises and wine commerce, and it was not until after the absorption of some of them by Parisian banks in the 1920s that funds became more readily available for the development of new industrial enterprises. These facts help explain the relative lack of interest by the Bordeaux municipality in building scientific and technical education and research.[5]

In the eighteenth century Bordeaux had been the site of a university, which was founded in 1441, and an academy that boasted Montesquieu as a member. In 1808, a Société Philomathique was founded as a reconstitution of the pre-Revolution Musée de Bordeaux. The Linnaean Society was established in 1818. One of Bordeaux's most flourishing nineteenth-century scientific organizations was the Société des Sciences Physiques et Naturelles, which originated in 1848 in a group meeting with Jean-François Laterrade, director of the Jardin des Plantes. Its members were to include Baudrimont, Paul Bert, Duhem, and other Science Faculty members such as the botanist P. M. A. Millardet and the chemist Ulysse Gayon.[6]

The Letters Faculty, founded under Napoleon, was suppressed in 1815, leaving Theology as the lone Bordeaux Faculty until the establishment of Sciences and the reestablishment of Letters in 1838. A lycée provided postbaccalaureate students for these institutions. In addition, the Philomathic Society adopted educational duties in 1825, creating courses in general physics, mechanics applied to the arts and industry, botany, mineralogy, geography, astronomy and geology, French history, and French literature.[7] The city thus had an educational infrastructure of state and local institutions encouraging scientific studies.

As was true in other cities, Bordeaux also had a network of Catholic educational organizations, including ten communal schools in 1870. Since Catholicism was seen by many as an antidote to the unrest that had resulted in the upheavals of 1848, the government reemphasized religion in lycées and collèges by means of the Falloux law of 1850. A network of Catholic activities was organized around parish schools, including adult courses, social centers, apprentice societies, and Catholic worker groups. After the suppression of the Paris Commune in 1871, the return to power of a papal-supported monarchy seemed possible, until the elections of 1876 gave republicans 340 of 495 seats in the National Assembly. Republicans then inaugurated a program of anticlericalism in education.

Before the republican victory, the law on "liberty of higher education" passed the assembly in 1875, leading to the establishment of Catholic "universities" in Angers, Lille, Lyon, Paris, and Toulouse. Subsequent republican measures deprived the Catholic universities of their title, however, so that they became "institutes," and by 1889 various Catholic institutions of higher education were teaching only 724 students, compared to 20,000 in the state system.[8] This was only one among many results of the relentless legislative program launched against clericalism in the last decades of the century in an attempt to eradicate church influence from education and politics in France.[9]

The city of Bordeaux was the site of a Faculty of Catholic Theology and the seat of an archbishopric. From 1873 until 1882 two men shared the administration of the Bordeaux bishopric: Msgr. Donnet and Msgr. de La Bouillerie. Donnet was a cardinal, a friend of Louis-Napoleon, and a sympathizer with the prominent views of F. A. P. Dupanloup, bishop of Orléans from 1849 to 1878, on distancing the French Catholic church from ultramontanism, or subservience to the papacy. In contrast, Msgr. de La Bouillerie was ultramontanist and legitimist in his political views, eschewing any collaboration between the church and the Republic. A strong advocate of the introduction of neo-Thomistic philosophy into Catholic education, La Bouillerie became a favorite of Pope Pius IX.[10]

Their successor was Msgr. Guilbert, who was sympathetic to democratic traditions and committed to the independence of the sciences from the jurisdiction of the church. This was important at a time when influential church figures argued that scientific teaching was undermining religious faith. To combat the influence of the legitimist Cercle Ozanam, Guilbert established the Cercle Fénelon for Catholic students who wanted to meet together outside the boundaries of a political ideology.[11] Msgr. Victor L. S. Lecot, bishop at Dijon and Guilbert's successor, had close links with the Carnot family and was equally well disposed toward republicanism. He welcomed Leo XIII's 1892 encyclical and *Letter to the French Cardinals* calling for Catholic unity and expressing the wish that French Catholics renounce all opposition to a republican regime.[12]

But two events were to contribute to the destruction of the emerging rapprochement between Catholicism and republicanism: the death in 1903 of Leo XIII and the French government's 1905 Law of Separation, under which the government confiscated the inventories of seminaries and episcopal palaces. In Bordeaux, Msgr. Pierre Paulin Andrieu succeeded Lecot in 1909, and the climate of peace among Bordeaux Catholics was succeeded by an atmosphere of struggle. The new cardinal aligned himself with Pope Pius X against the "errors" of "naturalism" and "modernism" and encouraged Catholics to resist the state policy of laicization. After demonstrations at the cathedral in June 1909, the correctional tribunal of Bordeaux condemned the cardinal to pay a 600-franc fine for counseling disobedience to the law.[13]

The struggle at Bordeaux between Catholic republicans and Catholic ultramontanists is a mirror of struggles occurring in cities and départements throughout France during the nineteenth century. But we mention it here to demonstrate the effect on perceptions of Bordeaux by government ministers and prospective Faculty members and students. Duhem was the best-known French scientist to take a militant conservative and ultramontanist Catholic position, espousing views closely associated with Msgr. Andrieu and the Cercle Ozanam. It was said that Duhem treated unfairly students who expressed opposition to his views. This may not be true, but the *perception* counted a great deal and influenced the fortunes of the Faculty. Let us look in more detail at the characteristics of this Bordeaux Sciences Faculty with which Duhem's name became intimately associated.

Founded in 1838, the Sciences Faculty originally had six professors: Victor Amedée LeBesgue in pure mathematics; Joseph C. C. Chenou in astronomy and rational mechanics; Jeremie J. B. Abria in physics; Auguste Laurent in chemistry; H. Provana de Collegno in botany, mineralogy, and geology; and Isidore Geoffroy Saint-Hilaire in zoology. Already an illustrious member of the Paris Academy of Sciences, Saint-Hilaire was named dean. However, within a year he resigned and returned to the Paris Sciences Faculty and the Museum of Natural History. He was succeeded by P. F. A. Bazin in the zoology chair and de Collegno in the deanship.[14]

Originally one of the largest provincial Sciences Faculties, Bordeaux experienced no expansion in its number of chairs until 1876, when Le Verrier's student, Georges Rayet, was appointed to a new chair of physical astronomy and two chairs were created from one: botany (for P. M. A. Millardet) and geology and mineralogy (kept by V. Raulin). In 1889, a second chair of chemistry was created (industrial chemistry for J. A. Joannis), and in 1895, the chair of physics was transformed into theoretical physics for Duhem at the same time that a new chair was created for J. A. F. Morisot in experimental physics.[15]

From all accounts, the Bordeaux Sciences Faculty was very badly housed until 1885. Placed in municipal buildings near the Hôtel de Ville, the Faculty had one amphitheater that was employed from 1839 to 1886 for public evening lectures. A physics laboratory opened on the court of the Hôtel de Ville, but did not connect with the Cabinet de Physique. The chemistry laboratory was illuminated by one window six meters above the floor, and Laurent transferred his work to his house.[16] In a letter to the mayor of Bordeaux, the Parisian chemist and minister L. J. Thenard had promised that the new Sciences Faculty would transform the city: "The teaching of the sciences will bring into being the taste for applications. The city of Bordeaux will thus become a manufacturing center and will acquire a taste of wealth."[17]

But the city did not rise to the bait. A municipal promise to contribute 6,000 francs for the establishment of instruments and collections actually resulted in only 3,000 francs. Sufficient funds were not available to pay heating bills, and the ministry decreased the already penurious budget for buying books in early 1841.[18] Hard decisions had to be made about purchases. Volumes 16 and 17 of the *Memoires* of the Academy of Sciences cost fifty francs, the same price as a thermometer for electrochemical experiments and half the cost of a pocket sextant.[19]

When the Bordeaux municipal council demanded the creation of a law faculty, Victor Duruy tried to strike a bargain: the construction of a proper building for the Sciences and Letters Faculties in exchange for a law faculty. Bordeaux's academy rector reported the failure of this approach to Duruy in a letter of March 8, 1869. The municipal council was outraged and had voted for "only one thing, the construction of a building for the Law Faculty" (a Faculty created in 1870).[20] In his

November 1869 report on the affairs of the Sciences Faculty, Dean Abria chided his public, saying that he was "far from asking that Bordeaux impose on itself the sacrifices that certain German cities do not hesitate to make." He continued, "I am far from asking that our Faculty be put on the footing of the analogous establishment of Heidelberg, a small city of 10,000 souls whose laboratories have been reorganized under the eminent physiologist Helmholtz."[21] With the Franco-Prussian War over and the French defeated, Rector Séguin rebuked the mayor and the municipal council in 1874, saying that Bordeaux surely was wealthy enough to pay for the honor of helping France to "vanquish other nations in all respects in intellectual development."[22] Predicting the passage of the law of liberty of higher education, Séguin correctly warned Bordeaux that it must take on the duty of building up higher education, lest its neighbor Toulouse develop higher education in the southwest at Bordeaux's expense.[23]

The city's immediate response was to buy three houses near the Hôtel de Ville and put them at the disposal of the Sciences Faculty. This favorable action may have been rendered likelier by the strong republican commitments of (Protestant) Emile Fourcand, who was elected deputy mayor and mayor in the 1870s.[24] Louis Liard, who taught philosophy in the Bordeaux Letters Faculty in 1878–79, helped formulate a project for building a university palais at a cost of 1.4 million francs, a project completed in 1885 on the site of the old lycée, which had been transferred to the collège from which Jules Ferry's ministry expelled the Jesuits in 1881.[25]

This period of municipal interest in higher education corresponded with the establishment of a new national salary scale for French professors in 1876, widespread talk of establishing six or seven provincial universities, the introduction of several hundred scholarships for licence and agrégation students, the doubling nationally of the number of Faculty positions (including the creation of lectureships), and an emphasis on research. When installed in its new quarters, the Faculty numbered eight professors, two chargés de cours complémentaires (mathematics), and three lecturers (chemistry, zoology, botany). There was much talk among Faculty members of more chairs, particularly in mathematics (applied mechanics) and chemistry (mineral, organic, or industrial chemistry). But at this time Bordeaux was granting only a handful of science licence degrees.[26]

For decades, Bordeaux's Sciences Faculty, like so many provincial

Faculties, taught very few university-level students, awarding two or three licences each year and hardly any doctorates. Grading the written part of the baccalaureate and conducting oral examinations took up a good deal of time each November, April, and August. In the 1880s, 170 to 260 lycée graduates wrote examinations each session, and three juries met each day for oral examinations, one in the morning and two in the afternoon, with each jury examining ten candidates per day. Faculty members regarded the examinations as a necessary evil, since they derived an average yearly income of 1,682 francs from examination fees during the period 1854–1875, when a provincial Faculty salary was only 4,000 francs.[27] Like other Faculties, Bordeaux professors also were unwilling to give up the power they exercised over selection of the elite through the grading of these qualifying examinations.[28]

Given the paucity of real university students, it is not surprising that the Bordeaux Faculty adopted the now-familiar strategy of offering evening lecture courses, in the hope of avoiding an empty classroom. In response to the ministry's criticism, the Faculty defended the quality of its public audiences, which were made up of industrialists, sailors, and medical doctors who were busy during the day.[29] When the ministry formally encouraged the teaching of applied science in 1854, the Bordeaux Faculty responded enthusiastically, mentioning wine and colonial enterprises, agriculture, forest products, shipbuilding, and railroad concerns. They promised that all needs for applied science courses could be met by the current staff, augmented only by a supervisor of graphic arts and a supervisor of laboratory work in physics and chemistry.[30] Again in 1864, Bordeaux, like other Faculties, was called on by the ministry to report on applied and evening courses. They assured the ministry that licence students, mostly lycée teachers, had followed course work entirely separate from the evening lectures, with the exception of infinitesimal calculus. But that did not mean evening courses were elementary. Indeed, Baudrimont defended the view that as far as chemical education was concerned, the evening courses dealing with applications were as important as those leading to the licence.[31]

The question of how to attract licence and especially agrégation students to Bordeaux remained. Typical was the occasion in 1869 when one student distinguished himself in physical science examinations and then moved to Strasbourg, where he could better complete graduate-level studies in physics and chemistry.[32] The loss of Strasbourg to Germany in 1870–71 was a major blow to this kind of strategy. From 1819 to 1869, fifty-two doctoral theses had been defended in the sciences at Strasbourg, considerably more than at any other French provincial Science Faculty.[33] In 1873, the Bordeaux Faculty supported the establishment in France of university-affiliated positions akin to the German Privatdozenten to encourage students to study beyond the licence and agrégation. To assure the development of science studies outside Paris, they suggested five measures: (1) the suppression of the privileges enjoyed by graduates of the Ecole Polytechnique for recruitment into the special engineering schools of Ponts et Chaussées, Mines, and so on; (2) requirement of the licence ès sciences for any professor teaching in Medical Faculties and schools; (3) the award of a 500-franc indemnity to collège professors possessing a licence; (4) control by the individual Faculty of the disposition of a certain number of scholarships; and (5) the grouping of provincial Faculties into fairly autonomous provincial universities.[34]

In 1878–79, the total enrollment in Bordeaux's four Faculties was 3,507, including 1,419 in Law, 1,140 in Letters, 859 in Medicine (a Faculty since 1874), and 89 in Sciences. By 1894, the Sciences enrollment was 134, increased as in other Science Faculties largely because of the PCN program. In 1900, it was 255.[35] By the academic year 1912–13, the Sciences Faculty enrollment was 316, but the total enrollment (at what was now the University of Bordeaux) had decreased since 1911, a result attributed to modification in the national law regarding military service. (The 1905 law had ended most university students' two years of exemption from a mandatory three years of military conscription.) The mood of the Sciences Faculty was not optimistic, and the yearly report stressed the lack of laboratory facilities for physics, botany, and other services at a time when facilities had just been completed for the Law Faculty.[36]

Like many of her provincial sisters, Bordeaux began establishing special institutions as appendages to the Faculty. An observatory had been built at Floirac in 1876. Now an Ecole de Chimie was created in 1891 by Baudrimont's successor in the chemistry chair, Gayon. By 1908, the School of Chemistry was awarding a university diploma in chemical engineering to foreign and French students in a three-year curriculum that was more strenuous than the original two-year program. From 1891 until 1908, 134 school diplomas were awarded; and from 1908 until 1921, 82 university engineering diplomas. The numbers of students in this program were not comparable, however, to rival provincial chemical-engineering schools, and Bordeaux did not attract a large foreign clientele. In 1920–21, only 15 foreign students were enrolled in the Sciences Faculty; in 1929–30, there were 99.[37]

Gayon, who had studied fermentation and wines with Pasteur, also directed a Station Agronomique et Oenologique, created by decree of the Ministry of Agriculture in 1880. Here laboratory analyses were carried out free of charge for local enterprises. Gayon also taught a course in agricultural chemistry, as Baudrimont had done before him. Collaborating with his colleague, the botanist Millardet, Gayon worked on remedies for the phylloxera disease that ravaged French vineyards, and Millardet received an Academy prize in 1894 for his research along these lines.[38] Maurice Vèzes began teaching a course in 1899 on the extraction and distillation of resins, in a teaching program that resulted in municipal, départemental, and private funding for a research laboratory on wood products, culminating in the creation in the early 1920s of the Institut du Pin.[39]

To these institutes was added the Ecole de Radiotélégraphie, shortly after a chair of industrial physics (which Duhem had always opposed) was created in 1920. In its second year, this school had thirty-eight students, most of them studying free of charge as students enrolled in the Ecole de Navigation de Bordeaux.[40] A difficulty for this school and the other institutes was that they did not bring in satisfactory revenues for their own programs and those of the Faculty. When Foch became director of the School of Radiotelegraphy in 1923, he informed the dean that student fees for laboratory work had not been raised since 1914. Whereas a 25-franc fee was allowed legally for certain programs, Bordeaux was only charging 10 francs. Similarly, the School of Chemistry found in 1920 that it was charging considerably less than Nancy for laboratory work.[41] So inadequate was the budget for industrial physics that its teaching program was reduced in 1923 and the chair was transformed in 1927, for lack of funds, into a chair of "physics" that was given to Mercier, then professeur sans chaire.[42]

It is striking that by 1930 at Bordeaux only one (industrial chemistry) of the seventeen disciplines formally tied to chaired professors and professeurs sans chaire was clearly concerned with applied science. The Faculty offered fifteen certificates of higher studies, only three of them in technical subjects (applied chemistry, physiological chemistry applied to agriculture, and radiotelegraphy).[43] We do not find a level of research activity in either applied or pure research comparable to the other four provincial universities.

One way to measure this activity is by the number and type of doctoral theses written at Bordeaux. At no time did the Bordeaux Sciences Faculty show the sudden surge of doctoral theses that is the outcome of a steady, ongoing, and substantial research program (table 4).[44] From 1810 to 1890, only five doctoral theses were defended at Bordeaux.[45] Duhem had few students, as we shall see.

Of course, traditionally, French students did not plan their studies around a scientist or research program, but around examinations. And

TABLE 4
DOCTORAL THESES DEFENDED IN THE SCIENCES FACULTIES

| | | 1896–97 | 1905–06 | 1909–10 | 1914 | 1925 | 1930 |
|---|---|---|---|---|---|---|---|
| Bordeaux | Total | 1 | 0 | 1 | 0 | 2 | 2 |
| | Univ. | 0 | 0 | 0 | 0 | 1 | 2 |
| | State | 1 | 0 | 1 | 0 | 1 | 0 |
| Grenoble | Total | 0 | 11 | 13 | 10 | 2 | 2 |
| | Univ. | 0 | 11 | 13 | 10 | 2 | 1 |
| | State | 1 | 0 | 0 | 0 | 0 | 1 |
| Lyon | Total | 4 | 2 | 1 | 2 | 7 | 17 |
| | Univ. | 0 | 1 | 1 | 1 | 4 | 11* |
| | State | 4 | 1 | 0 | 1 | 3 | 6 |
| Paris | Total | 36 | 28 | 46 | 36 | 60 | 81 |
| | Univ. | 0 | 4 | 9 | 10 | 8 | 20* |
| | State | 36 | 24 | 37 | 26 | 52 | 61 |
| Nancy | Total | 0 | 6 | 7 | 2 | 12 | 8 |
| | Univ. | 0 | 4 | 7 | 2 | 10 | 7 |
| | State | 0 | 2 | 0 | 0 | 2 | 1 |
| Toulouse | Total | 1 | 1 | 2 | 2 | 4 | 7 |
| | Univ. | 0 | 0 | 2 | 2 | 4 | 4 |
| | State | 1 | 1 | 0 | 0 | 0 | 3 |

NOTE: These University degrees include in each case one doctorate in engineering.

French scientists do not seem to have prided themselves on their research students until relatively recently. Further, Bordeaux did not attract many Faculty outsiders to vacant chairs; instead, the Science
Faculty frequently nominated their own alumni to staff positions and
promoted Faculty members from within their own ranks.[46]

One interesting and exceptional division within the Faculty in this
regard occurred over the question of hiring Albert Turpain, a former
student who completed a thesis on electric waves with Duhem in 1899.
(Duhem did not participate in the discussion.) The grounds for a negative vote (6–4) on Turpain focused on his recent polemical articles,
which were said to have enraged members of the Ecole Polytechnique
and the engineering corps of the grandes écoles. "Courteous relations,"
reads the report written by another of Duhem's former students, Marchis, must exist between members of the universities and the engineering bodies. This was not possible with Turpain, who had called education at the Ecole Polytechnique "superficial" and had recommended an
enlarged role for Sciences Faculties in engineering education. The Bordeaux Sciences Faculty did not want someone who would rock the
boat. It also was alleged that Turpain's recent experimental work was
merely a paraphrase of his thesis.[47]

In general, then, it can be said that the caliber of scientists at Bordeaux was not as high in the crucial 1875–1905 period as had been the
case in earlier years. Jacques Hadamard taught briefly at Bordeaux from
1893 to 1896, and then left the Bordeaux chair of astronomy and rational
mechanics to be chargé de cours at the Sorbonne.[48] Ernest Esclangon
was in charge of the Bordeaux Observatory before moving to Paris.
Paul Sabatier was lecturer in chemistry and physics before moving to
Toulouse in 1882, and the geologist Charles Jacob was a staff member
before moving to Toulouse. But few other luminaries passed through
Bordeaux in these years.[49]

This is a little surprising, given the distinguished reputations of Laurent and Baudrimont in chemistry and the presence of Duhem in physics. Laurent, the first chair holder in chemistry, left Bordeaux in 1847
for a frustrating career in Paris, where he first served as Jean-Baptiste
Dumas's suppléant at the Sorbonne, but never received a decent appointment because of misunderstandings with Dumas and other chemists regarding his theories of chemical substitution and types. Laurent's
collaboration with Charles Gerhardt is rightly considered to be important in the development of the modern theory of chemical valence, and

their working out of values for atomic weights became the basis for modern chemistry after the Karlsruhe Congress of 1860 (though not in France until the turn of the century).[50]

Laurent's successor in chemistry at Bordeaux, Baudrimont, was similarly prolific, innovative, and controversial in his work. He held the chemistry chair from 1849 until his death in 1880. His interests mirroring his diversified background in pharmacy, medicine, and the physical sciences, Baudrimont wrote more than 175 articles. He strongly supported Laurent's theory of types, that is, that the formulas of organic chemistry can be organized under four formulas of mineral chemistry, namely, $HCl$, $HOH$, $CO_2$, and $NH_3$. Baudrimont also worked out numerical and empirical relationships among the chemical elements, and their atomic and equivalent weights, leading him to an interesting periodic classification, although one that proved to be less fruitful than Mendeleev's. Baudrimont had his periodic table drawn up on a large scale for his students, using it as a teaching device in his lecture courses.[51]

Baudrimont attended the Karlsruhe Congress, where he supported the general atomistic views of Laurent, Gerhardt, and Wurtz. On the occasion of the 1877 debates in the Paris Academy of Sciences regarding the relative merits of the systems of atomic and equivalent weights, Baudrimont offered arguments for atomism on both experimental and methodological grounds, citing his earlier work in the 1844 *Traité de chimie*. Against the phenomenalist or positivist view that science must be empirical and avoid speculative theories like the discredited notions of heat "caloric" and light "corpuscles," he insisted that the speculative part of science is necessary in order to interrelate the parts of science and understand the whole.[52] He regarded chemical polymorphism (a substance can crystallize in two or more different forms) and allotropism (an element is found in different forms) as convincing evidence that atoms may be arranged in different ways to form chemical elements and molecules. He also favored the idea that the chemically active groups are not always the chemical elements, and he gave the general name *mérons* to particles of chemical activity,[53] which might be one or more atoms of like or unlike elements. Mérons included the so-called radicals like methyl or ammonium.

Baudrimont's successor in chemistry, Gayon, had primarily practical interests, reflected in publications in local Bordeaux or Gironde journals, that resulted in significant contributions to oenology. More akin to

Baudrimont in the possession of a philosophical and theoretical turn of mind was Pierre Duhem, with whom Baudrimont would have differed on most major issues in the epistemology and methodology of physical science.[54] Had the two served together in the Faculty (Duhem did not arrive until 1894), they most certainly would not have gotten along. It is Duhem who has become the centerpiece of the Science Faculty's history, for a number of reasons that illuminate both the history of science in Bordeaux and the history of science in France.

## PIERRE DUHEM AND SCIENTIFIC SKEPTICISM

Duhem came to Bordeaux as chargé de cours in physics, in fall 1894, when the chair holder, Joseph Eugène Pionchon, requested a transfer to Grenoble.[55] Duhem had previously taught at Lille (1887–1893) and at Rennes (1893–94). When he came to Bordeaux, he was thirty-three years old, recently widowed, and the father of one daughter, Hélène (another daughter had died in infancy). Duhem was born in Paris and had studied at the Catholic Collège Stanislas before entering the Ecole Normale Superiéure in 1882. A brilliant student, he took baccalaureate degrees both in letters (1878) and sciences (1879), and placed first in entrance examinations at the Ecole Normale and Ecole Polytechnique.[56]

When entering the Ecole Normale, Duhem already was well prepared in the physical sciences, particularly in thermodynamics, because of his course work with Jules Moutier at the Collège Stanislas. A theoretical physicist in orientation, Moutier was interested in applying the theorems of thermodynamics to problems of chemical dissociation. According to Duhem, Moutier preferred the kinds of mechanical explanations in physical theory associated with the traditions of Descartes and the atomists,[57] but he introduced Duhem to the newest non-mechanical approaches in thermodynamics. Duhem especially studied Georges Lemoine's description in *Etudes sur les équilibres chimiques* (1882) of J. W. Gibbs's work, as well as the first part of Hermann von Helmholtz's "Die Thermodynamik chemischer Vorgänge." Helmholtz's paper was concerned with the problem that François Raoult took up from the experimental standpoint: the distinction between chemical heat and voltaic heat in a battery. Helmholtz developed what later were called the Gibbs and Helmholtz "free energy" functions.[58]

Within three years after entering the Ecole Normale, before he had even received the licence, Duhem presented a thesis for the doctorate which developed the concept of thermodynamic potential in chemistry and physics, including criticism of Berthelot's principle of maximum work. Whereas other physicists and chemists, like van't Hoff, Arrhenius, and Ostwald, were to use osmotic pressure as a measure of chemical affinity, Duhem developed results for solutions, gas systems, and electrified systems using expressions for thermodynamic potentials, drawn from analogies between rational mechanics and thermodynamics.[59] This work was later published as *Le potentiel thermodynamique et ses applications à la mécanique chimique et à la théorie des Phénomènes électriques* (1886). The thesis was refused at the University of Paris. Duhem then prepared a second doctoral thesis, on the theory of magnetism treated from a thermodynamic point of view. Gaston Darboux and Edmond Bouty reported the positive reaction of the jury to the new thesis, which was submitted in mathematical sciences.[60] Duhem worked on the thesis while serving as an aide-agrégé in physics at the Ecole Normale and teaching at Lille.[61]

While lecturer in physics at Lille, Duhem published his second book, *Hydrodynamique, elasticité, acoustique* (1891), which was especially praised by Hadamard, his colleague and friend, who said it influenced significantly his own work. One of the surprising results in Duhem's study came in his conclusion for viscous fluids: only "quasi-waves," not true waves, are possible. This idea was important to the development of theorems on shock waves in air.[62] Duhem also began at Lille his long "Commentaire au principes de la thermodynamique," which he came to regard as one of his most important series of papers. Here he attempted to develop a mathematical treatment of equilibrium processes that is formally analogous to the mechanics of Lagrange.

Duhem intended this work to be logically continuous with the tradition of mechanics, yet include mechanics as a special case. He laid out fundamental postulates and principles in an axiomatic mathematical argument, from which the first and second thermodynamic laws could be deduced. From these laws he derived definitions of entropy and thermodynamic potential. From the notions of energy and work came an algebraic definition of heat, which he suggested will "perhaps scandalize some minds."[63] Within a few more years, after moving to Bordeaux, Duhem published *Traité élémentaire de mécanique chimique fondée sur la thermodynamique* (1897) and *Thermodynamique et chimie* (1902), in which he worked out more fully the thermodynamic program for chemistry within what he called the contemporary tradition of Horstmann, Mou-

tier, Gibbs, Helmholtz, and Sainte-Claire Deville. The science of mechanics or movement, he wrote,

> has ceased being the reigning doctrine from which all theories are
> reclaimed and has become only a branch—the simplest of all—of a
> more general science. This science embraces not only movement
> which displaces bodies in space but also every change of qualities,
> properties, physical state, chemical constitution. This science is con-
> temporary thermodynamics or, according to the word created by Ran-
> kine, Energetics.[64]

Duhem continued in the next decade to systematize physical science
within the framework of thermodynamics, although he was not successful in integrating electricity and magnetism.[65] He paid a great deal
of attention to problems in electricity and magnetism, but self-con-
sciously and programmatically steered clear of modern research using
corpuscular theories.[66] He preferred to work from Helmholtz's electromagnetic theory rather than Maxwell's or the newer theories of Lo-
rentz. In 1898, Duhem predicted that the corpuscular and kinetic theories associated with Clausius and Boltzmann were approaching
disaster, citing work on the law of phases as a decisive defeat for atomism. He had no sympathy for the synthetic treatment of electromag-
netism and atomism characterizing Einstein's work in the early 1900s.[67]

To be sure, Duhem regularly read the newest literature on Maxwell's
theory, problems in classical mechanics, and discoveries in corpuscular
and kinetic research. He wrote critical book reviews for the popular
Catholic scientific journal, *Revue des Questions Scientifiques*. In 1893 and
1894 he published in this journal his earliest papers on the relations
between physics and metaphysics, the proper aims of physical theory,
and the impossibility of a "crucial experiment." He began to distinguish his views from those of Henri Poincaré, whose ideas he thought
led to the skeptical conclusion that theoretical physics escapes the laws
of logic.[68] He repudiated the Comtian positivism associated with the
names of Emile Littré and Berthelot, as well as the new pragmatism of
Edouard Le Roy.[69] Duhem evinced an admiration for Ernst Mach,
whose *Die Mechanik in ihrer Entwicklung historischkritisch dargestellt* (1883)
he complimented highly: "What is lacking in this varied book, *sobre
vivant*, for seducing the French reader? Only to be written in
French!"[70]

To an important degree, Duhem owed his skepticism of mechanical
explanation and his concern with rigor and logic to his education at the

Ecole Normale as well as at the Collège Stanislas. At the Collège Stanislas he gained the conviction of the correctness of syllogistic reasoning, and at the Ecole Normale Jules Tannery was probably an important influence. Tannery emphasized logical rigor in mathematical and physical reasoning, with the intention of requiring his pupils to untangle the experiential and intuitional foundations of mathematics from logical foundations. He insisted that his students understand that mathematical science is the result of reasoning with signs and symbols.[71]

Tannery's pupils included Jean Perrin as well as Duhem—two men dramatically different in religious views, political commitment, and scientific orientation. Anticlerical, Dreyfusard, and socialist, Perrin became a convinced atomist and one of the firmest supporters in France of kinetic theory and the new corpuscular physics. Perrin also became the first chair holder in physical chemistry at the Sorbonne, firmly directing its teaching and research programs in an orientation very different from Duhem's. Perrin, for example, supported the "ionist" school of physical chemistry for which Duhem had little sympathy.[72] (One thus must be careful in drawing inferences about intellectual influences and their necessary outcomes.)

Among the provincial chemical scientists on whom we have concentrated (Raoult, Sabatier, Grignard), Duhem was alone in his virulent antiatomism. What cemented in Duhem the special antagonism against atomism was its identification with mechanical explanation and his conviction that the inevitable goal of any mechanical representation is to offer an explanation, however fragmentary, of the real, objective world in search of a unified world view. But Duhem took a unified worldview to be the province of religious belief and metaphysics, not of rational science. Here he was influenced by the neo-Thomist education of the Collège Stanislas and, like the philosopher Emile Boutroux, by the tradition of Pascal ("The heart has its reasons, which Reason does not know").[73]

In addition to these influences, Duhem claimed that teaching physics and chemistry at Lille convinced him that the so-called inductive method of Newton (eschewing hypotheses in favor of reasoning from experimental evidence alone) did not work as a teaching device. Explanations by mechanical hypotheses, he concluded, were *meaningless* because they could not be proved true or false. "Some of the students especially asked for an exposition of the principles of Thermodynamics. In attempting this, I began to understand that physical theory is neither a metaphysical explanation nor an ensemble of general laws for which experiment and induction have established the truth."[74]

Duhem concluded instead that physical theory is an artificial construction, fabricated by means of mathematical quantities; these quantities are signs or symbols of designated abstract notions, and a theory is a kind of synoptic tableau, which summarizes and classifies laws of observation. Rival physical theories may be equivalently accurate in the number and precision of physical laws effectively represented. In such cases, he suggested, a choice between theories is made on the basis of elegance, simplicity, commodity, and reasons of convenience that are essentially subjective, contingent, and variable with time, schools, and persons. As he was moving from Rennes to Bordeaux in 1894, Duhem published an elegant paper in the *Revue des Questions Scientifiques* which argues that once a theory is constructed, it is impossible to dissociate the parts from the theory and submit them individually to experimental verification. The theory must be taken as a whole.[75] While some of Duhem's ideas have clear affiliations with those of the Boutroux circle (the Tannerys, Boutroux, Poincaré), the latter conclusion is pure Duhem.

Arriving at Bordeaux in fall 1894, Duhem had an important body of work already behind him in both physics and philosophy. He seems to have made it clear that he thought his work superior to his current professional situation. His rector at Rennes reported to the ministry in summer 1894 that "Duhem would rather be at his place in a chair of the Collège de France, where he could expound only the results of his personal works and do pure science, rather than in a Faculty chair."[76] It is not clear whether the Bordeaux Faculty had a good sense of their new colleague's strong-willed, outspoken character, but the Faculty soon was to be radically affected by his presence. When he arrived he must have been under considerable strain, too, because of the recent death of his young wife and second daughter.

Duhem was appointed to Bordeaux at the rank of chargé de cours, a step up from his lectureship at Lille and Rennes and an ordinary appointment before the formal step into a recently vacant chair. At this time, there were two other physicists on the full-time staff. One was Morisot, a former Bordeaux lycée teacher who had published only fifteen articles in some thirty years (mostly in Bordeaux periodicals). The other was Emile Gossart, who, like Morisot, was a lecturer; he taught a popular course in industrial electricity.[77]

Two weeks after Duhem's appointment the Sciences Faculty council voted to transform Pionchon's chair of physics into a chair of experimental physics, to be confided to Morisot. The council also voted to request the ministry to create a new chair for physics for Duhem, with the title mathematical physics. The wording of the proposal for Duhem's chair makes it clear that he was considered a special case and a distinguished scientist. The wording conveys Duhem's personal style and point of view.

> Physics has made great progress in recent years and developed so that the number of people charged with presenting it has greatly increased. A deep schism has been produced between what one calls, on the one hand, experimental physics—which seeks the numerical properties of bodies—and on the other hand, theoretical physics, which attempts to encompass the ensemble of phenomena in laws or mathematical formulas.
>
> France has possessed and still possesses, experimentalists, but it appears that the tradition of Ampère, Fourier, Poisson and Lamé has for some time been neglected.
>
> Abroad, Helmholtz, Maxwell, Thomson [i.e., Kelvin], Tait and Gibbs represent an order of researches that our country has inaugurated.

And, the resolution went on, mathematical physics is represented at Paris in the teaching of Boussinesq and Poincaré. Circumstances now allow the organization of similar courses at Bordeaux's Faculty,

> which has as a chargé de cours M. Duhem, one of the young and brillant representatives of the School of Poisson. The Sciences Faculty Council then unanimously expresses the wish that a chair of mathematical physics be created at Bordeaux.[78]

Although the vote was unanimous, the situation would have been different if Duhem simply had been recommended to Pionchon's chair over Morisot's head. In a letter to Liard, Dean Rayet explained that a majority of the Faculty council would not want the current chair of physics, which in practice was oriented toward experimental physics, transformed into another kind of chair. In addition, Morisot, who was fifty-eight years old and almost ready to retire, would benefit from retirement at the rank of chair holder, with no additional funding necessary for his position since his salary was already as high as a chair holder's.

> If the chair is declared vacant, I and one or two others would vote for Duhem, but we have no certainty of a fourth vote that would make the

majority. In any case, such a vote would be morally deplorable for putting our differences into evidence. We would avoid all this by the creation of a chair of mathematical physics. Everyone would applaud Duhem's being named to it and be happy to see Morisot in the old chair because of his long service.[79]

Liard's subsequent recommendation, corresponding to the principal text of the resolution, was that Pionchon's chair be declared vacant, with the title of *theoretical* physics (and with the understanding that Duhem be appointed). The Faculty council made no comment on the change in title (from *mathematical* physics), but remained adamant about Morisot. As a consequence, Morisot's lectureship was transformed into a chair of experimental physics. Duhem's chair of theoretical physics would now be the only one in France and one of the few in Europe at a time when "theoretical physics" was just beginning to acquire definition as a discipline.[80]

Morisot died unexpectedly in 1896. To the dean's disappointment, there were few candidates for the vacancy in experimental physics. The paucity of candidates simplified matters for those who wanted to recommend their colleague Gossart.[81] But the combination of Gossart and Duhem proved explosive. A schism developed in the Faculty, rivaling and exemplifying the one Duhem perceived in the physics discipline as a whole. After decades of relative calm, one controversy succeeded another in the Faculty for the next ten to twelve years.

Dividing up credits and the curriculum for the physics program became a major problem. First of all, the budget was judged insufficient, in comparison to other Sciences Faculties. Bordeaux had only 2,550 francs for laboratory expenses in 1895, while Rennes had 3,000 and Grenoble 5,000. Instrument-collection credits also compared unfavorably, for example, with 4,100 francs at Bordeaux and 6,080 at Nancy.[82] Duhem had several students working with him, all requiring considerable outlays for equipment despite the fact that the title of his chair implied the need only for pencil and paper. In 1898, Lucien Marchis had already spent 1,800 francs for thermometers to do his studies of the deformation of glass and now required 1,900 francs for apparatus to study metals. Another student, Fernand Caubet, was working on the liquefaction of gas mixtures and needed 600 francs. Turpain, who wanted to study Hertzian waves, had abandoned the Faculty for industrial facilities, owned by M. Renons.[83]

Gossart began lodging complaints that he did not have proper access to instruments kept in the cabinets of the general physics laboratory associated with Duhem's chair, and the Faculty assembly in late spring 1900 took formal steps to ensure that Gossart had a key. Six years later this was still an issue, with Gossart alleging that Duhem had control of *all* the physics equipment and Duhem claiming this was false.[84] In summer 1906, one group within the Faculty (Gossart in experimental physics, Joseph Kunstler in anatomy and comparative embryology, Camille Sauvageau in botany, Pierre Cousin in mathematics, and Emile Vigoroux in industrial chemistry) signed a letter demanding a discussion of the distribution of credits and instruments assigned to the PCN program and experimental physics. Since Caubet and Morisot's son were in charge of the physics part of the PCN program, it can be surmised that Gossart again was claiming unfairness by Duhem and his group.[85]

Some of the younger researchers became involved in the hostilities between Duhem and Gossart. In 1900, Gossart lodged a complaint that young Morisot, who was an aide in physics, had been helping Duhem's former student Marchis in his courses on heat engines. The assembly voted that this was legitimate, since Morisot's services were supposed to be shared among the physics Faculty members.[86] When Gossart's assistant, Charles Henri Chevallier, was proposed by the Société des Amis de l'Université (which underwrote the course costs) to fill in for Gossart in industrial electricity, Duhem objected that Chevallier's doctorate, which he had directed, was not a state degree, that the defense of his thesis had been weak, and that papers he presented at the Société des Sciences Physiques et Naturelles lacked originality. When the council vote went against Duhem (6–2), he remarked that his judgment had now been opposed several times with regard to Chevallier. When Chevallier was brought up again in the fall, Duhem refused to take part in the discussion.[87]

A dramatic showdown between Duhem and Gossart came in summer 1906, when it was learned that Dean Gayon and the rector had made discreet inquiries of the ministry concerning the creation of a third chair of physics at Bordeaux, without consulting Gossart and some other Faculty members. The chair would be in general physics, corresponding to the certificate of general physics, and the chair holder would direct the general physics laboratory, which Duhem was no longer actively doing himself. It was clear to everyone that the inquiry had been made with Duhem's former student, Marchis, in mind. Marchis had ranked first in the national 1887 agrégation in physics, a

remarkable, rare achievement for a provincial student. His doctoral thesis, *Les modifications permanentes du verre et le déplacement du zéro des ther-
momètres,* was completed in 1898, and he had experience in teaching
which included industrial physics.[88] Gossart was furious, saying that
for the last ten years Duhem had taken over the physics laboratories
and had chased him away. He had been able to support that behavior
from Duhem, but he would not tolerate it from Marchis.[89]

At the next council meeting, after Duhem had praised Marchis's
qualifications for a chair, Gossart delivered a scathing attack on Duhem
and his student. Some points were well taken, others were vituperative
and ad hominem. Gossart had a reasonable argument in claiming that
there was no point in establishing a third chair of physics because there
were so few students then studying under Duhem and his two lecturers. During the period 1897–1905, there were awarded only seven certificates in general physics, seven in optics and mineralogy, and one
agrégation. Gossart argued that the only justification for a new chair
was a circumstance that did not exist at Bordeaux: "unparalleled merit
to be rewarded, or a new science to be developed—as was done for
Monsieur Curie." In addition, Gossart said, Bordeaux's Sciences Faculty was doing itself great harm by continuing to recruit from within the
Faculty for new chairs. Here at Bordeaux, "there has been a succession
of chair holders of perpetually decreasing quality as a consequence of
the amiable weakness that one feels for one colleague after another."
(Of course, this argument applied to Gossart as well.) As for Marchis's
qualifications, Gossart did not mince words.

> For twenty years Monsieur Marchis has done research under the eye
> of Monsieur Duhem: at Lille and at Bordeaux. His latest work, a large
> one of 400 pages, is still the work of a student, assigned, corrected,
> modified and finally approved by the master, in which there is found
> 100 pages inserted from the very lectures of the master.

Gossart concluded his remarks by exhorting his colleagues to free
Marchis from the tutelage of his master and invite him to do his own
work. "When the good hazards of nomination lead M. Duhem to Paris,
it is with pleasure and profit that we could choose Marchis as the
successor of his teacher." The council vote was a close one: six for
Marchis, five against, and one abstention.[90]

The next year or so continued to be tumultuous, with vote after vote
resulting in deadlock until Gossart retired in 1909.[91] When Gossart's

chair became vacant in 1909, Duhem did an about-face, demanding that the chair of experimental physics be transformed into a new chair of mathematics "because there is no place at Bordeaux for three chairs of physics. There is neither the space nor the instruments nor the necessary budget." This suggestion was approved unanimously in November 1909, but before it could be decided for sure, Marchis accepted a chair in aviation at the Sorbonne, and the Bordeaux mathematicians considered the idea of a transformation of the chair of general physics instead of the chair of experimental physics. In the end, both physics chairs were maintained and candidates from outside Bordeaux were appointed, with Duhem declining to serve on committees to help select them. Eventually, after the war, it was Duhem's chair that disappeared.[92]

As one would expect, Duhem had strong opinions about most questions, not just those directly affecting the Gossart struggle. However, his views on the development of applied science and engineering waxed and waned. He wrote a long article in 1899 in which he argued that municipalities needed to support the Sciences Faculties with laboratory space and instruments that were functional and numerous, instead of building sumptuous new palais for the institutes.[93] But despite the fact that he was a good friend of Dean Gayon, who was director of Bordeaux's Ecole de Chimie, Duhem did not support Gayon on all issues. On some occasions this was because Duhem wanted a stronger mathematics program at Bordeaux, especially in rational mechanics; both Duhem and his students did research that was as much mathematical as physical in nature.[94]

In 1900, Duhem opposed Gayon's suggestion to create a special lecture or conference in electrochemistry at the Ecole de Chimie. Duhem argued that there were not yet many electrochemical factories in the Pyrenees, and there was no need for this curriculum. His view failed to persuade the Faculty.[95] In contrast, Duhem proposed the creation of a course on industrial resins, which had to do with a traditional industry in the Bordeaux region. He argued in support of Gayon that teaching a course is a valuable way to recognize the right avenues for new research. This argument countered Rayet's unsuccessful objection that the Faculty's creation of courses complementary to its regular curriculum was depriving lecturers of time for research that would benefit both their own needs and the economic needs of the region.[96]

In 1901, after some discussion of establishing a chair of mineral chemistry, which would also bear the title physical chemistry, the ministry approved a transformation of Joannis's chair of industrial chemistry into one of mineral chemistry alone.[97] Physical chemistry was nevertheless taught under the rubric of mineral chemistry by Maurice Vèzes, as a "branche de la science, jusqu'à présent trop negligée en France."[98] Some students, for example, Charles Hugot, seem to have worked closely with both Vèzes and Duhem. Hugot wrote his thesis in 1900 and taught in mineral chemistry and physical chemistry at Bordeaux, where he was lecturer in the 1920s. As would be expected, Duhem's direction of his students' research in physical chemistry paralleled his own thermodynamic program for physical chemistry.

Once the industrial chemistry chair had been abolished, Duhem rallied to arguments by Rayet against reinstating it. Rayet defended science for science's sake against practical programs. Duhem insisted (against Gossart and the zoologist Perez) on the distinction between true Faculty students and those enrolled in the Ecole de Chimie who did not have the baccalaureate. Duhem and Rayet lost this vote in 1902, although Duhem was influential in having the title of the new chemistry chair read "applied" rather than "industrial."[99] A pessimistic note crept into this discussion with the observation by Vèzes that students graduating from the Bordeaux Ecole de Chimie found themselves at a disadvantage in competing for jobs with the better-trained students from "Paris, Lyon and Nancy."[100]

It appears that in Bordeaux during this period personal alliances and conflicts strongly determined the outcomes of curriculum and personnel decisions. For a crucial decade, a schism paralyzed the effectiveness of the Sciences Faculty, both in building viable new programs and in carrying out important research. Duhem was perceived as the troublemaker by higher-level administrators. In his 1899 report on Duhem, Bordeaux Rector Bizos enlarged Dean Brunel's comment that Duhem "has an obstinate and rather combative character" by saying, "This is a little indulgent toward a man who is violence itself."[101]

There is evidence that it was not simply the character of the Bordeaux Faculty (e.g., Gossart) that brought out this temperament in Duhem. According to the account of Demartres, Lille's dean, Duehm did not get along well with colleagues at Lille and made hostile remarks about them in front of students. Demartres almost came to blows with him. Duhem's transfer was requested by Demartres, and Duhem was ready to agree to it.[102] By 1900, the Bordeaux rector was blaming Duhem for fundamental problems in the Bordeaux Sciences Faculty: *"le grand mal* from which the Sciences Faculty suffers, is the recent influence of Duhem and his preponderance in administrative affairs."[103]

But it must be kept in mind that unfriendly opinions of Duhem were influenced by Duhem's militant Catholicism and antirepublicanism. Most academy rectors were chosen by the education ministry for their dependable republican sympathies and anticlericalism. These considerations moved Bizos to inform the ministry in June 1899 that Duhem had presided over a meeting of the Association Amicale des Anciens Elèves de l'Ecole et de l'Institution Sainte-Marie, a "rival institution to our lycées and collèges."[104] In the next years the rector continued to describe Duhem as a political and religious fanatic and a "perpetual mischief maker." He reported that Duhem declaimed against his colleagues and against students not belonging to the antirepublican and Catholic coterie of the Cercle Ozanam. Duhem's antirepublicanism has been given as the reason for his refusing nomination to the Legion of Honor in 1908.[105]

Of course, as we have seen, Duhem had loyal and admiring students, several of whom taught at Bordeaux (Marchis, Caubet, Manville, Hugot), some at the Catholic Institute at Lille (E. Monnet, E. Lenoble), and others at the Universities of Lille (H. Pélabon) and Poitiers (Turpain). But since Duhem's opposition to Maxwellian physics, atomistic physics, and ionist physical chemistry did not lead to research in areas that turned out to be the forefront of modern physics and chemistry, his research school had little distinction in the long run. Nor did the recruitment of his former students at Bordeaux become an asset to the Faculty in terms of innovative or important research programs.[106]

It was in the early 1900s that Duhem's interests turned decisively to the history of science, with his historical research flowing naturally out of his interest in epistemology and the foundations of his discipline, much as was the case for Berthelot in chemistry and Mach in physics. But Duhem trained no students at Bordeaux in the history of science, and it is said that he turned down an opportunity for doing so at the Collège de France. He declined to pose his candidacy for a history of science chair on the grounds that he was a physicist.[107]

At Bordeaux, Duhem's formal teaching duties were to prepare students for examinations for the licence and agrégation (in general physics, until Marchis's appointment). To this end, he offered an ordinary series of lectures. In addition, in 1897–98, he began giving a second series of lectures which did not lead to any examination. In this series, he chose a different subject each year corresponding to his current interests. In 1897–98, for example, he lectured on hysteresis from a thermodynamic standpoint; in 1899–1900, on the theories of Maxwell and the experiments of Hertz; in 1902–03, on stability and small movements. In 1903–04, the title of his second lecture series was "Theoretical Physics: Its Object and Structure," the same title as his book published in 1906. From 1904 to 1909, he gave lectures on various aspects of "energetics" (which became part of the 1911 *Traité d'energetique*), and in 1909–10, on the history of physical theories, in particular, on the formation of the astronomical system of Copernicus.[108] Whereas in 1908–09 there were only six auditors in his lectures on the conduction of heat and thermodynamic stability, the number of his students jumped to around sixty in 1909–10 when he began teaching the history of science.[109]

Many scholars have suggested that Duhem's interest in the history of science was a direct consequence of his militant Catholicism, since he developed a historical thesis on the debt of Leonardo da Vinci and Galileo Galilei to medieval church fathers, especially those at the University of Paris. Thus modern science, for Duhem, is rooted in the achievements and traditions of the church, and modern scientific epistemology owes fundamental debts to Thomist theology, on the one hand, and medieval nominalism, on the other. But there is good evidence that the discovery by Duhem of medieval work in mechanics was unanticipated, and, as it turned out, it was a discovery that did not displease him.[110]

Duhem's studies in the history of science required long periods of time away from Bordeaux, as is evidenced in comments in *Les origines de la statique* (1905–06) about his laborious analysis of manuscripts at the Bibliothèque Nationale and Bibliothèque Mazarine in Paris.[111] His *Evolution of Mechanics* (1903) appeared as a series of articles in the *Revue Générale des Sciences* from January to April 1903 and *Les origines de la statique* appeared in the *Revue des Questions Scientifiques* from 1903 to 1906. The latter journal had a tradition of publishing articles in the history of science,[112] many of them written by Catholic priests like Ignace Carbonelle and Julien Thirion. Duhem developed an account of the historical evolution of the sciences, demonstrating the gradual achievements of a kind of disciplinary knowledge which he claimed was autonomous from metaphysics or other systems of belief. For Duhem, the celestial mechanics of Greek and medieval astronomy was the only part of pre-modern scientific work that had developed into true science "to save the phenomena" by formally representing phenomena without at-tempting causal explanation.[113]

Aristotle's physics, hardly considered seriously by scientists since the Renaissance, was now revived by Duhem for physicists, as it recently had been revived by Emile Boutroux for philosophers. Duhem claimed Aristotelian physics was on the right track, although not yet freed from metaphysics, inspiring the evolution of science in the Middle Ages.

> It is necessary now that science cease being consecrated only to the study of local movement, in order to embrace general laws of all trans-formation of material things. . . . It is necessary that it treat not only change of place in space but also all movement of alteration, of genera-tion and of corruption. . . . Thermodynamics is the new science, whose construction appears to be the great work of physicists of the nineteenth century, for it embraces all these kinds of changes . . . it is truly the physics for which Aristotle sketched general outlines, but it is the physics of Aristotle developed and made precise by efforts of ex-perimentalists and mathematicians.[114]

In a thorough critique of Duhem's philosophical and historical writ-ings up to 1904, the philosopher Abel Rey praised Duhem's originality, but suggested that Duhem had not himself avoided injecting a meta-physical point of view into his physics. Duhem's system "is the scien-tific philosophy of a believer," Rey concluded.[115]

In response to Rey, Duhem wrote a remarkable essay, later printed as an Appendix to editions of La théorie physique and entitled "Physique de croyant." It was first published in 1905 in the Annales de Philosophie Chrétienne. He began by saying that he had failed in his purpose if one must be a believer to accept his view of the formation of physical theo-ries, and he nicely explained the relationship between his normative view on the way theories should be constructed and his historical con-clusions on the way they have been constructed. Positivist science has led to skepticism, he said, about the validity and unity of physical theo-ries, many of which have fallen into disuse or have been shown to be logically incompatible with one another. Given the representational character of physical theories, there is no logical reason to reject the simultaneous use of two theories having incompatible first principles. But physicists intuitively sense that logical unity is an ideal for science. The history of science, Duhem concluded, does not demonstrate this scientific ideal to be utopian, for old theories are not ruined by new facts, they are merely rendered insufficient or restricted. Historically, theories have progressed and have arranged experimental laws in an order that is more and more analogous (but never identical) to the transcendent order of things.[116]

Of course, Duhem was correct in claiming that one need not be a believer to adopt his general epistemological point of view; this is demonstrated in Duhem's seminal influence in contemporary philosophy of science and in the history of science.[117] But his historical studies moved Duhem even further from the mainstream of new research in physical science and cemented his conviction that this research was on the wrong track. With the onset of the First World War, he turned, too, to criticizing German influences in science, distinguishing the German from the French and English "esprits."[118]

If Duhem believed that "positive science," the byword and slogan of the Third Republic, had endangered the scientific enterprise, he nevertheless believed that it was the French esprit that would put science back on the right track. Emile Picard said of Duhem that one reason he wanted an appointment in Paris was that it would be the only means of having an effect on the orientation of current research in physics and chemistry.[119] It was because the education ministry did not welcome his political and ideological orientations, regardless of his physics, that he was not called to Parisian physics. In addition, his criticism of the older Berthelot school of chemistry, on the one hand, and of the younger ionist school of physical chemistry, on the other, guaranteed that many Parisian physical scientists would not welcome him.

In the long run, Duhem's name has added brilliance to the reputation of the University of Bordeaux, but in the short run his presence was a significant factor in the limited expansion and modest research program of the Sciences Faculty. Duhem never became a Faculty leader at Bordeaux, perhaps because he lacked interest in administration and in the entrepreneurial activities it entails. But, more to the point, his combative personality and ultramontanist reputation made it virtually impossible for him to be appointed to any leadership role. Further, his methodological approach placed him and his students outside the mainstream of scientific research in his fields of expertise in physics and physical chemistry.

The weakness of Bordeaux's sciences program cannot, however, be blamed on one individual. In comparing the history of Bordeaux's

Sciences Faculty with other Faculties, I have found striking what I would call a *lack of enthusiasm* at Bordeaux. This, of course, is a purely subjective judgment based on the reading of Faculty minutes and published speeches. It is tempting to speculate that Duhem's philosophical critique of modern science had a demoralizing effect on his colleagues. I suspect, however, that Duhem's influence in this regard was personal rather than intellectual. Enthusiasm at other Sciences Faculties was rooted in new monies for laboratories and lecture halls, expanding and changing personnel and staff, increasing enrollments, and local support that was both financial and moral. The Bordeaux Sciences Faculty did not enjoy these benefits, because of its economic, geographical, and social milieu, which was agricultural and conservative, until the development of petrochemical and newer industries in recent decades. Except for agricultural chemistry, especially oenology, close links between "fundamental" science and "applied" science developed later at Bordeaux than we have found elsewhere. The result was a comparative stagnation in contrast to the efflorescence of scientific teaching and research at other major provincial universities.

# 7

# CONCLUSION: THE CHARACTER AND ACHIEVEMENTS OF FRENCH PROVINCIAL SCIENCE

## THE CENTRALIZATION ISSUE

Traditional wisdom is that centralization and lack of competition between rival institutions adversely affected French science. Joseph Ben-David has made this argument forcefully in *The Scientist's Role in Society*. Ben-David's thesis is questioned by Shinn, who argues that alternating periods of centralization and decentralization in French university science do not correlate with alternating periods of scientific decline and scientific productivity. However, Shinn's characterization of the period 1876–1914 as one of effective decentralization accompanied by increased productivity is problematic. Decentralization lasted only from roughly 1876 to 1900, as we have seen, and even then the movement was controlled carefully by the Parisian educational bureaucracy in a process qualified as "déconcentration" by Robert Fox and George Weisz.[1]

Ben-David compares the institutional system of French science unfavorably with nineteenth-century German science and twentieth-century American science. But, in fact, his own chronological account of the social organization of science, beginning with seventeenth-century England and ending with twentieth-century America, lends support to the conclusion that the historical evolution of scientific institutions has led to stronger control by national bureaucracies. By 1900, local German initiatives in new scientific fields were addressed routinely to the central government, and in both Germany and the United States, federal government support of science by the mid-twentieth century altered the conditions under which the sciences had flourished previously. Increasing centralization occurs inevitably because of the progressive identification of scientific education and research with economic growth and military preparedness as well as with the certification and power of professional elites.[2]

This process of scientific centralization has been initiated to a considerable extent by scientists' demands for government funding and by scientists' growing self-identity as a civic and political elite. These characteristics are demonstrated clearly for French science in Charles Gillispie's *Science and Polity in France at the End of the Old Regime* and in recent essays in the collection *Professions and the French State, 1700–1900*.[3]

In his analysis, Ben-David astutely focuses on immobility within the French system as a source of frustration for scientists and science. A principal and continued problem in nineteenth- and early twentieth-century France, in contrast to Germany, was the lack of incentive for scientists to move laterally from one provincial university to another. Superimposed on the egalitarian mediocrity of provincial universities were the superior Parisian institutions. Clearly, this superiority was not merely a matter of perceived prestige, but of institutional and economic arrangements brought into force during the Napoleonic period.

We have seen how this system worked in practice. The education ministry offered no benefits to scientists interested in remaining in the provinces and frowned on individually negotiated local arrangements that might entice a scientist to move from Lyon to Grenoble, or from Bordeaux to Toulouse. In contrast to German scientists, French scientists often resigned higher-ranking provincial positions for lower-ranking "central" ones at no financial disadvantage. We need only remember the careers of Elie Cartan, Aimé Cotton, and Jacques Hadamard. Quantitative data on physics around 1900 indicate that of 106 nineteenth-century physics professors at university-level institutions in France, seventeen moved from a provincial university to Paris, with eight accepting a position lower than professor. In contrast, of 169 professors in Germany, only sixteen moved from a "provincial" university to Berlin, and of these, one accepted a lower rank.[4]

Another difficulty for provincial scientists was that the 1896 law on universities established a larger number of regional universities than the central government could effectively support. The government continued to exert authority over Faculty appointments, curricula, and the awarding of state degrees. But by decreasing its support to the regional universities in the 1890s, just at the time when enrollments and research activities in some of the provincial universities were really growing, the central government presented severe obstacles to any regional university becoming an influential counterweight to Paris as a scientific or cultural center.

Especially devastating in this respect was the administrative reform of the Ecole Normale Supérieure in 1903. Decisions about the agrégation in the next few years decisively programmed the provincial Faculties for failure in higher-level teaching and research. It became clear that the national education ministry intended the provincial Faculties to cater only to a clientele of would-be secondary teachers, medical students, and engineers, none of whom required scientific class work or laboratory work beyond the elementary university level.[5] Given this emphasis, the continued advances of provincial science are astonishing.

Placed in an embarrassing situation by the award of the 1912 Nobel Prize to provincial scientists, the Academy of Sciences created a nonresident section in 1913. Still, Parisians did not readily relinquish traditional privileges. It was not until 1950 that nonresident members could vote on any issue other than elections to their own section, and it was not until 1964 that the obligation for ordinary Academy members to reside in the Paris region was suppressed altogether.[6] In honorary institutions, as in education and research, the power and prestige of the Parisian elite continued to be protected against incursions from rival elites in the provinces.

A source of stagnation for French science not sufficiently emphasized by Ben-David and other scholars is the educational system's closed-door policy for foreign scientists. If any system of national scientific institutions is to be vigorous and innovative, the scientific community must maintain an international character among its students and its fully qualified professional colleagues. In France, as illustrated in the history of provincial science, the universities did not employ scientists who were not French citizens. This chauvinism was a major source of inertia and weakness for French science.

To be sure, foreign students invigorated provincial universities from the 1890s on. Many foreign students wrote provincial doctoral theses (see table 4). The training of foreign students enhanced French prestige abroad, especially in Eastern Europe, North Africa, and the Far East. As with the Saint-Simonian movement a century earlier, science and engineering became exports of French culture. Measured against German universities, it even appears that there was a higher proportion of foreign students in French provincial universities than in German higher education after 1900, particularly in scientific and engineering studies.[7]

Had some of these graduates taken up posts in French universities, as was customary in Great Britain, Germany, and the United States, French science might have shown greater vigor in the first half of the twentieth century. American scientific advances in the twentieth century should not be attributed solely to the post-Hitler emigration of Europeans to the United States. By 1930, there was already one transplanted European teaching theoretical physics in the United States for every two or three native-born Americans. In contrast, during the 1930s even distinguished refugee scientists most often did not find permanent employment in France. W. M. Elsasser, Fritz London, Laszlo Tisza, L. W. Nordheim, and Felix Bloch were among those who worked briefly in French scientific laboratories and then emigrated to the United States or Great Britain.[8]

A final difficulty with Ben-David's emphasis on centralization and with Shils's center-periphery model is the underestimation of the changes that have occurred in France since the early nineteenth century and the failure to acknowledge that much of the impetus for change *originated in the provinces*, or on the "periphery." Two important examples illustrate the evolution of new forms of centralized administration of science *as a result of* the development of strong, competitive programs in provincial science. One is the creation in 1939 of the CNRS, the umbrella organization in France sponsoring research in pure and applied science. The second example is the establishment in 1948 of the ENSI, the unified system of prestige engineering schools. Let us look briefly at these two examples.

The First World War marked a watershed for relations between science and government in the major Western nations, including France. For scientists, the war demonstrated the potential advantages of new forms of government and industrial support for scientific research, separate from educational programs. University scientists in France worked on the manufacture and detection of explosives, poison gases, submarines, and other weapons, either as members of the armed forces or under the Direction des Inventions. Victor Grignard did war research in several settings: a navy laboratory, his own university laboratory, and with a team project in Paris. Manufacture of nitrogen products, liquid chlorine, and aluminum alloys was carried out with electrical furnaces in Moissan's laboratory in Paris and Georges Flusin's laboratory in Grenoble.[9] Gradually, it was recognized that there was a need for coordination of nationwide facilities and activities.

There were no sudden and dramatic changes in the organization of scientific research at the war's end.[10] But in 1924, the Chamber of Deputies levied a tax on industries, the *taxe d'apprentissage*, earmarking some of these funds for support of scientific research in laboratories of higher education.[11] When construction began on the multibillion franc Maginot Line in the 1930s, Jean Perrin persuaded Edouard Herriot to sponsor a national defense bill amendment that would provide funds for a "research corps." A "superior council of research" was established, intended by Perrin and his closest colleagues to serve as a counterweight to the clique of older professors traditionally dominating the Ministry of Public Instruction and the Academy of Sciences.[12] Perrin's aim was to create a new organizational base for scientific research, separate from educational institutions, and to staff it with young, active researchers. The Centre Nationale des Recherches Scientifiques developed around this nucleus, with divisions for both basic and applied research. Henri Longchambon, dean at Lyon, became director of the applied sciences division at its founding in 1939.[13]

Initially, many provincial scientists criticized the new organization as another instance of centralization. The control in Paris of funds that previously had been administered by individual laboratories throughout France was fought by some scientists in the 1930s. However, the new organization also was praised for its coordination of research activities, its provision of new funds, and its emphasis on hiring younger scientists as full-time researchers.[14]

Just as the CNRS was the result of efforts to incorporate provincial laboratories and observatories into a centralized organization, so the ENSI was the result of efforts to coordinate and standardize the separate educational programs of the applied science institutes. As described earlier, the state education ministry in 1924 drew up a master list of specialized provincial programs, with the aim of imposing order on the mosaic of independently developed applied science programs. Although interrupted by the Second World War, the movement was under way by 1930 and resulted in the establishment of the unified system of the Ecoles Nationales Supérieures d'Ingénieurs. The 1948 reform authorized a national concours examination controlling entry to the ENSI schools, an examination similar to that for the grandes écoles. About thirty schools were designated for the ENSI system, including selected provincial institutes and the Ecole Supérieure de Physique et de Chimie and the Ecole Supérieure d'Electricité in Paris. Directors of the schools were required to be Sciences Faculty professors, and many members of the schools' teaching personnel were Faculty professors, as in the past.

What must be emphasized about the ENSI is that it was not a completely new creation. Its establishment constituted both a recognition and a new means of control of the independent successes of major provincial Sciences Faculties and their institutes. Reciprocal tensions and competition developed between Parisian and provincial Faculties, both in fundamental and applied science, as provincial scientists and administrators successfully developed new research and educational programs. In response, Parisian scientists and administrators planned new centralized umbrella organizations that could incorporate provincial achievements and institutions. The history of provincial science is not a demonstration of central planning and Parisian initiatives. Rather, when the state moved decisively in the modern period, as often as not, it was the result of a trend already under way at the scientific periphery. That the direction state activity took was of a centralizing kind should surprise no student of modern French history. Recognizing these developments is very different from viewing scientific initiatives and innovation as originating exclusively in Paris.

## THE ROLE OF APPLIED SCIENCE

There is no general agreement about the role played in scientific advance by practical and technological imperatives. Shinn is among those who have suggested that provincial science was not as successful as it might have been because it was too closely linked with local industries. Testimony from provincial scientists (e.g., the minutes of Faculty meetings at Toulouse in 1927) seems to confirm this view. But if the provincial universities became overly "specialized" and "sluggish" after 1900,[15] it cannot be due simply to close alliance with industry. Indeed, as we have seen, insofar as Bordeaux only narrowly maintained its traditional reputation as a scientific center during this period, it was in part because the Sciences Faculty lacked strong industrial connections, not because it had them. Paul argues that the provincial and applied science institutes "created genuine faculties of science in the provinces."[16] This generalization is apt; what success the provincial faculties enjoyed was achieved because of local industrial interests, not despite them.

In comparing French universities with their British and German counterparts in the late nineteenth century, we noted earlier that educational reformers demanded "modern" education and the availability of specialized and technical preparation. In Great Britain, technical schools multiplied faster than university facilities for science, and by 1900, enrollments in university physics courses were stagnating or declining because of competition from the technical curriculum.[17] Local commerce and industry largely founded and funded English provincial or "civic" universities after 1870, and spokesmen for English colleges and universities made clear statements that one of their most important functions was to serve local industry. These circumstances are similar to the regional pressures exerted on French legislators in the 1890s to ensure the creation of "universities" in their cities. Regional rivalries were characteristic of both countries. In this connection, Sanderson writes of provincial English universities, "It is probably fair to say that competition between the Victorian cities themselves was a more important dynamic in urging them to form universities than any fear of competition they might all have felt collectively from the Germans."[18] Again, we are reminded of the importance of focusing on tensions and relations within the national scientific community, rather than too exclusively on international rivalries.

In both England and France, technically oriented and university-affiliated programs were well established by the beginning of the twentieth century. The English, French, and German systems began to strongly resemble one another, as university institutions became multifunctional and the separation between science and technology diminished.[19] The Germans themselves, like Von Treitschke in his ridiculing of French Alsatian schools as "bread-and-butter" institutions, claimed German universities were bastions of pure scholarship and learning.

However, there were close links between Germany's model research institute, the PTR, and both the University of Berlin and industrially sponsored research (chap. 1). We also have noted the granting to the Berlin Technische Hochschule in 1899 of university status and the right to award doctoral degrees. For this occasion, the kaiser donned the uniform of an officer of engineers.[20] Historians now are reevaluating the collaboration among university scientists, government officials, and industrialists in Germany.[21]

Perhaps Germany, England, and France were not so different after all around 1900. In France, industry provided provincial Sciences Faculties with new chairs and new facilities in industrial chemistry, electrotechnology, electrometallurgy, and industrial physics. Industrially oriented Faculty scientists often provided personal vigor to science programs, as in the cases of Grenoble's Louis Barbillon in industrial electricity and Lyon's Léon Vignon in industrial chemistry. In the late nineteenth century the *expectation* developed of a practical research role for university scientists.

The latter effect has been underestimated. Recent writers on French science have emphasized that French universities were transformed from teaching schools to research institutions during the nineteenth century.[22] It has not been stressed enough, however, that demands for research as an expected duty of university professors came not only from a central government interested in emulating German patterns of success but also from regional industrialists and agriculturalists who required assistance in pulling out of the long economic depression that lasted from the late 1870s through the late 1890s. This depression coincided with scientific breakthroughs in electrical studies, metallurgy, and organic chemistry which promised rapid technical application and industrial expansion. In France, applied science demands were allowed full expression after the mid-1870s by a new republican government attempting to establish a strong base of support among the entrepreneurial bourgeoisie. When industrialists were allowed to aid research, it is not surprising that they demanded some payoff in return.

The promise of professional and specialized jobs for university graduates was an important result of the evolution of science-related industries. A symptom of the stagnant environment at Bordeaux's Sciences Faculty was the inability of the School of Chemistry to place its graduates in the face of competition from students trained at Lyon, Nancy, and Paris. In contrast to the situation at Bordeaux, Grenoble engineers never lacked jobs. As the job market compressed in the 1920s and 1930s, the Grenoble Sciences Faculty decreased the number of school admissions, so that it could guarantee jobs to its graduates.

At Nancy, graduates of the Ecole Supérieure de la Métallurgie et de l'Industrie des Mines always found jobs in the 1920s (chap. 2). Other newly established applied science schools at Nancy were successful in placing graduates either because the students were sons and daughters of the owners and managers of local industries or because of steady expansion in the new science-related industries of metals, engineering, mining, food, drink, chemicals, and public utilities.[23]

Industrial rather than academic placement became one of the best career hopes of university graduates in France, because the national birthrate and secondary school enrollments were holding steady after 1870. Thus, new positions in secondary education were not feasible.[24] At the university level, the French national educational system still valued the scientific generalist, rather than the specialist, so that the education ministry created few new state chairs in response to evolving interests in new specialized fields. The resistance at the Sorbonne in the early 1900s to creation of a chair of physical chemistry is one indication of this bias. Until the founding of the CNRS, there were few openings for young, specialized scientific researchers in France except in applied science fields, and even these positions closed up in the interwar period, with a few exceptions, such as Toulouse.

Increases in enrollments in the universities following the First World War were especially great in the technical institutes. The numbers were swelled by the return of foreign students in the mid-1920s and by the matriculation of increasing numbers of women. While the number of students doubled, however, the number of university professors in France actually declined between 1914 and 1925.[25] The dependence of provincial research funds on income from the institutes encouraged the institutes to enroll more and more students. Predictably, this led to a decline in the quality of education. In Paris, Georges Urbain characterized the burgeoning enrollments as a deplorable palliative for a serious problem. In Lyon, Victor Grignard regretted that the schools of applied chemistry had opened their doors "in order to get student fees," with the result that "there is an apparent plethora of chemists, but of bad chemists."[26]

Thus, it appears that there was a natural limit to the contributions provincial Sciences Faculties could make in education and research when drawing largely on their own resources. Without incentives and financial support from Paris, the institutes gradually found their ability to provide a superior scientific education overwhelmed by students with a largely practical orientation. It follows that the quality of provincial Faculty members' scientific contributions might have declined and their professional aspirations might have languished.

At the same time this was occurring, the Parisian grandes écoles were becoming more elitist in their admissions policies. While the number of candidates for all the grandes écoles increased by 20 percent during the period 1900–1932, the number admitted actually declined by 15 percent.[27] By the end of the 1920s, some provincial institutes committed to quality and prestige began to talk seriously about limiting their own admissions, despite the disadvantageous implications for their budgets. As noted earlier, Grenoble was a leader in this movement, along with Nancy's Institut Electrotechnique. In the long run, the decision to limit enrollments enhanced the standing of these provincial institutions. However, by the mid-1920s, the Parisian degree had garnered even greater prestige than in the 1900s. Thus, on the face of it, the decentralization movement in science had failed by 1930.

Why, if industrial and practical interests had proven so beneficial to scientific expansion and to regional reinvigoration from the 1800s to the First World War, did some scientists claim that practical concerns were stifling pure science? We have seen this claim made at Toulouse, to a lesser extent at Lyon, and most vociferously in Paris. We need not look only at French institutions for this kind of claim, for we find it also in England and Germany.

As discussed earlier (chap. 1), mathematics and the classics were the basis for leadership training in European countries and in Great Britain throughout the nineteenth century. Despite its name, France's Ecole Polytechnique schooled "generalists" rather than "specialists," as demonstrated in Shinn's *Savoir scientifique et pouvoir sociale*, and Weiss's *The Making of Technological Man*. Ringer has stressed the strong commitment in modern Europe to Latin and classical studies for the making of culture and the gentleman, and Pyenson more recently has made the same argument for the role of mathematics. The aim of training gentlemen was expressed succinctly in 1867 by John Stuart Mill: "There is a tolerably general agreement about what a university is not. It is not a place of professional education."[28]

The role of mathematics in modern science has obscured the fact that higher mathematics was valued not for its usefulness but for its inaccessibility and esotericism. Similarly, the study of scientific *method* was valued in secondary and higher education as part of the tradition of logic and rhetoric. In contrast, the *practice* of laboratory science was regarded as applied or *specialized*, professional knowledge.[29] The relationship between pure science and applied science in French universities was a fundamental aspect of debates over the proper norms for training the leadership elite.

There is some evidence that the social status of scientists declined during the course of the nineteenth century. According to Weiss, the section for physical sciences at the Paris Academy became more diverse in social origins during this period. Zwerling's analysis of scientists educated at the Ecole Polytechnique and Ecole Normale Supérieure suggests that scientists' social standing was shifting downward from 1800 to 1900.[30] As scientific education developed more emphasis on laboratory research (some of it industry or agriculture related), scientists may have suffered in attributed social status, as well as in real social status determined by family origins. These developments likely affected the willingness of some academically superior secondary students to consider scientific careers. Diminished social status was less a source of concern for university professors in Germany who generally enjoyed higher social status than their counterparts in France.[31]

Nonetheless, the wave of the future lay in close links between industry and the university. For example, Frédéric Joliot and Lew Kowarski, both of whom were leaders in French fission research, were trained in industrial chemistry, Joliot at the Ecole Supérieure de Physique et de Chimie in Paris (where Langevin taught) and Kowarski at Lyon's Ecole de Chimie (where Grignard taught). Kowarski was an emigré from Russia, and it was through the new structure of the CNRS, and later the French Atomic Energy Commission and the international research agency CERN, that he was able to become a full-fledged member of the French scientific community. After he completed his doctoral thesis in 1930 at Marie Curie's Radium Institute, Joliot initially could find no academic position. He would have taken a job as an engineer in industry but for Jean Perrin's obtaining for him a research scholarship with the Caisse Nationale des Sciences, the predecessor of the CNRS.[32]

It was in research disciplines related to growth industries that provincial scientists were to excel. After the First World War, local economies prospered in those cities with flourishing Sciences Faculties. Electricity was the most important of the new science-based industries, but the synthesis and marketing of organic chemicals also prospered. Figures show that from 1905 to 1930 French economic growth in production per capita was unmatched by its nearest competitors. Steel mills in northern and eastern France enlarged, electrometallurgic plants in the Alps consolidated into major groups, and dyestuffs thrived, as did luxury and technically oriented manufactures like photography and automobiles. Lyon's diversified economy continued to justify its nickname, the "French Manchester," and Grenoble led the country in patent applications during the period 1895-1940.[33]

After the Second World War, the potential latent in the institutional infrastructure constructed from 1880 to 1930 began to be more fully realized in provincial cities. The long-lived generation of 1890, which had grown old during decades of population stagnation, was gone. In technical and scientific fields, four of the provincial Sciences Faculties emerged among the leaders in French science: Grenoble, Toulouse, and Nancy in electronics and electrotechnology; Grenoble and Toulouse in computer science; and Nancy in chemistry, with the institutes and schools of Lyon not far behind. In a 1982 education survey, only the Ecole Supérieure d'Electricité in Paris outranks Grenoble's Institut Nationale Polytechnique and Toulouse's Ecole Nationale Supérieure d'Electronique, d'Electrotechnique, d'Informatique et d'Hydraulique (ENSEEIH) in electronics. Nancy's Ecole Nationale Supérieure des Industries Chimiques ranks third behind the Ecole de Physique et de Chimie of Paris and the Université de Paris I in chemistry. The Université de Grenoble I and Grenoble's Ecole Nationale Supérieure d'Informatique et de Mathématiques Appliquées are the leaders in France in computer science.[34] It is difficult in these fields to separate "basic" and "applied" science, and it is not by accident that such hybrid fields originally achieved eminence in the provinces. The history of these institutions demonstrates the crucial role of provincial scientific communities and provincial leadership in modern France.

## THE QUESTION OF GERMANY AND
## NATIONAL SCIENTIFIC PRODUCTIVITY

Nothing succeeds like success. During the period 1880-1930, the reputation of German scientific institutions outstripped its reality, as did the French reputation for backwardness. The assumption that German scientific and technological education was far superior to that of France and England must come under serious challenge, despite the long-standing belief that this was the case.

The Englishman S. H. Higgins claimed in the early 1900s that stories of German superiority were greatly exaggerated and that Charlotten-burg's PTR "suffers by comparison" with the University of Manchester. Sir William Siemens claimed that the German Technische Hochschulen taught too many practical procedures, which their students took for unchanging principles.[35] American postgraduate study in physical chemistry at German scientific centers began to decline after 1908, and the American chemist Harry Clary Jones observed in 1904 that many changes had taken place in Germany in the previous ten years: "all in all, Germany has lost much of her prestige in chemistry. In physics the loss is still greater. . . . I am fully convinced that there is more good work being done in America in physics, physical chemistry, and inorganic chemistry than in Germany."[36]

If we attempt to work out quantitative evaluations of relative national scientific productivity, the results are inconclusive, as demonstrated in calculations by Paul Forman, John Heilbron, and Spencer Weart for physics in *Physics circa 1900*. Relating figures for gross productivity of physics papers to estimates of the numbers of physicists doing the work produces very similar rates of individual productivity throughout northern Europe. Indicators of scientific activity such as the numbers of academic posts, personal incomes, laboratory and new plant expenditures, and so on, show that physics productivity around 1900 in England, Germany, and France was remarkably uniform in relation to national population and national income, though Forman et al. estimate French productivity was declining.[37]

These authors make no judgments about the "relative contribution of each paper to the progress of physics, or about the existence and significance of national gifts and styles." The use of words like "discovery" or "innovation" invokes a qualitative judgment of meritorious scientific work which often is not simple to make. As a consequence, sociologists and historians sometimes use frequency of citation to indicate the influence of a paper, or they attempt to weight the prestige of journals or scientific societies with which scientists are associated. The Nobel Prize or other prizes (like the Le Conte Prize awarded to Blondlot) may also be used to evaluate the significance of research results and research schools. However, comparison of the numbers of

Nobel Prizes awarded to French and German scientists in the early decades of the twentieth century does not result in any clear verdict of relative national scientific rank. While German scientists won a higher percentage of Nobel Prizes than their French counterparts, Germany had a larger population and proportionally higher enrollments in education, including scientific education.

Paradoxically, when scandals like the N-ray episode occur, they seem to carry more negative weight for national scientific reputation than the positive weight associated with achievements like Henri Becquerel's discovery of radioactivity. Blondlot hurt favorable perceptions of French national scientific stature more than Becquerel helped. In addition, rates of achievement for the different sciences need to be distinguished from one another rather than lumped together. Physics may not be a representative discipline among the sciences. Insofar as comparisons of national scientific stature have been made, we can only conclude that the qualitative and quantitative productivities of France, England, and Germany were close enough overall in the nineteenth and twentieth centuries that no unequivocal statement about national superiority and inferiority is possible.[38]

Where the Germans continued to excel was in sheer numbers of scientists and engineers. This was in large part the result of the expanding German population. In contrast to the French population, which registered a negative birthrate in 1890 and in five more years before 1914, the German population continued to expand steadily. In 1914, the German population was 67.8 million and the French population only 39.5 million. Nor did the French population increase significantly in the years after the First World War.[39]

Simply on the basis of an expanding youthful population, Germany had a vigor that France did not. For scientific training, this meant there was a greater opportunity for employment in secondary education, and correspondingly in higher education. German industries, more open to organizing research laboratories than their French and English counterparts, absorbed a large number of scientists and engineers. In addition, as noted by Grignard, German scientists and engineers had financial incentives for industry-related research not found in France. The 1844 French patent law favored protection of products rather than procedures (chap. 4). The German law, favoring procedure, rewarded scientists for innovative methods of making an old product in a new way.[40]

Nonetheless, good results were obtained in France after the 1870s on the basis of moderate expansions in numbers of new scientists and moderate amounts of investment in instructional and research facilities. The opening up of secondary education to a wider sector of society, limited decontrol of university Faculties by the education ministry, and small amounts of financial support in the 1870s benefited a generation of young scientists who took their places in lecture halls and laboratories in the twenty-five years or so after 1877. Quantitative data for physics demonstrate that the number of publications required before a physics professor received his first appointment rose precipitously from 1880 to 1900, and per capita professorial productivity rose above that of Germany during the period 1875–1890.[41]

Contemporaries and later historians and sociologists have spoken of a "renaissance" in French science around 1900,[42] and it was perceived that any recent decline in quality or quantity of French scientific work had been reversed. If gradual and moderate increases in the number of Faculty members and resources had continued, with only moderate increases in student enrollments, the fact and perception of scientific revitalization in France likely would have become well established. If, under these conditions, lateral immobility and national chauvinism had significantly lessened, even more might have been achieved. In any event, the differences among Great Britain, Germany, and France have been exaggerated. All remained great powers in the sciences and all enacted substantial and similar reforms in the organization of scientific education and research from 1860 to 1930.


## THE EVALUATION OF SCIENTIFIC RESEARCH PROGRAMS

Much has been written by historians and sociologists of science regarding the organization of research schools and the conditions under which they flourish. Drawing in part on J. B. Morrell's comparative study of Liebig's school of chemistry at Giessen with that of Thomas Thomson at Glasgow, Gerald Geison has constructed a checklist for characteristics of the ideal research school. These characteristics include a charismatic director, a distinctive research approach, simple research techniques, a pool of talented recruits, control over publications, and financial support.[43]

Sociologists have analyzed "invisible colleges" and influence networks, investigating their roles in the structure of the scientific community, the development of new subdisciplines, and the appearance of innovation and discovery.[44] The notions of research schools and influence networks are helpful in assessing the importance of provincial scientific centers relative to one another and to Paris.

Conventional wisdom has it that the only notable influence network in France radiated from Paris, that the best students studied in the institutions of the capital, and that research trends originated in Paris and spread from there to the hinterlands. We have demonstrated that this portrayal of French science must be radically qualified.

While the Paris-centered influence networks were crucial, other networks existed and performed important functions. The case of Nancy has shown the importance of local family networks like the Bichats and the Blondlots who exercised control and influence not only in university science but in local political elites. The friendships of these families with other families, like the Poincarés, in turn linked the councils and circles of Nancy to the innermost power cliques in Paris. Local industrial and mercantile circles similarly had influence both regionally and nationally. In Lyon, the industrialist Joseph Gillet was a leader in marshaling moral and financial resources for the benefit of local scientific institutions. In Grenoble, Casimir Brenier played a similar role.

Student and faculty friendships and allegiances at Lyon provide an example of the importance and potential of rival school loyalties to those of the Parisian grandes écoles. The youthful comradeships at the Ecole Normale d'Enseignement Spéciale among Grignard, Meunier, and Wiernsberger resulted in an esprit de corps similar to that inculcated in students in institutions like the Ecole Polytechnique and Ecole Normale Supérieure. There is every reason to think that, had it not been eliminated by the education ministry, the Cluny school would eventually have had an influence similar to the rival grandes écoles.

Research schools in the provinces formed collaborative groups of Sciences Faculty members and networks of masters and students. Striking examples of the former are the Lyon and Nancy Faculty groups in organic chemistry and electrical physics. Examples of master-pupil groups include Sabatier and his students Alphonse Mailhe, Marcel Murat, Léo Espil, and Georges Gaudion, and Pierre Duhem and his students Lucien Marchis, Fernand Caubet, Octave Manville, and Albert Turpain.

As described by sociologists, the good scientific maître is often charismatic and forceful. The scientific leader has a knack for evoking excel-

lence in others and imparting a sense of taste and judgment in recognizing and solving problems. Sabatier appears to be a good example of the
more authoritarian of this type, and was outspoken about the need for
a powerful research director on what he called the German model. In
contrast, Grignard seems to have employed a more collaborative approach, winning the affection of colleagues and students. Both developed straightforward and characteristic research techniques that could
be applied to fundamental problems in chemical synthesis, with subsets
of related problems for students and collaborators to develop successive-
ly.

The social environment at Lyon, especially in the 1890s and 1900s,
provided students with the mutual psychological support important in
early years of scientific training. Grignard, Meunier, and Wiernsberger
arrived together from Cluny to pursue studies under Bouveault,
Barbier, and Gouy, whom they found to be abreast of new ideas and
adept in experimental techniques. When Grignard began to pursue single mindedly his new method of organic synthesis, his colleague Tissier
was quick to offer aid, as did Meunier. Old friendships brought Grignard back to Lyon in 1920 after his intermediate years in Nancy and
Paris.

Blondlot and his colleagues at Nancy exhibit similar psychological
and social characteristics to the group at Lyon. Work in electrical physics and N-rays was carried out in mutual collaboration by the good
friends Bichat and Blondlot and by Bichat's godson Camille Gutton.
Colleagues at the Medical Faculty kept up with the physicists' work
and addressed the physiological implications of N-rays. It was the overall strength of the Nancy programs, in combination with the tenacity of
regional and university loyalties, that ensured the successes of Nancy's
sciences programs after the N-ray fiasco. Thus, Nancy established the
first physical chemistry chair in France in 1899, and when Grignard
joined the Faculty in 1909, he enjoyed the collegial relationships and
research interests not only of Müller in physical chemistry but also of
Guntz, Minguin, Guyot, Wahl, and his former lycée teacher, Petit.

Especially because attracting research students remained a problem
in the provinces, peer relationships and comradeship were crucial for
scientific enthusiasm and innovation. It is scarcely surprising that the
aging of this Faculty in the 1920s and 1930s, with few replacements and
hardly any expansion in the Faculties, weighed heavily against new
scientific advance and contributed to an enervating nostalgia among old

comrades-in-arms. However, it must be noted that the gerontocracy problem affected Parisian professors as well during the same period.

The grievous difficulty of establishing systematic provincial research programs with a ready supply of talented recruits is no better demonstrated than in the case of Raoult. Raoult worked out a set of ideas combining physics and chemistry which resolved old problems and helped found a new discipline. Yet he worked largely alone and trained no students. As judged by national examinations, the best French students continued to migrate to Paris.

Given the exodus to Paris, it is somewhat paradoxical that lectures and laboratory work in the provinces frequently were more innovative and avant-garde than in Paris. This clearly was true in chemistry around 1900. In his recent study of the French physics community in the 1920s and 1930s, Dominique Pestre argues that this was specifically the case for physics. Not infrequently, more modern lectures were given in the provinces than in Paris. Pestre mentions the courses of young maîtres de conférences like Albert Kastler at Bordeaux, Georges Bruhat at Lille, Jean Cabannes at Montpellier, and Gustave Ribaud at Strasbourg. Especially at the beginning ranks of the professoriate, the provinces often had younger, more dynamic Faculty members than were to be found in Paris, where many maîtres de conférences were forty or fifty years old.[45]

Whereas most provincial university scientists unhesitatingly moved on to Paris once given the opportunity, this was not the case for the scientists on whose careers we have focused. Unusual people are often mavericks, and in general, the French educational system mitigated against the development and influence of mavericks and "outsiders." We may speculate that for some provincial scientists, it was their experience as social or psychological outsiders from the mainstream of Paris-based research schools which helps account for their creativity.[46]

Of course, those provincial scientists whose careers we have followed in detail do not fully exemplify the features of provincial scientific life. Since they were scientific notables, they are by definition atypical of the majority of their colleagues in important respects. Nonetheless, their biographies provide us with valuable insights into the intellectual environments in which they worked and the scientific communities they helped establish. Their unusual achievements and personal reputations were instrumental in making their provincial scientific communities prosperous and influential. Whereas regional interests encouraged the development of different specialties at different university Faculties in France, individuals determined the relative creativity within those specialties.

In conclusion, the persistent diversity among provincial scientific communities historically has been essential to scientific growth and progress in France. For too long, provincial science has been ignored in discussions of modern French science. As a consequence, it is usually assumed that a scientist of the stature of Victor Gignard must have been Parisian. That mistake should not be made again.

# ABBREVIATIONS

## UNPUBLISHED SOURCES

ADI, FdR:     Archives Départementales d'Isère et de l'Ancienne Province de Dauphiné. Fonds du Rectorat.
AN:     Archives Nationales de France.
NI:     Notice Individuelle in AN dossier.
RC:     Renseignement Confidentiel in AN dossier.
AN $F^{17}$ 21574:     François Raoult dossier.
AN $F^{17}$ 22028:     Emile Boutroux dossier.
AN $F^{17}$ 22097:     René Blondlot dossier.
AN $F^{17}$ 23295:     Pierre Duhem dossier.
AN $F^{17}$ 23735:     Benjamin Baillaud dossier.
AN $F^{17}$ 23767:     Georges Gouy dossier.
AN $F^{17}$ 24106:     Paul Sabatier dossier.
AN $F^{17}$ 26752:     Victor Grignard dossier.

### From Archives of Université de Bordeaux I

Bordeaux, Fac. Sci., PV (I): Bordeaux. Faculté des Sciences. Délibérations. 1839–1894.
Bordeaux, Fac. Sci., PV Assem. (I) and (II): Bordeaux. Faculté des Sciences. Procès-Verbaux de l'Assemblée. 1895–1904; and 25 nov. 1904–1916.
Bordeaux, Fac. Sci., PV Conseil (I), (II), (III), and (IV): Bordeaux. Conseil de la Faculté des Sciences. Registre des Délibérations. 1886–1901; 1901–1909; 1909 au 11 juin 1920; and 24 juin 1920 au 10 fév. 1938.

### From Archives Départementales d'Isère (ADI, FdR)

Grenoble, Fac. Sci., PV (I), (II), (III), and (IV): Grenoble. Délibérations du Conseil et de l'Assemblée de la Faculté des Sciences. 1811–1896; 1897–1918; 1918–1923; and 1923–1933.

### From Archives Départementales du Rhône

Lyon, Fac. Sci., PV: Lyon. Registres des Procès-Verbaux des Séances du Conseil de la Faculté des Sciences de Lyon. 6 janv. 1886–3 mai 1897; 24 mai

1897–3 fév. 1903; 28 fév. 1903–16 déc. 1912; and 14 mars 1913–19 mars 1932.

## From Archives of Université de Nancy I

Nancy, Fac. Sci., PV (I): Nancy. Faculté des Sciences. Procès-Verbaux des Séances. Décrets. Arrêtes Ministères. 1854–1906.

Nancy, Fac. Sci., PV (II): Nancy. Procès-Verbaux des Séances de l'Assemblée de la Faculté des Sciences de Nancy, 1872–1945.

Nancy, Fac. Sci., PV (III) and (IV): Nancy. Procès-Verbaux des Réunions du Conseil de la Faculté des Sciences de Nancy. Déc. 1888–11 janv. 1917; and Commencé le 16 juin 1917.

## From Archives of Université de Paris I

Paris, Fac. Sci., PV Conseil: Paris. Faculté des Sciences. Procès-Verbaux des Séances du Conseil. 1873–1902; 1902–1919.

Paris, Fac. Sci., PV Conseil, Pièces-Annexes: Paris. Faculté des Sciences. Pièces-Annexes au Procès-Verbaux des Séances du Conseil. 1883–1909 (pp. 47–370); 1909–1920 (pp. 1–27).

Paris, Fac. Sci., PV Assem.: Paris. Faculté des Sciences. Procès-Verbaux des Séances de l'Assemblée. 1886–1925.

## From Archives of Université de Toulouse I
## (Université Paul Sabatier)

Toulouse, Fac. Sci., PV Assem.: Toulouse. Faculté des Sciences. Registre des Délibérations de l'Assemblée. 1886–1909; 1909–1922; 1922–1936.

Toulouse, Fac. Sci., PV Conseil: Toulouse. Faculté des Sciences. Procès-Verbaux du Conseil. Registre des Délibérations. 1886–1909; 1907–1921; and 1921–1932.

## PUBLISHED SOURCES

*Amer. Chem. Jl.: American Chemical Journal*
*Ann. Chem.: Annalen der Chemie*
*Ann. Chim. Phys.: Annales de Chimie et de Physique*
*Ber. Deut. Chem. Gesell.: Berichte der Deutsche Chemische Gesellschaft*
*Bordeaux, Livret Etud.: Annuaire de l'Université de Bordeaux. Livret Etudiant pour l'Année Scolaire*
*Bordeaux, Rentrée Facs.: Bordeaux. Séance Annuelle de Rentrée des Facultés de l'Université*
*Bordeaux, Univ. Conseil: Bordeaux. Rapport Annuel du Conseil de l'Université*
*Bull. Soc. Chim.: Bulletin de la Société Chimique de France*
*CR: Comptes Rendus Hebdomadaires de l'Académie des Sciences*
*Dépêche: La Dépêche de Toulouse*
*DSB: Dictionary of Scientific Biography*

*Enquêtes Enseignt.: Enquêtes et Documents relatifs à l'Enseignement Supérieure*
*Grenoble, Livret Etud.: Annuaire de l'Université de Grenoble. Livret Etudiant pour*
    *l'Année Scolaire*
*Grenoble, Rentrée Facs.: Grenoble. Séance Annuelle de Rentrée des Facultés de l'Uni-*
    *versité*
*Grenoble, Univ. Conseil: Grenoble. Rapport Annuel du Conseil de l'Université*
*HSPS: Historical Studies in the Physical Sciences*
*Jl. Amer. Chem. Soc.: Journal of the American Chemical Society*
*Jl. Chem. Soc.: Journal of the Chemical Society* (London)
*Lyon, Livret Etud.: Annuaire de l'Université de Lyon. Livret de l'Etudiant pour l'An-*
    *née Scolaire*
*Lyon, Rentrée Facs.: Lyon. Séance Annuelle de Rentrée des Facultés*
*Lyon, Univ. Conseil: Lyon. Rapport Annuel du Conseil de l'Université*
*Nancy, Livret Etud.: Annuaire de l'Université de Nancy. Livret de l'Etudiant pour*
    *l'Année Scolaire*
*Nancy, Rentrée Facs.: Nancy. Séance Annuelle de Rentrée des Facultés de l'Université*
*Phil. Mag.: Philosophical Magazine*
*Phys. Zeit.: Physikalische Zeitschrift*
*RdQS: Revue des Questions Scientifiques*
*Rev. Fr. Soc.: Revue Française de Sociologie*
*Rev. Gén. Sci.: Revue Générale des Sciences*
*RIE: Revue Internationale de l'Enseignement*
*RS: Revue Scientifique*
*Statistique Générale: France. Statistique Générale de la France. Annuaire Statistique*
*Toulouse, Livret Etud.: Annuaire de l'Université de Toulouse. Livret de l'Etudiant pour*
    *l'Année Scolaire*
*Toulouse, Rentrée Facs.: Toulouse. Séance Annuelle de Rentrée des Facultés de l'Uni-*
    *versité*
*Toulouse, Univ. Conseil: Toulouse. Rapport Annuel du Conseil de l'Université*
*Zeit. phys. Chem.: Zeitschrift für physikalische Chemie*

# NOTES

## INTRODUCTION: THE PARIS-PROVINCES DICHOTOMY

1. See Jared Wenger, *The Province and the Provinces in the Work of Honoré Balzac* (Ph.D. diss., Princeton University, 1937), p. 31.

2. Gustave Flaubert, *Madame Bovary*, trans. Paul de Man (New York: Norton, 1965), pp. 50–51.

3. Ibid., p. 57.

4. Ibid., p. 41.

5. François Mauriac, *La province* (Paris: Hachette, 1926), p. 17.

6. Ibid., p. 51.

7. Edward Shils, "Metropolis and Province," in *The Intellectuals and the Powers and Other Essays* (Chicago: University of Chicago Press, 1972), pp. 335–371.

8. Edward Shils, "British Intellectuals in the Mid-Twentieth Century," in *The Intellectuals and the Powers*, pp. 135–153, esp. p. 145.

9. Quoted in Arthur Schuster, *Biographical Fragments*, p. 201, and cited in Robert Kargon, *Science in Victorian Manchester: Enterprise and Expertise* (Baltimore: Johns Hopkins University Press, 1977), p. 50. See recent essays on British provincial science in Ian Inkster and Jack Morrell, eds., *Metropolis and Province: Science in British Culture 1780–1850* (Philadelphia: University of Pennsylvania Press, 1983).

10. See Jean-François Gravier, *Décentralisation et progrès technique* (Paris: Flammarion, 1953), p. 36.

11. John Merz, *A History of European Thought in the Nineteenth Century*, Vol. I, Pt. I, *Scientific Thought* (1904; reprint, New York: Dover, 1965), p. 227.

12. Maurice Caullery, *La science française depuis le xvii<sup>e</sup> siècle*, 2d. rev. ed. (Paris: Armand Colin, 1948; 1st ed., 1933), p. 199.

13. On these institutions before the Revolution, see Charles C. Gillispie, *Science and Polity in France at the End of the Old Regime* (Princeton, N.J.: Princeton University Press, 1980).

14. Terry Shinn, "The French Science Faculty System 1808–1914: Institutional Change and Research Potential," *HSPS*, 10 (1979), 271–332; Harry W. Paul, "Science and the Catholic Institutes in Nineteenth-Century France," *Societas*, 1 (1971), 271–285, and "Apollo Courts the Vulcans: The Applied Science Institutes in Nineteenth-Century French Science Faculties," in Robert Fox and

George Weisz, eds., *The Organization of Science and Technology in France 1808–1914* (Cambridge, England: Cambridge University Press, 1980), pp. 151–181. Robert Fox, "The *Savant* Confronts his Peers: Scientific Societies in France, 1815–1914," ibid., pp. 241–282; also, Fox, "Learning, Politics and Polite Culture in Provincial France: The Sociétés Savantes in the Nineteenth Century," in Donald N. Baker and Patrick J. Harrigan, eds., *The Making of Frenchmen: Current Directions in the History of Education in France, 1679–1979*, special issue of *Réflexions Historiques*, 7 (1980), 543–564; and Fox, "Presidential Address: Science, Industry, and the Social Order in Mulhouse, 1798–1871," *British Journal for the History of Science*, 17 (1984), 127–168.

15. Fritz Ringer, *Education and Society in Modern Europe* (Bloomington: Indiana University Press, 1978); Theodore Zeldin, "Higher Education in France, 1848–1940," *Journal of Contemporary History*, 2 (1967), 53–80, and *France, 1848–1945*, 2 vols. (Oxford: Oxford University Press, 1973–1977); George Weisz, *The Emergence of Modern Universities in France, 1863–1914* (Princeton, N.J.: Princeton University Press, 1983), and "The French Universities and Education for the New Professions, 1885–1914: An Episode in French University Reform," *Minerva*, 17 (1979), 98–128; John M. Burney, *The University of Toulouse in the Nineteenth Century: Faculties and Students in Provincial France* (Ph.D. diss., University of Kansas, 1982); and John E. Craig, *Scholarship and Nation Building. The Universities of Strasbourg and Alsatian Society, 1870–1939* (Chicago: University of Chicago Press, 1984). Although not focused on the universities, another valuable recent study is Robert Gildea, *Education in Provincial France, 1800–1914: A Study of Three Departments* (New York: Clarendon Press, 1983).

Also see William Bruneau, *The French Faculties and Universities, 1870–1902* (Ph.D. diss., University of Toronto, 1977); and for the grandes écoles, Robert J. Smith, *The Ecole Normale Supérieure and the Third Republic* (Albany: SUNY Press, 1982); Craig Zwerling, "The Emergence of the Ecole Normale Supérieure as a Centre of Scientific Education in the Nineteenth Century," in Fox and Weisz, *The Organization of Science and Technology*, pp. 31–60; Terry Shinn, *Savoir scientifique et pouvoir sociale: L'Ecole Polytechnique, 1794–1914* (Paris: Presses de la fondation nationale des sciences politiques, 1980), and "Des sciences industrielles aux sciences fondamentales: La mutation de l'Ecole Supérieure de Physique et de Chimie (1882–1970)," *Rev. Fr. Soc.*, 22 (1981), 167–182. Also, John H. Weiss, *The Making of Technological Man: The Social Origins of French Engineering Education* (Cambridge: MIT Press, 1982); and A. Brunot and R. Coquand, *Le corps des Ponts et Chaussées* (Paris: Editions du CNRS, 1982).

16. On decline, see Mary Jo Nye, "Scientific Decline: Is Quantitative Evaluation Enough?" *Isis*, 75 (1984), 697–708; on general characteristics of French science in relation to the content of twentieth-century physics, see Dominique Pestre, *Physique et physiciens en France, 1918–1940* (Paris: Editions des archives contemporaines, 1984). Also, more generally, Gerald Geison, *Michael Foster and the Cambridge School of Physiology: The Scientific Enterprise in Late Victorian Society* (Princeton, N.J.: Princeton University Press, 1978), and "Scientific Change, Emerging Specialties, and Research Schools," *History of Science*, 19 (1981), 20–40; also see Robert K. Merton, "The Sociology of Science: An Episodic Mem-

oir," in Robert K. Merton and Jerry Gaston, eds., *The Sociology of Science in Europe* (Carbondale: Southern Illinois University Press, 1977), pp. 31–141, esp. pp. 75–76.

17. Claude Bernard, *Rapport sur les progrès et la marche de la physiologie générale en France* (Paris, 1867); Louis Pasteur, *Le budget de la science* (Paris, 1868); and Adolphe Wurtz, *Les hautes études pratiques dans les universités allemandes* (Paris, 1870). And Emile Aglave, "Les Universités allemands," *Revue des Cours Scientifiques*, 6 (1869), 753–754, on p. 753.

18. See the report of Maurras's remarks in Maurice Ajam, "La décentralisation—où en sommes nous," *Revue Politique et Parlementaire*, 57 (1908), 31–50, on pp. 35–36.

19. See Robert Scott Root-Bernstein, *The Ionists: Founding Physical Chemistry, 1871–1890*, 2 vols. (Ph.D. diss., Princeton University, 1980), p. v; and John William Servos, *Physical Chemistry in America, 1890–1933: Origins, Growth, and Definition* (Ph.D. diss., Johns Hopkins University, 1979), esp. pp. 34, 36.

20. François Mauriac, *La province*, p. 17. My translation.

21. Quoted in Lucien Babonneau, "Paul Sabatier," in *Génies occitans de la science* (Toulouse: Privat, 1947), pp. 167–189.

## 1: FRENCH UNIVERSITY SCIENCE BEFORE THE FIRST WORLD WAR

1. Barrett Wendell, "Impressions of Contemporary France, I: The Universities," *Scribner's Magazine*, 41 (March 1907), 314–326.

2. On modern French education, see Antoine Prost, *Histoire de l'enseignement en France 1800–1967* (Paris: Armand Colin, 1968); Félix Ponteil, *Histoire de l'enseignement en France: Les grandes étapes, 1789–1964* (Paris: Sirey, 1966); George Weisz, *The Emergence of Modern Universities in France, 1863–1914* (Princeton, N.J.: Princeton University Press, 1983); and William Bruneau, *The French Faculties and Universities, 1870–1902* (Ph.D. diss., University of Toronto, 1977). For an analysis of the recent reforms, see Habiba S. Cohen, *Elusive Reform: The French Universities, 1968–1978* (Boulder, Colo.: Westview Press, 1978).

3. On the role of the academies and societies in the eighteenth and nineteenth centuries, see Daniel Roche, *Le siècle des lumières en province: académies et académiciens provinciaux, 1680–1789*, 2 vols. (Paris: Mouton, 1978); and Robert Fox, "The *Savant* Confronts his Peers: Scientific Societies in France, 1815–1914," in Robert Fox and George Weisz, eds., *The Organization of Science and Technology in France, 1808–1914* (Cambridge: Cambridge University Press, 1980), pp. 241–282, and "Learning, Politics and Polite Culture in Provincial France: The Sociétés Savantes in the Nineteenth Century," *Réflexions Historiques*, 7 (1980), 543–564.

4. George Weisz emphasizes new functions undertaken by the late-nineteenth-century universities, namely, a stronger research orientation and technical and commercial training. Republican educational ministries, particularly under the leadership of Jules Ferry and Louis Liard, were to emphasize also a role

for universities, supplementary to primary and secondary education, in promulgating a scientific and secular ideology consistent with republican political strategy. See Weisz, *Emergence of Modern Universities*, esp. pp. 3–4, 115–116, 129, 285. Valuable sources on technical education are Michel Bouillé, *Enseignement technique et idéologies au 19ème siècle* (Ecole des Hautes Etudes en Sciences Sociales thèse de 3ème cycle, 1972), and Frederick B. Artz, *The Development of Technical Education in France, 1500–1850* (Cambridge: MIT Press, 1966).

5. On the history of the universities before 1800, see Simone Guenée, *Bibliographie de l'histoire des universités françaises des origines à la révolution*, Vol. I, *Généralités: Université de Paris;* Vol. II, *D'Aix-en-Province à Valence et académies protestants* (Paris: Picard, 1978–1981). This includes titles up to 1975. On the establishment of the Catholic universities, see René Aigrain, *Les universités catholiques* (Paris: A. Picard, 1935).

6. See Louis Liard, *L'Enseignement supérieur en France*, 2 vols. (Paris: Armand Colin, 1888–1894), Vol. I, *1789–1889*, pp. 1–2, 11–15. (Vol. II, *1889–1893.*)

7. For a short time, the écoles centrales taught a modern curriculum of French grammar, drawing, modern history, and the sciences. See L. Pearce Williams, "Science, Education and the French Revolution," *Isis*, 44 (1953), 311–330; John E. Talbott, *The Politics of Educational Reform in France, 1918–1940* (Princeton, N.J.: Princeton University Press, 1969), pp. 6–10; and Theodore Zeldin, *France: 1848–1945*, Vol. II, *Intellect, Taste, and Anxiety* (Oxford: Oxford University Press, 1977), pp. 244–245. On the grandes écoles parisiennes and engineering education, see Terry Shinn, *Savoir scientifique et pouvoir social: L'Ecole Polytechnique, 1794–1914* (Paris: Presses de la fondation nationale des sciences politiques, 1980); John H. Weiss, *The Making of Technological Man: The Social Origins of French Engineering Education* (Cambridge: MIT Press, 1982); and A. Brunot and R. Coquand, *Le corps des Ponts et Chaussées* (Paris: Editions du CNRS, 1982). On institutions before the Revolution, see Charles C. Gillispie, *Science and Polity in France at the End of the Old Regime* (Princeton, N.J.: Princeton University Press, 1980).

8. Talbott, p. 9.

9. See W. R. Fraser, *Education and Society in Modern France* (London: International Library of Sociology and Social Reconstruction, 1963), pp. 23–24.

10. There were forty academies in 1808, including some in Belgium, Switzerland, and Italy. After Napoleon's defeat the number varied, though it was sixteen or seventeen during much of the modern period. Ponteil, pp. 123–135.

11. Ponteil, pp. 158–159.

12. I have in part followed here Shinn's summary of the requirements of the 1808 decree. He notes that from 1835 to 1840 at Lyon and Montpellier, 50 percent of the staff worked simultaneously in the secondary-school system, and at Besançon between 1840 and 1845, 67 percent. These people taught in the lycées fifteen to twenty-five hours per week. See Terry Shinn, "The French Science Faculty System, 1808–1914: Institutional Change and Research Potential in Mathematics and the Physical Sciences," *HSPS*, 10 (1979), 271–332.

13. J. B. Piobetta, *Le baccalauréat* (Paris: J. B. Baillière, 1937), p. 69.

14. Ponteil, p. 162.

15. Ibid., p. 163.

16. André Gain, "L'enseignement supérieur à Nancy de 1789–1896," in Louis Bruntz, Charles Adam, M. E. Husson et al., *L'Université de Nancy (1572–1934)* (Nancy, 1934), pp. 25–42.

17. See Victor Duruy, *Notes et souvenirs (1811–1894)*, 2d ed., 2 vols. (Paris: Hachette, 1902), Vol. I, p. 197.

18. R. D. Anderson, *Education in France, 1848–1870* (Oxford: Clarendon Press, 1975), pp. 61, 68, 78–79. On Fortoul's earlier attempt to give science new prominence in secondary education, see Nicole Hulin, "A propos de l'enseignement scientifique: une réforme de l'enseignement secondaire sous le second empire, la 'bifurcation' (1852–1864)," *Revue d'Histoire des Sciences*, 35 (1982), 217–245.

19. Anderson, p. 207.

20. See Roger Grignard, *Centenaire de la naissance de Victor Grignard, 1871–1971* (Lyon: Audin, 1972), pp. 24–25. See also Rod Day, "Education, Technology and Social Change in France: The Short, Unhappy Life of the Cluny School, 1866–1891," *French Historical Studies*, 8 (1974), 427–444; and for a contemporary account, Matthew Arnold's description of "special secondary education" in Matthew Arnold, *Works*, Vol. XII, *A French Eton. Higher Schools and Universities in France. Higher Schools and Universities in Germany* (London: Macmillan, 1904), pp. 249–256.

21. Grenoble, Fac. Sci., PV (I), 26 August 1855, and Fortoul's letter, dated 17 July 1855, following the PV for this date; also see Shinn, in *HSPS*, p. 295. Shinn points out that the ministry long ignored requests from Lyon for chairs in applied mechanics and applied chemistry in support of the regional textile industry. The Lyon Sciences Faculty in 1840 employed some cunning in getting permission for a chair of applied mathematics and then recruiting someone who could teach applied mechanics as well (Shinn, pp. 288–290). On Fortoul's certificate, see Weisz, *Emergence of Modern Universities*, pp. 43–44.

22. Claude Digeon, *La crise allemande de la pensée française (1870–1914)* (Paris: Presses Universitaires de France, 1959), p. 364. See "Les universités allemandes," *Revue des Cours Scientifiques* [later *RS*], 6 (30 October 1869), 753–754, and (13 November 1869), 785; and Ernest Renan, "L'Enseignement supérieure en France, son histoire et son avenir," *Revue des Deux Mondes*, 51 (1864), 73–95.

23. In 1865, for example, Matthew Arnold was traveling in France, charged by the British Schools Enquiry Commission with investigating the system of education for the upper and middle classes in France, Italy, Germany, and Switzerland. See S. J. Curtis and M. E. A. Boltwood, *An Introductory History of English Education since 1800* (London, 1960), p. 84.

24. See William Keylor, *Academy and Community: The Foundation of the French Historical Profession* (Cambridge: Harvard University Press, 1975), p. 21; and Anderson, *Education in France*, pp. 219–220.

25. On German university education, see Friedrich Paulsen, *Geschichte der gelehrten Unterrichtsauf den deutschen Schulen und Universitäten vom Ausgang des Mittelalters bis zur Gegenwart*, 3d ed., Rudolf Lehmann, ed. (Berlin, 1921); Fritz K. Ringer, *Education and Society in Modern Europe* (Bloomington: Indiana University Press, 1978); and R. Steven Turner, "The Growth of Professorial Research in Prussia, 1818–1848—Causes and Context," *HSPS*, 3 (1971), 137–182.

26. Ringer, p. 39.

27. Seminars first were established in philology and history (after 1810) as programlike institutions in which ten to twenty students were selected to enroll for a two- or three-year program, receiving intensive training from the professor in charge. Their members received subsidies from the state in the form of scholarships, prizes, and exemptions from state tests. See Steven Turner, in *HSPS*, p. 145, n. 23.

28. It was Justus von Liebig who introduced this innovation. See Christa Jungnickel, "Teaching and Research in Physical Sciences and Mathematics in Saxony, 1820–1850," *HSPS*, 10 (1979), 1–47, esp. pp. 18–19, 27.

29. Arnold, *Works*, Vol. XII, p. 384.

30. In reports in the Archives Nationales, cited in Terry Shinn, *HSPS*, p. 300.

31. Letters from Emile Boutroux to the minister of public instruction, 28 January and 3 March 1869, AN F$^{17}$ 22028 (Emile Boutroux); and Emile Boutroux's remarks in *Revue des Cours Littéraires de la France et de l'Etranger* (later, *Revue Bleue*), 1871.

32. Duruy, *Notes et souvenirs*, p. 301; *Statistique de l'enseignement supérieur*, 4 vols. (Paris: Imprimerie Impériale, 1868–1898).

33. Duruy, *Notes et souvenirs*, pp. 313–314; Ponteil, p. 249. There were five research sections in the EPHE: mathematics, physical sciences, natural sciences, historical and philological sciences, and economic sciences. No diplomas were given. See "L'Ecole Pratique des Hautes Etudes," *Revue des Cours Scientifiques*, 5 (15 August 1868), 585–592.

34. Shinn, *HSPS*, p. 301.

35. Another indication of the new emphasis on research was Jules Ferry's interest in publishing a yearly report of the work done by all professors in the Sciences Faculties to "help show that professors of the state take part in the progress of science in our country and the world." Circular from Ministry of Public Instruction to academy rectors, dated 18 March 1880, ADI, FdR, no. 200. This plan was realized in the publication, after 1902, of the *Bibliographie scientifique française: Recueil mensuel*. See Weisz, *Emergence of Modern Universities*, p. 200.

36. Robert A. Nye, *Crime, Madness, and Politics in Modern France: The Medical Concept of National Decline* (Princeton, N.J.: Princeton University Press, 1984), esp. chap. 5.

37. See Weisz, *Emergence*, pp. 106–107, for the figures. Weisz's book also has a good deal of information on funding for the university system.

38. Shinn, *HSPS*, p. 308.

39. Keylor, pp. 26–63.

40. Lyon, Fac. Sci., PV, 27 March 1896 and 10 April 1920. Another decree of 1920 organized the new category of *professeur sans chaire*, a position that might be requested for a senior staff member who had been teaching courses for which there was no established chair. Usually, the creation of the professeur sans chaire was personal and awarded additional salary; the new professor's former position was abolished. See Lyon, Fac. Sci., PV, 28 January 1921 and 18 February 1921.

41. Prost, pp. 239–240; and Weisz, *Emergence*, pp. 5, 72–83, 131.

42. Ponteil, p. 312.

43. See Jean Delsarte, *Les activités scientifiques et techniques de la Faculté des Sciences de Nancy: Le problème des Ecoles Nationales Supérieures d'Ingénieurs*, Les Conférences du Palais de la Découverte, Series A, no. 222 (26 May 1956), pp. 7–8.

44. The Société de l'Enseignement Supérieur was founded in 1878 and had over 500 members by 1880, drawn largely from provincial teachers. See Weisz, "The Anatomy of University Reform, 1863–1914," *Réflexions Historiques*, 7 (1980), 363–380, p. 367.

45. On the Ferry laws, see Talbott, p. 24, and Frederic Ernest Farrington, *French Secondary Schools* (London, 1915), p. 82. Farrington notes (p. 83) that illiteracy in France decreased from 25 percent for men and 37 percent for women in 1870 to 4.7 percent and 7.2 percent, respectively, in 1898. Also see Louis Liard, *Universités et facultés* (Paris, 1890), p. 229; and Prost, pp. 237–238. Women's education took place mostly in convent schools and private boarding schools before the 1880s. Girls' schools were not allowed to teach Latin, which made it practically impossible for girls to take the baccalaureate. In 1862, Julie-Victoire Daubié, a teacher and writer on girls' education, became the first woman to take the baccalaureate, after Rouland could find no legal justification for refusing permission. In the 1860s, women first made an appearance in the Medical Faculties, and by 1882 the Paris Faculty had given 19 doctorates to women, only five of whom were French. See Anderson, pp. 189–190.

46. Liard, *Universités et facultés*, pp. 142–149, 199; and Prost, p. 236.

47. On the changing functions of the Faculty, see Paul Sabatier, "La Faculté des Sciences et ses instituts techniques en 1919," in *Documents sur Toulouse et sa région* (Toulouse: Privat, 1910), Vol. I, p. 167; and *Toulouse, Univ. Conseil, 1882–1883*, pp. 54–55.

48. On changes in the baccalaureate grading system, see Weisz, *Emergence of Modern Universities*, pp. 210–211.

49. An opinion confided to the rector of the Toulouse academy, Claude Perroud; cited in Louis Rascol, *Claude-Marie Perroud* (Paris: Didier, 1941), pp. 246–247.

50. See letters from Benjamin Baillaud to Claude-Marie Perroud and to the ministry, 24 January 1882, AN F[17] 24106 (Paul Sabatier). On this point, see Steven Turner, in *HSPS*, pp. 172–173. Turner argues that it was ministerial policy, not individual university initiatives, that built the German universities into research institutions. The ministry emphasized creativity and originality in publication as a basis for appointments rather than fraternal, pedagogical, and collegial values, which had been stressed in the eighteenth century.

51. Prost, p. 233.

52. See *Toulouse, Univ. Conseil, 1887–1888*, p. 83, and *1887–1888*, p. 75; and Toulouse, Fac. Sci., PV Assem., 27 January 1894, 28 April 1896, 28 November 1900.

53. Weisz, in *Réflexions Historiques*, p. 370, including n. 18; Shinn, *HSPS*, p. 305; and "La nouvelle Sorbonne," *RIE*, 29 (1895), 209–244, 326–357, 401–422.

54. Shinn, *HSPS*, p. 312; and Weisz, *Emergence*, pp. 164–166, 170–171. See

also Paul Forman, John Heilbron, and Spencer Weart, *Physics circa 1900: Personnel, Funding, and Productivity of the Academic Establishments*, HSPS, 5 (1975), pp. 66–67. In Germany during the period 1891–1900, 9 of 27 towns possessing higher schools made money contributions to the universities or Technische Hochschulen in addition to grants of land. In England, the percentage of total university income deriving from annual grants from local authorities averaged 15 percent (ibid., pp. 78–79).

55. See Shinn's graphs, esp. fig. 4, p. 331 in *HSPS*; also pp. 305–306.

56. Prost, p. 230.

57. See Ringer, Table XI, p. 335. He also gives the numbers of students in other Faculties, and per population and age group.

58. Terry Shinn, "From 'corps' to 'profession': the emergence and definition of industrial engineering in modern France," in Fox and Weisz, eds., *Organization of Science and Technology*, pp. 183–210. The French-language version of the article is in *Rev. Fr. Soc.*, 19 (1978), 39–71. Also see Ezra N. Suleiman, *Elites in French Society: The Politics of Survival* (Princeton, N.J.: Princeton University Press, 1978).

59. They also expressed doubt about the ability of the congress to reach trustworthy decisions, according to David Cahan, *The Physikalisch-Technische Reichsanstalt: A Study in the Relations of Science, Technology and Industry in Imperial Germany* (Ph.D. diss., Johns Hopkins University, 1980), pp. 63–72.

60. Quoted in Cahan, p. 76, n. 53; and p. 88.

61. Ibid., pp. 213–302, esp. 293–302, and p. 359; the quotation is from *Annalen der Physik*, 292 (1895), 451–456, p. 455.

62. See Lewis Pyenson, *The Goettingen Reception of Einstein's General Theory of Relativity* (Ph.D. diss., Johns Hopkins University, 1974), pp. 130, 164–167. Göttingen became the first German university to appoint professors in applied mathematics and applied physics (the latter in electrotechnology and applied mechanics), according to Pyenson, p. 126.

63. Ferdinand Lot, "De la situation faite à l'enseignement supérieur en France," *Cahiers de la Quinzaine*, series vii, 9 (1906), 1–107, and 11 (1906), 109–237, pp. 15, 55, 64. A statement of Kaiser Wilhelm's often quoted in popular journals was: "We have won, and we will win on the fields of battle of commerce and industry." See Antoine Léon, *Histoire de l'education technique* (Paris: Presses Universitaires de France, 1961; 2d ed., 1968), p. 87. Also see Maurice Caullery, "L'Evolution de notre enseignement supérieur scientifique," *Revue du Mois*, 4 (1907), 513–535, esp. p. 522; and J. Gosselet, "L'Enseignement des sciences appliquées dans les universités," *RIE*, 37 (1899), 97–106.

64. The Ecole Polytechnique was administered by the Ministry of Defense, not the Ministry of Public Instruction. Shinn has estimated that of the 14,000 polytechniciens who graduated between 1808 and 1880, fewer than 100 devoted themselves afterward to pure research and less than 1,000 to the applied sciences. See *Savoir scientifique*, pp. 180–181.

65. See C. Rod Day, "Education for the industrial world: technical and modern instruction in France under the Third Republic 1870–1914," in Fox and Weisz, eds., *Organization of Science and Technology*, pp. 127–153, esp. p. 149.

66. See Anderson, *Education in France*, pp. 98–101. On Mulhouse, Zeldin, *France*, Vol. II, pp. 210, 245–247.

67. On Manchester especially, see Arnold Thackray, "Natural Knowledge in Cultural Context: The Manchester Model," *American Historical Review*, 79 (1974), 672–709; and Robert Kargon, *Science in Victorian Manchester: Enterprise and Expertise* (Baltimore: Johns Hopkins University Press, 1977). Cardwell suggests that industrial demand played no more part in the specialization of the various science degrees than in the specialization of the Classics Tripos or the London B.A. In D. S. L. Cardwell, *The Organization of Science in England* (London: Heinemann, 1972), p. 246.

68. On this point, see Delsarte, p. 8.

69. André Gain, *L'Enseignement supérieur à Nancy de 1789 à 1896* (Nancy: Imprimerie Berger-Levrault, 1934), p. 64.

70. Auguste Ehrard, *L'Université de Lyon* (Lyon: A. Rey, 1919), pp. 209–210.

71. See Paris, Fac. Sci., PV Conseil, p. 306. In 1887, the salary ranges were: maître de conférence, 1,800 francs; chargé de cours, 2,800–5,200 francs, with those in the provinces making 400–500 francs less; and professeur, 6,000–11,000 francs in the provinces and 12,000–15,000 francs in Paris. See Prost, pp. 75, 234, 358.

72. See Weisz, *Emergence of Modern Universities*, p. 166.

73. See Paris, Fac. Sci., PV Conseil, 21 October 1897, pp. 244–245, and 4 July 1899, pp. 287–288. Moissan's remark is in the PV for 10 November 1900, p. 325.

74. Toulouse, Fac. Sci., PV Conseil, 17 June 1887 and 16 March 1888.

75. Paris, Fac. Sci., PV Conseil, 15 May 1907, pp. 58–61.

76. Ibid., 10 November 1900, p. 326.

77. For the 1896 decree, see *Nancy, Livret Etud., 1896–1897*; Henry J. Furber, Jr., "Concerning the Attendance of Americans at the Universities of France," in *The Universities of France. A Guide for American Students. Published by the Franco-American Committee* (Paris: Lahure, 1899), pp. 2–5, p. 2; and Michel Bréal, "Opinions of an American Professor on the Universities of France," in ibid., pp. 5–9, p. 7, originally published in the *Journal des Débats*, 7 June 1895. On the 1895 decree, see Weisz, *Emergence of Modern Universities*, p. 260.

78. Bréal, "Opinion," p. 9.

79. See Paris, Fac. Sci., PV Conseil (1896), p. 218, and 23 June 1900, p. 317.

80. F. Lot, in *Cahiers de la Quinzaine*, p. 110; also see pp. 110–133.

81. A recent source is Weisz, *Emergence of Modern Universities*, pp. 252–268. See Albert Dubos, "Les études d'origine étrangère à Toulouse (1895–1910)," in *Documents sur Toulouse et sa région*, Vol. I, p. 215; *Toulouse, Univ. Conseil, 1909–1910*, pp. 15–18, 19–20; and ibid., *1928–1929*, pp. 72–73; also Paul Sabatier, "Institut de Chimie," in *L'Université de Toulouse: Son passé—son présent* (Toulouse, 1929), p. 299. Dubos suggested the appropriateness of a rewording of an "old cento" (p. 215):

> Paris pour voir,
> Lyon pour avoir,
> Bordeaux pour dispendre,
> TOULOUSE pour apprendre.

82. See table 2; also Weisz, pp. 241–252, including the tables, pp. 246 and 295.

83. *Toulouse, Univ. Conseil, 1903–1904*, p. 14.

84. See L. Clédat's discussion of the reforms. Clédat was dean of the Lyon Letters Faculty. "La réforme de l'Ecole Normale Supérieure et les universités de province," *RIE* (1906), 46–62; and the defense of the prerogatives and curriculum of normaliens by an anonymous student, "A propos de l'Ecole Normale et de la réorganisation des Facultés de province," *RIE, 53* (1907), 230–240.

85. Forman, Heilbron, and Weart, p. 33. See Victor Karady, "L'Accès aux grades et leurs fonctions universitaires dans les Facultés des Sciences au 19$^e$ siècle: examen d'une mutation," in *Réflexions Historiques, 7* (1980), 397–414. The preparation of Sorbonne students of course resulted in heavier teaching work loads for professors at the Ecole Normale. See the remarks by Jules Tannery, in Paris, Fac. Sci., PV Conseil, 23 December 1909, pp. 113–114.

86. Clédat, p. 52.

87. Ibid., p. 49. Bouglé moved to the Sorbonne in 1901. From 1935 to 1940, he was director of the Ecole Normale Supérieure (Zeldin, *France*, Vol. II, p. 338).

88. Clédat, pp. 51–52. In 1904, the scholarships for the licence and agrégation were distributed in the sciences as follows: 12 licence and 6 agrégation scholarships to Paris; 49 licence and 9 agrégation scholarships to the 17 provincial Sciences Faculties (p. 58).

89. Anonymous student, *RIE* (1907), p. 234. For a detailed discussion of the curriculum, examinations, degrees, and daily life at the Ecole Normale around 1908, see Farrington, *French Secondary Schools*, pp. 345–377. The written examination for the licence scholarships and Ecole Normale Supérieure was given at each academy seat. Those who were selected to go on to the oral examination took the oral part in Paris. Science candidates chose between two groups of requirements for the written part of the examination (see accompanying table). For the *agrégation des sciences mathématiques* or *physiques*, it was required to obtain three licence-qualifying certificates (the certificats d'études supérieures in special fields) and one *diplôme d'études supérieures* in either mathematics or physical sciences. Then, competitive examinations had to be taken. In the physical sciences, there were three compositions to write, each seven hours in duration. This was followed by oral proofs before a jury, where the candidate was required to explain the program of study for a lycée lesson on a topic chosen by the jury and to carry out the lesson. In addition, the candidate was required to demonstrate practical chemical laboratory work and give both physics and chemistry lessons, with experimental demonstrations. In *Lyon, Livret Etud., 1910–1911*, pp. 161–162, 223–225.

90. See Suleiman, p. 48; and Shinn, *Savoir scientifique*, pp. 101–139.

91. Grenoble, Fac. Sci., PV (II), 23 May 1907.

92. Victor Karady, Table C, p. 407. Also see *Etat numérique des grades, 1795–1885*, no. 21 in *Enquêtes Enseignt.* (Paris, 1886). According to Weisz, in *Emergence of Modern Universities*, pp. 238–241, 40 percent of students preparing for scientific teaching careers did so in Paris in 1876. This would include nonuniversity students. By 1914, only 23 percent of all science students were in Paris, but

QUALIFYING EXAMINATION FOR THE ECOLE NORMALE SUPÉRIEURE
AND LICENCE SCHOLARSHIPS CA. 1908

| Group I | Duration | Points |
|---|---|---|
| A composition of special mathematics | 6 hours | 6 |
| A composition of mathematics on the parts of the special mathematics program, to be determined by a ministerial decree | 4 hours | 4 |
| A composition of physics (program of the class of special mathematics) | 6 hours | 7 |
| A French composition | 3 hours | 1 |
| Two translations chosen among three texts in Latin, German, and English | 2 hours | 2 |
| | | $\overline{20}$ |

| Group II | Duration | Points |
|---|---|---|
| A composition of general mathematics, according to a program determined by ministerial decree | 4 hours | 4 |
| A composition of physics (program of certificat d'études physiques, chimiques et naturelles) | 6 hours | 5 |
| A composition of chemistry (same program) | 4 hours | 4 |
| A composition of natural sciences (same program) | 4 hours | 4 |
| A French composition (joint test with Group I) | 3 hours | 1 |
| Two translations chosen among three texts in Latin, German, and English (joint test with Group I) | 2 hours | 2 |
| | | $\overline{20}$ |

most of the 76 percent in the provinces were not intending to be teachers in French universities or lycées.

93. Charles Bruneau, "Le passé et l'avenir de l'Université de Nancy," in *L'Université de Nancy*, pp. 189–194, p. 190.

## 2: NANCY

1. Quoted in Phillipe Garner, *Emile Gallé* (New York: Rizzoli, 1976), p. 11, from Henri Frantz, in *The Studio*, 28 (1903), 108–117. The latter part of this chapter closely follows my previously published article "N-Rays: An Episode in the History and Psychology of Science," *Historical Studies in the Physical Sciences* 11 (1980), 125–156.

2. Ibid., p. 29.

3. Gaston Floquet, *L'Exposition retrospective lorraine des sciences* (Nancy: Imprimerie Berger-Levrault, 1912), p. 7; Eugène Martin, "L'Université de Pont-à-Mousson," in *L'Université de Nancy (1572–1934)*, Louis Bruntz, Charles Adam, E. Husson et al., eds., pp. 1–24, p. 16; and *Vosges, Lorraine, Alsace: Les guides bleus* (Paris: Hachette, 1922), pp. [2], 11–18.

4. See John E. Craig, *A Mission for German Learning: The University of Strasbourg and Alsatian Society, 1870–1918* (Ph.D. diss., Stanford University, 1972), p. 81. I am grateful to Steven Turner for calling this study to my attention. Craig's dissertation has been incorporated into his more recent *Scholarship and Nation Building: The Universities of Strasbourg and Alsatian Society, 1870–1939* (Chicago: University of Chicago Press, 1984).

5. Ibid., pp. 32–33. Paul Appell's boyhood experiences were typical of many Alsatians. Sermons at the church were preached in Alsatian, and while Appell learned French at school in the 1860s, his mother had been sent from her home in Klingenthal to a family in Saulxures in the Vosges to learn French, while a Saulxures boy came to Klingenthal in her place to learn German. See Paul Appell, *Souvenirs d'un alsacien, 1858–1922* (Paris: Payot, 1923), pp. 56–57.

6. Jean-Claude Bonnefont, ed., *Histoire de la Lorraine de 1900 à nos jours* (Toulouse: Privat, 1979), pp. 28, 37.

7. Appell, *Souvenirs*, p. 117.

8. Bonnefont, p. 80.

9. Garner, p. 50.

10. The 1788 building later became the municipal library. See Eugène Martin, pp. 16–20.

11. André Gain, "L'Enseignement supérieur à Nancy de 1789–1896," pp. 25–42 in *L'Université de Nancy (1572–1934)*, on pp. 26–31.

12. See André Gain, *L'Enseignement supérieur à Nancy de 1789 à 1896* (Nancy: Imprimerie Berger-Levrault, 1934), p. 37. This is a longer version of the essay published in the commemorative volume edited by Bruntz et al.

13. Gain, "L'Enseignement supérieur," pp. 32–36; and Jacques Aubry, "Les précurseurs," ms., 13 pp., p. 2.

14. Discussed in John Craig, *Mission for German Learning*, pp. 108–109.

15. Ibid., pp. 94–95, 315.

16. Quoted in ibid., p. 118, from *Annales de l'Assemblée Nationale*, 3 (1871), annexe no. 272, 30 May 1871, p. 197.

17. Craig, *Mission*, pp. 248–250; and Ferdinand Lot, "Deux universités: Strasbourg et Nancy," *RIE*, 32 (1896), 138–141.

18. See Nancy, Fac. Sci., PV (I), pp. 1, 57.

19. E. Husson, "La Faculté des Sciences de Nancy," in *L'Université de Nancy (1572–1934)*, pp. 83–122, p. 84.

20. Garner, *Emile Gallé*, p. 44. That Gallé's interest in natural history was scholarly as well as artistic is demonstrated in his active role around 1880 in the Société Lorraine d'Horticulture, including editing the society's journal. Ibid., p. 44.

21. Appell, *Souvenirs*, pp. 141–143.

22. Theodore Zeldin, *France: 1848–1945*, Vol. II, *Intellect, Taste, and Anxiety* (Oxford: Oxford University Press, 1977), pp. 275–277.

23. Aubry, "Les précurseurs," p. 3; and Nancy, Fac. Sci., PV (I), pp. 64–71.

24. *Nancy, Rentrée Facs., 1879–1880.*

25. Nancy, Fac. Sci., PV (II), 22 January 1881.

26. Ibid., 19 July 1882, 7 January 1896, 9 April 1897.

27. I enjoyed conversations with Professor Aubry and Dean Depaix on these and other questions, June 12 and 14, 1979, at the Université de Nancy I in Villers-Vandoeuvre.

28. Jean Delsarte, *Les activités scientifiques et techniques de la Faculté des Sciences de Nancy: Le problème des Ecoles Nationales Supérieures d'Ingénieurs*, Les Conférences du Palais de la Découverte, Ser. A, no. 222 (26 May 1956), pp. 7–8.

29. Ministère de l'Instruction Publique et des Cultes, "Instruction sur l'exécution du règlement du 26 décembre 1854 relatif à l'enseignement des sciences appliquées" (Paris: Imprimerie Nationale, 1855), pp. 3–4. In ADI, FdR.

30. Ibid., p. 5.

31. Quoted in E. Husson, "La Faculté des Sciences de Nancy," p. 83.

32. Quoted in E. Bichat, "L'Enseignement des sciences appliquées à la Faculté des Sciences de Nancy," *RIE*, 35 (1898), 299–307, p. 300.

33. J. Nicklès, for example, taught inorganic chemistry as a Faculty course and metallurgy and exploitation of mines in the Ecoles des Sciences Appliquées. Jules M. A. Chautard taught acoustics and electricity to licence students and electrochemistry and its applications to applied students. Nancy, Fac. Sci., PV (I), 24 March 1857, pp. 22–24.

34. Ibid., 25 November 1864, pp. 44–45.

35. Ibid., 25 January 1858, pp. 23–24; and E. Husson, "La Faculté des Sciences de Nancy," p. 83.

36. Nancy, Fac. Sci., PV (I), January 1868, p. 49, and November 1868, p. 49; 21 September 1868, pp. 50–53, and 9 December 1871, p. 70. The Station Agronomique founded by Grandeau in 1867 was the first in France.

37. Aubry, "Les précurseurs," p. 4. His son became a mayor of Lunéville and his grandson a deputy in the National Assembly. (Conversation with Professor Aubry and Dean Depaix, at Nancy, June 14, 1979.)

38. Bichat, "L'Enseignement des sciences appliquées," p. 307; and Husson, "La Faculté des Sciences de Nancy," photograph on p. 122.

39. See Aubry, "Les précurseurs," p. 4; and Haller's obituary notice in *Bull. Soc. Chim.*, 39 (1926), 1037–1092. During the period 1901–1915, Haller was nominated for the Nobel Prize in chemistry in 1907 by Lyon's Léo Vignon; in 1911, by Bordeaux's Maurice Vèzes and by Geneva's Philippe Guye; in 1912, by a Nancy group, including J. Minguin, A. Guntz, Paul Petit, and Paul-Thiébaud Müller; in 1914, by Vignon, Barbier, Le Chatelier, Ostwald, and Ramsay; and in 1915, by Grignard. Nobel Prize Archives, Royal Swedish Academy of Sciences, courtesy of Elisabeth Crawford and Wilhelm Odelberg. On the system of nomination and selection of Nobel laureates, see Elisabeth Crawford, *The Beginnings of the Nobel Institution: The Science Prizes, 1901–1915* (Cambridge: Cambridge University Press, 1984).

40. See chap. 5 for a more detailed discussion on this topic.

41. See François Leprieur and Pierre Papon, "Synthetic Dyestuffs: The Relations between Academic Chemistry and the Chemical Industry in Nineteenth-Century France," *Minerva*, 17 (1979), 197–224, esp. 205, 210–211; and Paul M. Hohenberg, *Chemicals in Western Europe: 1850–1914, An Economic Study of Technical Change* (Chicago: Rand McNally, 1967), pp. 27–28, 40.

42. Husson, p. 87; and Gain, "L'Enseignement supérieur," p. 39.

43. Bichat, "L'Enseignement des sciences appliquées," p. 302; and Delsarte, p. 15.

44. Nancy, Livret Etud., 1903–1904, p. 180.

45. Quoted from municipal council proceedings, 27 May 1886, in Husson, p. 87.

46. Garner, Emile Gallé, p. 11.

47. Jean Diez, Le régionalisme économique: Organisation de la région économique de l'est (Nancy) (Paris: Rousseau, 1925), pp. 20–22. Like the area around Lyon-Savoie-Dauphin, Lorraine and the north of France had a strong banking system that made possible industrial development in the direction of heavy industry (Hohenberg, p. 46).

48. Aubry, "Les précurseurs," p. 3; and Aaron Ihde, The Development of Modern Chemistry (New York: Harper and Row, 1964), p. 466.

49. Alsace possessed petrol and potassium. See Husson, p. 117.

50. See Hohenberg, pp. 23–24. The trick of Solvay's process lay in his 80-foot towers into which $CO_2$ was injected at the bottom, while ammoniated brine flowed downward from the top, ensuring a quick and complete reaction from which 95 to 99 percent of the ammonia could be recovered. See W. J. Reader, Imperial Chemical Industries: A History, Vol. I, The Forerunners, 1870–1926 (Oxford: Oxford University Press, 1970), pp. 45–46.

51. Solvay et Cie, Exposition Internationale: Nancy 1909, Solvay et Cie (1909), pp. 3–6.

52. In my conversation with Messrs. Aubry and Depaix, on June 14, 1979, they suggested that Solvay was not well disposed toward the University of Brussels because he identified it with a socialist and left-wing political orientation.

53. Nancy, Fac. Sci., PV (II), 21 October 1885.

54. Husson, p. 100.

55. Nancy, Fac. Sci., PV (II), 12 March 1885. Guntz did research mostly in thermochemistry and metallurgical and mineral chemistry. He died in 1935. See Marcel Delépine, "Notice nécrologique sur M. Antoine-Nicolas Guntz," CR, 201 (1935), 461–463.

56. Husson, p. 102.

57. "Rapport de M. Bichat, doyen de la Faculté des Sciences sur la situation et les travaux de la Faculté pendant l'année scolaire 1899–1900," p. 142.

58. Nancy, Fac. Sci., PV (II), 13 November 1893; Nancy, Fac. Sci., PV (III), 4 November 1899; Husson, pp. 100–102; and Nancy, Fac. Sci., PV (II), 22 July 1909.

59. Husson makes the claim, p. 102.

60. Nancy, Fac. Sci., PV (II), 20 February 1899.

61. See Michel Bauer and Elie Cohen, "Politiques d'enseignement et coalitions industrialo-universitaires. L'exemple de deux 'grandes écoles' de chimie, 1882–1976," Rev. Fr. Soc., 22 (1981), 183–203, pp. 190–191.

62. Bichat, "L'Enseignement des sciences appliquées," p. 303.

63. Husson, p. 104; and Aubry, "Les précurseurs," p. 5.

64. Husson, pp. 88, 96.

65. Ibid., p. 96.

66. Delsarte, p. 12.

67. Nancy, Fac. Sci., PV (I), p. 194.

68. The latter institute was supported by the patronage of the Comité de Forges de France, the Comité Central des Houillères de France, and various chambers of commerce. The Institute of Geology received major funding for construction from the Société Industrielle de l'Est; and the Agricultural and Colonial Institute received an annual subvention from the Indo-Chinese government. See Charles Adam, preface to *L'Université de Nancy (1572–1934)*, p. viii.

69. Nancy, Livret Etud., 1903–1904, p. 82.

70. Nancy, Rentrée Facs., 1904–1905, p. 113; ibid., 1889–1890, p. 4; and ibid., 1879–1880, p. 106. In 1903–04 the Sciences Faculty offered eighteen certificates, three of which could be combined for a licence; these were integral and differential calculus, rational mechanics, astronomy, superior analysis, superior algebra, superior geometry, general physics, applied physics, general chemistry, applied chemistry, mineralogy, zoology and general physiology, botany, geology, physical chemistry and electrochemistry, agricultural chemistry and geology, agricultural biology, agricultural zoology. In Nancy, Livret Etud., 1903–1904, pp. 84–97.

71. Ibid., 1896–1897, pp. 349–352.

72. Ibid., 1903–1904, pp. 180–181. The university authorities often mentioned the sports facilities as an important resource for foreign and French students: "Physical culture is an adjunct of intellectual culture." M. Auerbach, "Rapport annuel du conseil de l'Université de Nancy pour l'année scolaire 1912–1913," in Enquêtes Enseignt., no. 108 (Paris, 1914), 338–357, p. 347.

73. Nancy, Univ. Conseil, 1929–1930, pp. 1, 122; ibid., 1920–1921, pp. 125–126; and ibid., 1910–1911, pp. 42–45.

74. Gaston Floquet, "Rapport de M. Floquet, doyen" (1913–14), pp. 79–81.

75. Nancy, Fac. Sci., PV (II), 22 July 1909.

76. Husson, p. 94; Charles Adam, "Rapport," in Nancy, Univ. Conseil, 1925–1926, pp. 1–24, pp. 5–6; and Delsarte, pp. 10–11.

77. Adam, preface, p. vii.

78. Aubry, "Les précurseurs," p. 8; and Husson, p. 106.

79. Gain, L'Enseignement supérieur, p. 80.

80. Paris, Fac. Sci., PV Conseil, 10 February 1900, p. 306. Robert Fox, "Scientific Enterprise and the Patronage of Research in France, 1800–1870," Minerva, 9 (1973), 442–473; and Université de Nancy: Vingt-cinq années d'une université nouvelle 1897–1922 (Nancy: Coubé et Fils, 1923), p. 14.

81. Quoted in Vingt-cinq années, p. 10.

82. Fig. 1 is taken from Paul Petit, "Rapport du conseil de l'Université de Nancy," in Nancy, Rentrée Facs., 1912–1913, pp. 17–39.

83. See chap. 1.

84. Nancy, Fac. Sci., PV (III), 19 May 1913, 16 October 1913; and Gaston Floquet, "Rapport de M. Floquet, doyen" (1913–14), pp. 73–74.

85. Nancy, Fac. Sci., PV (III), 27 April 1909.

86. Bonnefont, *La Lorraine de 1900 à nos jours*, pp. 99–100; and *Vingt-cinq années*, pp. 23–24, 34.

87. See Paul Petit, "Rapport de M. Petit, doyen," in *Nancy, Univ. Conseil, 1920–1921*, pp. 115–152, on pp. 120–121, 129.

88. Nancy, Fac. Sci., PV (IV), 19 May 1925 and 5 June 1925.

89. See Charles Adam, "Rapport général sur la situation de l'Université, 1926–1927," in *Nancy, Univ. Conseil, 1926–1927*, pp. 1–12, on pp. 10–11. At this time, Nancy did not charge examination fees. The Faculty now proposed such fees, plus an increase in enrollment fees for university (not state) degrees from 60 to 240 francs.

90. *Nancy, Univ. Conseil, 1920–1921*, p. 129.

91. E. Husson, "Rapport du conseil de l'Université de Nancy," in *Nancy, Univ. Conseil, 1929–1930*, pp. 1–12, on pp. 2–3; and Nancy, Fac. Sci., PV (II), 23 June 1922; (IV), 19 May 1925 and 23 October 1925.

92. Adam, preface, p. vii.

93. Nancy, Fac. Sci., PV (IV), 20 May 1924 and 4 November 1924.

94. Ibid., 24 October 1928; and Aubry, "Les précurseurs," p. 9. Also Delsarte, p. 16. Maurice Letort, who directed the Chemical Institute from 1946–1956, also quit the Faculty for industrial research, taking on the directorship of the Laboratoire de Recherches des Charbonnages de France, according to Delsarte.

95. Nancy, Fac. Sci., PV (IV), 12 January 1926; and Husson, p. 98.

96. Husson, p. 118. In 1922 the Nancy Sciences Faculty assembly discussed the news that the Ecole Centrale in Paris had obtained permission for its concours to be considered equivalent to a prelicence certificate in general mathematics. Like other Faculties, Nancy's regarded this as a setback to the prerogatives of the university. In Nancy, Fac. Sci., PV (II), 21 February 1922.

97. Bauer and Cohen, "Politiques d'enseignement."

98. Delsarte, pp. 29–30.

99. Ibid., pp. 34–35.

100. Adam, preface, p. viii.

101. *Vingt-cinq années*, p. 13.

102. I am indebted to Aubry for calling the coat of arms to my attention.

103. Irving Langmuir, "Pathological Science" (Colloquium, Knolls Research Laboratory, 18 December 1953, ed. R. H. Hall), Langmuir Papers, Library of Congress, courtesy of Stanley Babb; Ian Firth, "N-Rays: Ghost of Scandal Past," *New Scientist*, 44 (1969), 642–643; Derek J. de Solla Price, *Science since Babylon* (New Haven: Yale University Press, 1961), pp. 153–160; Jean Rostand, *Error and Deception in Science* (New York: Basic Books, 1960); H. Piéron, "Grandeur et décadence des rayons N," *Année Psychologique*, 13 (1907), 143–169; G. F. Stradling, "A Resumé of the Literature of the N-Rays, the N' Rays, the Physiological Rays and the Heavy Emission," *Franklin Institute Journal*, 164 (1907), 57–74, 113–130, 177–199. On the rate of publication of N-ray papers, see Price, "Networks of Scientific Papers," *Science*, 149 (1965), 510–515.

104. For a more balanced view of the episode, see Jean Rosmorduc, "Une erreur scientifique au début du siècle: 'Les rayons N,'" *Revue d'Histoire des*

*Sciences*, 25 (1972), 13–25; Robert T. Lagemann, "New Light on Old Rays: N-Rays," *American Journal of Physics*, 45 (1977), 281–284; Spencer Weart, "A Little More Light on N-Rays," ibid., 46 (1978), 306; and, especially, Pierre Thuillier, "La triste histoire des rayons N," *La Recherche*, 9 (1978), 1093–1101, and Irving M. Klotz, "The N-Ray Affair," *Scientific American*, 242 (May 1980), 168–175.

105. Blondlot, "Sur la vitesse de propagation des rayons X," *CR*, 135 (1902), 666–670; "Sur l'égalité de la vitesse de propagation des rayons X et de la vitesse de la lumière dans l'air," ibid., 721–724; "Observations et expériences complémentaires relatives à la détermination de la vitesse des rayons X," ibid., 763–766. The first paper was translated by Erich Marx in *Phys. Zeit.*, 4 (1903), 310–314. On the invalidity of Blondlot's results, see Marx, "Die Geschwindigkeit der Röntgenstrahlen," ibid., 6 (1905), 768–777.

106. As Bruce Wheaton has pointed out, J. J. Thomson argued that the unidirectional cathode beam would give partially polarized X rays with an intensity in the line of the cathode beam twice that at right angles to it. Thomson, "Theory on the Connection between Cathode and Roentgen Rays," *Phil. Mag.*, 45 (1898), 172–183; Bruce Wheaton, *On the Nature of X and Gamma Rays: Attitudes toward Localization of Energy in the 'New Radiations,' 1896–1922* (Ph.D. diss., Princeton University, 1978), p. 47. See also Wheaton's *The Tiger and the Shark. Empirical Roots of Wave-Particle Dualism* (Cambridge: Cambridge University Press, 1983); and Blondlot, "Observation," *CR* (1902), 763–766.

107. Blondlot, "Sur la polarisation des rayons X," *CR*, 136 (1903), 284–286, reprinted in Blondlot, *Rayons 'N': recueil des communications faites à l'Académie des Sciences* (Paris, 1904), pp. 1–6; Cf. C. G. Abbott, "The N-Rays of M. Blondlot," *Smithsonian Institution, Annual Reports*, 58 (1903), 207–214.

108. Blondlot, "Action des rayons X sur de très petites étincelles électriques," *CR*, 134 (1902), 1559–1560. Blondlot's experiments bear a striking similarity to Hertz's investigations in 1887 of the influence of ultraviolet light on the sparking distance of an electric discharge. Heinrich Hertz, *Electric Waves* (New York, 1893), pp. 64–79.

109. Blondlot, "Sur une nouvelle espèce de lumière," *CR*, 136 (1903), 735–738, in *Rayons 'N'*, pp. 7–11.

110. Blondlot, "Sur des nouvelles sources de radiations," *CR*, 136 (1903), 1227–1229, in *Rayons 'N'*, pp. 16–21.

111. AN F$^{17}$ 22097 (René Blondlot); Léon Lecornu, "Nécrologie [de Blondlot]," *CR*, 191 (1930), 1033–1034; Henri Poincaré, "Rapport sur le prix Leconte, attribué à M. Blondlot," *RS*, 2 (1904), 843–844.

112. Blondlot, *Recherches expérimentales sur la capacité de polarisation voltaique* (Paris, 1881); AN F$^{17}$ 22097.

113. Blondlot, "Détermination de la vitesse de propagation des ondes électromagnétiques," *CR*, 113 (1891), 628–631; "Détermination de la vitesse de propagation d'une perturbation électrique le long d'un fil de cuivre," ibid., 117 (1893), 543–546; Blondlot and Gutton, "Vitesse de propagation des ondulations électromagnétiques," in *Congrès International de Physique, Rapports*, pp. 268–283, Ch. Ed. Guillaume and Lucien Poincaré, eds., 4 vols. (Paris, 1900–01).

114. AN F$^{17}$ 22097; Poincaré, "Rapport sur le prix Leconte," p. 844; and J. J.

Thomson, "Electric Waves," in *Encyclopedia Britannica*, 11th ed. (1910–11), on p. 208.

115. Blondlot, "Egalité des vitesses de propagation des rayons X et de la lumière dans l'air," *Archives des Sciences Physiques et Naturelles*, 15 (1903), 5–29, on pp. 25–29; *Rayons 'N'*, avertissement.

116. J. J. Thomson, "Roentgen Rays," in *Encyclopedia Britannica*, 11th ed. (1910–11); and Wheaton, *On the Nature of X and Gamma Rays*, pp. 45–48.

117. Blondlot, "Sur l'existence, dans les radiations émises par un bec Auer, de rayons traversant les métaux, le bois, etc.," *CR*, 136 (1903), 1120–1123, in *Rayons 'N'*, pp. 12–26.

118. "Lettre de M. H. Rubens," in *RS*, 2 (1904), 753–754. Rubens and Hagen, "Sur les rapports entre les qualités optiques et électriques des métaux," *Ann. Chim. Phys.*, 2 (1904), 185–214, p. 212, calculated that a metal plate 0.01 mm thick absorbs the infrared spectrum up to Hertzian waves.

119. Rubens, "Recherches sur le spectre infrarouge," *Rev Gén. Sci.*, 11 (1900), 7–13; cf. Hans Kangro, "Ultrarotstrahlung bis zur Grenze elektrisch erzeugter Wellen: Das Lebenswerk von Heinrich Rubens," *Annals of Science*, 26 (1970), 235–259, and 27 (1971), 165–170; also his "Heinrich Rubens," *DSB*, pp. 581–585.

120. Sagnac held that Blondlot's N-rays with refractive indexes 2.62, 2.436, and 2.29 were all diffracted images of the principal image of N-rays with index 2.942; Sagnac, "La longeur d'onde des rayons N déterminée par la diffraction," *CR*, 136 (1903), 1435–1437.

121. Blondlot, "Nouvelles sources," *CR* (1903).

122. Pierre Curie, "Radium," *RS*, 13 February 1904, translated in *Smithsonian Institution, Annual Report*, 58 (1903), 187–198.

123. Blondlot, "Sur une nouvelle action produite par les rayons N," *CR*, 137 (1903), 166–169, in *Rayons 'N'*, pp. 24–29; and "Nouvelles sources," *CR* (1903), p. 20.

124. Blondlot, "Nouvelles sources," p. 19; "Instruction pour confectionner des écrans phosphorescents propres à l'observation des rayons N," *Rayons 'N'*, pp. 73–76.

125. E. Bose, review of *Rayons 'N'* in *Phys. Zeit.*, 5 (1904), p. 560.

126. *Rayons 'N'*, pp. 11, 23; "Sur de nouvelles actions produites par les rayons N," *CR*, 137 (1903), 684–686, in *Rayons 'N'*, pp. 29–33; "Sur l'emmagasinement des rayons N par certain corps," *CR*, 137 (1903), 729–731, in *Rayons 'N'*, pp. 33–37.

127. Blondlot, "Sur le renforcement qu'éprouve l'action exercée sur l'oeil," *CR*, 137 (1903), 831–833, in *Rayons 'N'*, pp. 37–41; Curie, "Radium," pp. 194–195.

128. Blondlot, "Sur la propriété d'émettre des rayons N, que la compression confère à certains corps," *CR*, 137 (1903), 962–964, in *Rayons 'N'*, pp. 41–45.

129. The papers of Charpentier, Lambert, Meyer, and Gutton are summarized in *Nature*, 69 (1903–04), 182, 239–240, 264, 335, 359.

130. Arsène d'Arsonval, "Les radiations N," *Institut Général Psychologique, Bulletin*, 3 (1903), 25–30, on p. 28.

131. Discussion of the Groupe d'Etudes des Phénomènes Psychiques, in ibid., 4 (1904), 275–320, on pp. 276–282.

132. Ibid., 149–163, on p. 157.

133. D'Arsonval, "Les radiations N," p. 29.

134. Discussion of Groupe d'Etudes des Phénomènes Psychiques, p. 154.

135. Cf. Piéron, "Grandeur et décadence des rayons N," p. 144; memoirs of H. Baraduc, Carl Huter, and Darget, cited in CR, 138 (1904), 34, 189; d'Arsonval, "Remarques à propos des communications de M. A. Charpentier et des revendications de priorité auxquelles elles ont donné lieu," CR, 138 (1904), 884–885.

136. Rayons 'N', p. 32.

137. Cf. S. G. Brown and A. A. Campbell Swinton, letters of 23 January and 23 February 1904, Nature, 69 (1903–04), 296, 412, respectively; and the notes of H. Dufour, Piéron, and Perrin, 10 September, 29 October, and 12 November 1904, RS, 2 (1904), 338–339, 548, 622–623.

138. O. Lummer, "M. Blondlot's N-Ray Experiments," Nature, 69 (1904), 378–380; "N-Strahlen Diskussionen," Phys. Zeit., 5 (1904), 606–607; 674–677; cf. Stradling, "Résumé," p. 184.

139. Stradling, "Résumé," p. 183.

140. Burke and Schenck, letters of 8 February and 10 March 1904, Nature, 69 (1903–04), 486–487, respectively; M. Salvioni, "La question de la nature des rayons N," RS, 2 (1904), 73–78; ibid., 152–153, 658–659.

141. D'Arsonval, "Radiations N," p. 28.

142. Charpentier, La vision avec les diverses parties de la rétine (Ph.D. diss., University of Paris, 1877) and "Sur les phénomènes rétiniens," in Congrès International de Physique, Rapports, Vol. III, pp. 523–546.

143. Discussion of the Groupe d'Etudes des Phénomènes Psychiques, pp. 277–278.

144. Burke, letter of 21 June 1904, Nature, 70 (1904), 198; Langevin, "Enquête," RS, 2 (1904), 590–591; Blondlot, Rayons 'N', p. 5.

145. E. Rothé, "Essai d'une méthode photographique," CR, 138 (1904), 1589. Rothé later became a faculty member at Nancy. See "Les expériences de M. Bordier: prouvent elles l'existence des rayons N?" RS, 2 (1904), 783–785. D'Arsonval, who also supported Blondlot in this controversy, presented Bordier's note.

146. Blondlot, "Enregistrement, au moyen de la photographie, de l'action produite par les rayons N sur une petite étincelle électrique," CR, 138 (1904), 453–457, Rayons 'N', pp. 53–59.

147. William Seabrook, Doctor Wood (New York: Harcourt Brace, 1941), p. 237.

148. Blondlot to Director, Ministry of Public Instruction, 8 April 1903, AN F[17] 22097. Cf. Weart, "A Little More Light," p. 306.

149. "Déclaration de M. Blondlot," RS, 2 (1904), 620–622.

150. Nancy, Fac. Sci., PV (III), 19 March 1912.

151. Blondlot, "Sur la dispersion des rayons N et sur leur longeur d'onde," CR, 138 (1904), 125–129, in Rayons 'N', pp. 45–52. Blondlot further confused

matters by announcing in February 1904 the discovery of another new set of radiations N' mixed with N-rays, which diminished the brightness of N-ray detectors and decreased visual acuity. Their effect was noted by looking at a luminous surface tangentially rather than normally. See *Rayons 'N'*, pp. 59–62.

152. *Figaro*, 3 February 1904. This clipping, made by Gustave LeBon, is part of an extensive collection belonging to M. Pierre-Sadi Carnot and his sister, Lucie Carnot.

153. D'Arsonval, "Vibrations et radiations," *Institut Général Psychologique, Bulletin*, 4 (1904), 127–148, on pp. 142–143.

154. Schenck, letter of 10 March 1904, *Nature*, 69 (1903–04), 486–487.

155. "Opinion de M. Perrin," *RS*, 2 (1904), 622–623. Cf. Seabrook, *Doctor Wood*, p. 238.

156. Salvioni, "Question," pp. 76, 152–153.

157. *RS*, 2 (1904), 548, 590–591, 622–625, 656–657, 682, 719, 753.

158. Jean Becquerel likened N-rays to rays from radioactive substances and investigated N-ray sources for radioactivity. On Bouty, see d'Arsonval and Mascart, *RS*, 2 (1904), 591, 638.

159. J. Meyer, "Berichte aus den naturwissenschaftlichen Abteilungen der 76. Versammlung deutscher Naturforscher und Ärtze zu Breslau 1904," *Naturwissenschaftliche Rundschau*, 19 (1904), 569–570; "Discussion of N-Rays," *British Association for the Advancement of Science, Reports* (1904), 467–468.

160. "Les rayons-N existent-ils?" *RS*, 2 (1904), 545–552.

161. Seabrook, *Doctor Wood*, p. 236; "Opinion de M. Berget," *RS*, 2 (1904), 658; A. Berget, *La radium et les nouvelles radiations* (Paris, 1904); "Lettre de M. H. Rubens," *RS*, 2 (1904), 753–754.

162. Seabrook, *Doctor Wood*, p. 237.

163. Ibid., p. 238.

164. R. W. Wood, "La question de l'existence des rayons N," *RS*, 2 (1904), 536–538; cf. English original, "The N-Rays," *Nature*, 70 (1904), 530–531, and German translation, "Die N-Strahlen," *Phys. Zeit.*, 5 (1904), 789–791.

165. Moissan, "Enquête," p. 657.

166. "Déclaration de M. Blondlot," *RS*, 2 (1904), 620–622; "Déclaration de M. Julien Meyer," ibid., 718–719; Blondlot, "Nouvelles expériences sur l'enregistrement photographique de l'action que les rayons N exercent sur une petite étincelle électrique," *CR*, 139 (1904), 843–846, reprinted in *RS*, 2 (1904), 731–732; "Déclaration de M. E. Meyer," *RS*, 2 (1904), 753; Blondlot, "Nouvelles expériences sur l'enregistrement au moyen de la photographie de l'action exercée par les rayons N sur une étincelle électrique," *Rev. Gén. Sci.*, 16 (1905), 727.

167. Gutton, "Expériences photographiques sur l'action des rayons N sur une étincelle oscillante," *RS*, 5 (1906), 118–120; Mascart, "Sur les rayons N," ibid., 117–118, presented to the Academy of Sciences, January 15, 1906.

168. Aimé Cotton, "La question des rayons N," *Revue du Mois*, 1 (1906), 503–506; discussed in [J. D.], "A propos des rayons N," *RS*, 5 (1906), 376, and more recently in Rosmorduc, "Erreur," pp. 23–24; Jean Perrin, "A propos des rayons N," *Revue du Mois*, 1 (1906), 254–256.

169. A. Turpain, "Les rayons N et les expériments de contrôle," *RS*, 5 (1906), 491–495. Gutton's experiments were also questioned in a letter to *Nature* by A. Campbell Swinton, translated in *RS*, 5 (1906), 31.

170. *RS*, 6 (1906), 30–31, favored Turpain over Gutton. Cf. Emile Pierret, "Un moment de l'Ecole de physique de Nancy: les rayons N et N', réalités ou mirages," *Académie et Société Lorraines des Sciences, Bulletin*, 7 (1968), 240–257.

171. *RS*, 2 (1904), 548, 621, 537, 591. A typical plea for relevant details is found in H. Zahn, "Zu den Versuchen des Herrn Blondlot über N-Strahlen," *Phys. Zeit.*, 4 (1903), 868–870.

172. Alan Gauld, *The Founders of Psychical Research* (New York: Schocken, 1968), p. 140; David B. Wilson, "The Thought of Late Victorian Physicists: Oliver Lodge's Ethereal Body," *Victorian Studies*, 15 (1971), 29–48, on pp. 38–41.

173. *Vie Mondaine*, 6 January 1904; *Illustration*, 23 January 1904 (excerpts in Carnot Collection); C. G. Abbot, "N-Rays," p. 214; [A. R.], review of Charles de Reichenbach's *Les Phénomènes odiques*, trans. Ernest Lacoste (Paris, 1904), in *Annales des Sciences Psychiques*, 14 (1904), 186–187; Jules Regnault, "Phénomènes odiques nouvelles," ibid., 15 (1905), 167–177.

174. Joseph Maxwell, *Les phénomènes psychiques* (Paris, 1903), pp. 152–153.

175. *Illustration*, 23 January 1904, and an untitled clipping dated 19 May 1904 in the Carnot Collection; "Chronique scientifique: Les effluves humaines (rayons N)," *Le Nouvelliste de Bordeaux*, 12 (1904), preserved in Blondlot dossier at the Académie des Sciences.

176. Blondlot and J. Becquerel suggested that material particles ("émission pesante") might accompany N-rays in the ether. See Rosmorduc, "Erreur," and Lagemann, "New Light," p. 281.

177. Charles Adam, preface, *L'Université de Nancy (1572–1934)*, p. ix; Gaston Michel, "La Faculté de Médecine de Nancy," in ibid., 63–82; Dominique Barrucand, "L'Ecole de Nancy," in his *Histoire de l'hypnose en France* (Paris, 1967), pp. 102–134; Ernest Jones, *The Life and Work of Sigmund Freud*, 3 vols. (New York, 1953–1957), Vol. I, pp. 181–238.

178. John Ambrose Fleming, "Electrokinetics," *Encyclopedia Britannica*, 11th ed. (1910–1911), p. 216; Daniel Kevles, "Rowland," in *DSB*, 577–579; Victor Crémieu and Harold Pender, "On the Magnetic Effect of Electrical Convection," *Phil. Mag.*, 6 (1903), 442–444; *Nature*, 68 (28 May 1903); Wood, "Question," p. 538.

179. *RS*, 2 (1904), 552, 657; Seabrook, *Doctor Wood*, p. 239.

180. See Jacques Hadamard's report in Paris, Fac. Sci., PV Conseil, 21 and 24 January 1902, pp. 348–360. Also Harry W. Paul, *The Sorcerer's Apprentice: The French Scientist's Image of German Science, 1840–1919* (Gainesville: University of Florida Press, 1972).

181. Marjorie Malley, "The Discovery of the Beta Particle," *American Journal of Physics*, 39 (1971), 1454–1561, on p. 1459, n. 5. Ever anxious to claim priority, Le Bon said in 1903 that N-rays were identical with his "black light." In Blondlot, *Rayons 'N'*, p. 24.

182. Blondlot and Curie, *Sur un electromètre astatique pouvant servir comme wattmètre* (Nancy, 1889).

183. Jean Becquerel, *Notice sur les travaux scientifiques de M. Jean Becquerel* (Paris, 1934), pp. 89–90; also, *CR*, *139* (1904), 40–42, and ibid., 264–267.

184. Ernest Lavisse, *Questions d'enseignement national* (Paris, 1885), pp. 94–95, cited in William R. Keylor, *Academy and Community: The Foundation of the French Historical Profession* (Cambridge: Harvard University Press, 1975), p. 70.

185. On this group, see Mary Jo Nye, "Science and Socialism: The Case of Jean Perrin in the Third Republic," *French Historical Studies*, 9 (1975), 141–169.

186. These circumstances are explained in an unsigned letter to M. Lecornu, 27 November 1930, in the Archives of the Académie des Sciences.

187. Poincaré, "Rapport," pp. 843–844.

188. Undated note by Blondlot, Archives of the Académie des Sciences.

189. E. Bouty, *Radiations, électricité, ionisation*, 3d suppl. to J. Jamin, *Cours de physique de l'Ecole Polytechnique*, 4th rev. ed. (Paris, 1906), pp. 40–41.

190. Blondlot to Minister, 23 July 1909, AN F[17] 22097; Pierret, "Moment de l'Ecole de physique de Nancy," p. 254.

191. "Rapport du conseil de l'Université de Nancy pour l'année 1929–1930," in *Nancy, Univ. Conseil, 1929–1930*, pp. 1–12, on p. 5.

192. Gaston Floquet, *L'Exposition retrospective lorraine des sciences*, p. 18.

193. Blondlot's candidacy was supported by the dean of Nancy's Sciences Faculty and by Bichat, Jamin, and Berthelot. He was thought to be superior to the situation, but nonetheless eager to return home. Notes of 2, 8, and 12 February 1882, AN F[17] 22097. Also, Nancy, Fac. Sci., PV (II), 4 February 1882.

194. Blondlot to the Director, 29 December 1882, AN F[17] 22097. At this time, the Vosgien Jules Ferry was minister of public instruction.

195. My count, from Stradling, "Résumé," and Poggendorff's *Biographisch-Literarisches Handwörterbuch*.

196. On polywater, see the recent book by Felix Franks, *Polywater* (Cambridge: MIT Press, 1981). Before polywater studies ended in 1973, 500 publications related to polywater appeared (1963–1974), about half of them in research journals, reported by some 400 scientists. The polywater turned out to be dirty water—a solution of impurities in water.

197. Ezra Suleiman, *Elites in French Society: The Politics of Survival* (Princeton, N.J.: Princeton University Press, 1978), p. 4. According to Charles Adam, colleagues in the Academy of Sciences proposed to Blondlot, after the Academy instituted the class of nonresident members in 1913, that he allow his name to be brought up for membership. He declined. (In *Vingt-cinq années*, pp. 11–12.) An article in the newspaper *Le Temps* spread the rumor that Blondlot would be one of the first scientists elected to the new nonresident section. It described Blondlot as "the modest and conscientious scientist of Nancy, who already fifteen years ago, refused to pose his candidacy for a seat for Academy member, because he did not want to leave his province." On microfilm of *Le Temps*, 21 February 1913, from collection of Nobel Prize Archives, Royal Swedish Academy of Sciences, courtesy of Elisabeth Crawford and Wilhelm Odelberg.

## 3: GRENOBLE

1. The remains of the early settlement, Cularo, were discovered in 1974 when work was being done on an underground parking garage at the Maison du Tourisme in the rue de la République. See Vital Chomel, ed., *Histoire de Grenoble* (Toulouse: Privat, 1976), pp. 25–27.

2. *Université de Grenoble* (Grenoble: Dardelet et Cie, 1948), pp. 13–14; Robert Latouche, "L'Université de Grenoble au moyen âge et sous l'ancien régime," in *Université de Grenoble 1339–1939* (Grenoble: Imprimerie Allier Père et Fils, 1939), pp. 11–14; and W. A. B. Coolidge, "Grenoble," in *Encyclopedia Britannica*, 11th ed. (1910–11), pp. 579–580.

3. Latouche, pp. 15–20.

4. Chomel, p. 193; and Ed. Esmonin, "L'Université de Grenoble depuis la révolution," in *Université de Grenoble 1339–1939*, pp. 21–63, on pp. 21–22.

5. In the 1860s this new group held public meetings at which the Sciences Faculty physics professor Jean-Marie Seguin demonstrated cathode-ray tubes, and his colleague Valson discussed molecular action in capillary tubes. On the Académie and the Société de Statistiques des Sciences Naturelles et des Arts Industrielles du Département de l'Isère, see Jean-Guy Daigle, *La culture en partage: Grenoble et son élite au milieu du 19e siècle* (Presses Universitaires de Grenoble and Editions de l'Université d'Ottawa, 1977), pp. 128–154. Also Chomel, pp. 274–275.

6. Stendhal, *Life of Henri Brulard*, trans. Catherine Alison Phillips (New York: Alfred Knopf, 1939), pp. 125–126.

7. Ibid., p. 252.

8. Ibid., pp. 120, 29.

9. Ibid., pp. 266, 289.

10. Chomel, p. 248; Daigle, p. 26.

11. Chomel, pp. 254, 257.

12. AN F$^{17}$ 21574 (François Raoult).

13. See Chomel, pp. 278–279.

14. Ibid., p. 284.

15. Ed. Esmonin, pp. 45, 62; Grenoble, Univ. Conseil, 1928–1929, p. 8.

16. Chomel, pp. 263, 275. Fourier was prefect from 1802 to 1815.

17. *Université de Grenoble. Organisation de l'enseignement. Statistique des étudiants. Ressources scientifiques et matérielles* (Grenoble: Imprimerie Allier Frères, 1900), pp. 9, 14–17.

18. Daigle, p. 31.

19. It enrolled 105 students in 1804, 300 in 1848, and 700 in 1880. Chomel, p. 276.

20. Ed. Esmonin, pp. 32–33.

21. *Life of Henri Brulard*, p. 186.

22. Ibid., p. 247.

23. Ibid., p. 67.

24. From *L'Impartial Dauphinois*, 14 December 1862, quoted in Daigle, p. 112.

25. Quoted in Esmonin, p. 35.

26. J. Collet, "Notice historique," in "Centenaire de la Faculté des Sciences 1811–1911," pp. 260–279; in *Annales de l'Université de Grenoble*, 24 (1912), pp. 249–321, on p. 265.

27. At the public library, interest in science and the industrial arts increased from 8 percent of book demand in 1841 to 19 percent in 1877. In Daigle, pp. 117–119.

28. Quoted from *Barbe Bleue*, 13 April 1857, in Daigle, p. 114.

29. Grenoble, Fac. Sci., PV (I), 6 January 1848, p. 25*b*.

30. The assay laboratory was supported by some funds from the départemental general council. See P. Léon, *La naissance de la grande industrie en Dauphiné (fin du xviiᵉ siècle–1869)* (Paris: Presses Universitaires de France, 1954), pp. 485–501; Chomel, *Histoire de Grenoble*, p. 276; Daigle, pp. 155–156; Wilfrid Kilian, "Trois doyens de la Faculté des Sciences de Grenoble," in *Annales de l'Université de Grenoble* (1912), pp. 280–321, on pp. 283–287; and *Bulletin de la Société des Statistiques de l'Isère*, 5 (1876), 229–243, on Gueymard.

31. Kilian, "Trois doyens," p. 284.

32. Quoted in Daigle, p. 156.

33. Ministère de l'Instruction Publique et des Cultes, "Instruction sur l'exécution du règlement du 26 décembre 1854 relatif à l'enseignement des sciences appliquées" (Paris: Imprimerie Nationale, 1855), pp. 5–9, ADI, FdR.

34. Grenoble, Fac. Sci., PV (I), 26 August 1855, pp. 41*a*–42*b*, and documents transcribed on pp. 43*a*–46*b*.

35. J. Collet, "Notice historique," p. 267; and Grenoble, Fac. Sci., PV (I), 20 January 1880, p. 59*a*. Raoult, Joseph Carlet (botany and zoology), Jean Collet (pure mathematics), and Musset (botany) were the staff members affected.

36. AN F¹⁷ 21574, NI for 1875.

37. Collet, "Notice historique," p. 270.

38. "Rapport de M. Lory, doyen de la Faculté des Sciences," in *Grenoble, Rentrée Facs., 1880–1881*, pp. 24–31. And, circulars from the Ministry of Public Instruction to the rectors, dated 18 March 1880 and 21 January 1881, in ADI, FdR, no. 200.

39. *Université de Grenoble 1339–1939*, pp. 112–115.

40. Kilian, "Trois doyens," p. 290.

41. Ibid., pp. 289–299.

42. AN F¹⁷ 21574, RC for 1875.

43. Remarks of the rector, in Raoult's RC for 1899, AN F¹⁷ 21574.

44. Kilian, "Trois doyens," p. 308.

45. Grenoble, Fac. Sci., PV (I), 9 June 1891, pp. 70*a*–70*b*; and 4 July 1891, pp. 70*b*–71*a*. Paul Janet, who was chargé de cours at Grenoble since 1886, was appointed to the physics chair immediately after his thirtieth birthday in 1893. A graduate of the Ecole Normale Supérieure, his work was mathematical more than experimental, but he was an ardent supporter of industrial and applied science both at Grenoble and, later, at the Sorbonne. On his controversial report recommending the teaching of industrial design and practical mechanics at the Sorbonne, see Paris, Fac. Sci., PV Conseil, 15 May 1907, pp. 58–62. There is a discussion of his credentials in the Pièces-Annexes to ibid., pp. 86–87.

46. Kilian, "Trois doyens," p. 308; Grenoble, Fac. Sci., PV (II), 23 May 1900, pp. 46–47; 18 May 1901, pp. 62–64; 3 June 1902, pp. 94–95. And *Université de Grenoble 1339–1939*, p. 114.

47. Chomel, pp. 294–295. Keller was an engineer who, after the First World War, was president of the Grenoble Chamber of Commerce, president of the Comité Electrométallurgique de France, and vice president of the Chambre Syndicale des Forces Hydrauliques. See Chomel, p. 316.

48. Grenoble, Fac. Sci., PV (I), 16 November 1892, pp. 76*b*–77*b*.

49. The list is in Grenoble, Fac. Sci., PV (I), following minutes for 25 April 1896, unnumbered pages.

50. From an extract of the minutes of the 5 August 1892 meeting of the Grenoble Chamber of Commerce, in Grenoble, Fac. Sci., PV (I), following minutes for 25 April 1896, unnumbered pages.

51. M. Berger, "Rapport du conseil de l'Université de Grenoble pour l'année scolaire 1895–1896," in *Enquêtes Enseignt.*, no. 65 (Paris, 1897), pp. 106–118, on p. 110.

52. Grenoble, Fac. Sci., PV (II), 11 December 1897, p. 11.

53. *Université de Grenoble . . . Statistique* (1900), pp. 40–43. And *L'Institut Electrotechnique de l'Université de Grenoble* (Grenoble: Imprimerie Générale, 1908), pp. 3–7.

54. Quoted by Charles Petit-Dutaillis,"Allocution du recteur," in *Annales de l'Université de Grenoble* (1912), p. 259.

55. E. Boirac, "Discours du recteur," in *Grenoble, Rentrée Facs., 1900–1901*, pp. 5–28.

56. Grenoble, Fac. Sci., PV (II), 25 October 1904 and 1 December 1904, pp. 144–145.

57. In addition to these donations, the university borrowed 125,000 francs, the city contributed 200,000 francs, and the state 150,000 francs for construction and relocation costs for the Sciences Faculty. Of this total sum, 350,000 francs was designated for the Institut Polytechnique—Institut Brenier. See Pegoud, "Rapport du conseil de l'Université de Grenoble pour l'année scolaire 1906–1907," in *Enquêtes Enseignt*, no. 95 (Paris, 1908), pp. 135–156, p. 137. M. Michoud, "Rapport annuel du conseil de l'Université de Grenoble pour l'année scolaire 1907–1908," in ibid., no. 97 (Paris, 1909), pp. 149–184, on pp. 152–154, 161.

58. Michoud, p. 161. Barbillon held the title of director of the Institut Electrotechnique/Polytechnique from 1904 until after 1939. See Grenoble, Fac. Sci., PV (II), 25 October 1904, pp. 143–144.

59. Enrollment figures in ADI, FdR, no. 196.

60. *L'Institut Electrotechnique de l'Université de Grenoble* (1908), pp. 7–8, 11. In 1908–09, the curriculum was divided into elementary and higher divisions. The elementary division entrance examination included material corresponding to instruction in the Ecole Primaires Supérieures, the Ecoles Pratiques d'Industrie de Garçons, or the $1^{re}$ cycle of lycée classes. The higher division examination included material required for admission to the Ecole Centrale in Paris or the lycée classes in higher mathematics. See *Grenoble, Livret Etud., 1908–1909*, p. 109.

61. Pegoud, "Rapport annuel" (1906–1907), p. 140.

62. Chomel, p. 298.

63. "Rapport de M. Gosse, doyen de la Faculté des Sciences," in *Grenoble, Univ. Conseil, 1928–1929*, pp. 24–33, on p. 30. Chomel, p. 298.

64. P. 19 in A. Rosier, "Du chômage intellectual. De l'encombrement des professions libérales." An extract from *L'Enseignement Public* (Paris: Delagrave, 1934). Page reference is to manuscript of 78 pages, dated 31 March 1933, in ADI, FdR, no. 95.

65. Pegoud, p. 138; and Grenoble, Fac. Sci., PV (II), 30 October 1907, p. 200.

66. *Université de Grenoble 1339–1939*, pp. 112–115.

67. Institutes and programs were also associated with Ostwald in Leipzig, Arrhenius in Stockholm, and van't Hoff in Amsterdam. See Erwin N. Hiebert, "Nernst and Electrochemistry," in George Dubpernell and J. H. Westbrook, eds., *Proceedings of the Symposium on Selected Topics in the History of Electrochemistry* (Princeton, N.J.: Electrochemical Society, 1978), pp. 180–200.

68. Paris, Fac. Sci., PV Conseil, 23 June 1900, pp. 318–320.

69. Chomel, pp. 298–299. Also, letter from the rector Henri Guy, of the University of Grenoble, to the Conseiller Général, March 1924, ADI, FdR, no. 196. And information sheet for the period 1900–1924, signed by Emile Gau, and addressed to the academy rector, dated 13 March 1925, ADI, FdR, no. 196. On funds during the war, Grenoble, Fac. Sci., PV (II), 15 May 1915, p. 337; 2 December 1915, p. 340; 29 November 1917, p. 356; and *Grenoble, Univ. Conseil, 1919–1920*, p. 7.

70. Grenoble, Fac. Sci., PV (III), on 21 January 1919 and 3 November 1919, unnumbered pages; and 6 January 1921. The Ecole des Ingénieurs Hydrauliques became part of the Institut Polytechnique in 1928.

71. Grenoble, Fac. Sci., PV (III), 6 January 1921; and *Université de Grenoble 1339–1939*, p. 58.

72. Grenoble, Fac. Sci., PV (IV), 10 May 1930. All of the requirements are typewritten, with the French composition requirement penned in by hand.

73. Grenoble, Fac. Sci., PV (III), 27 January 1921.

74. Esmonin, pp. 53–54; and *Université de Grenoble . . . Statistique* (1900), pp. 31–35.

75. *Université de Grenoble . . . Statistique* (1900), p. 34; Chomel, p. 337; and Esmonin, p. 55. A Grenoble languages institute was established at Florence in 1911 and another at Naples in 1931. Esmonin, pp. 54–55.

76. *Université de Grenoble . . . Statistique* (1900), p. 35.

77. Petit-Dutaillis, "Allocution du recteur," p. 256; M. Balleydier, "Rapport annuel du conseil de l'Université de Grenoble pour l'année scolaire 1912–1913," pp. 223–253 in *Enquêtes Enseignt.*, no. 108 (Paris, 1914), pp. 243–244, 249–250. And enrollment figures for the Institut Polytechnique, ADI, FdR, no. 196.

78. See *Grenoble, Univ. Conseil, 1919–1920*, p. 4; and ibid., *1928–1929*, p. 34.

79. *Statistique Générale*, 45 (1929), p. 32.

80. Chomel, p. 338.

81. Petit-Dutaillis, p. 257.

82. "Université de Grenoble. Procès-verbal de la séance d'inauguration de

l'Institut Electrotechnique tenue sous la présidence de M. Liard, directeur de l'enseignement supérieur, le 11 mars 1901" (Grenoble: Imprimerie-Librarie de l'Université, 1901), p. 20.

83. Ibid., pp. 21–22.

84. Chomel, pp. 299–301.

85. Ibid., p. 420.

86. Ibid., pp. 438–440.

87. *Grenoble, Univ. Conseil, 1928–1929*, p. 35; and *Université de Grenoble . . . Statistique* (1900), p. 21. The first source shows that in the academic year 1928–29 Grenoble offered twelve certificates of higher studies in the sciences. The certificates, followed by numbers awarded in that year, were: higher mathematics (46), rational mechanics (18), geology and applied mineralogy (1), zoology (0), botany (7), general mathematics (98), industrial mechanics (10), general chemistry (72), general physics (22), electrotechnology (11), electrochemistry and metallurgy (31), and PCN (21).

88. Information supplied by Emile Gau to the rector, 13 March 1925, ADI, FdR, no. 196.

89. Grenoble, Fac. Sci., PV (I), 2 April 1892, pp. 72*b*–73*b*.

90. Ibid. (II), 7 June 1901, p. 67.

91. Ibid. (III), 21 January 1919.

92. Ibid. (III), 28 October 1920.

93. Ibid. (III), 12 November 1921; and (IV), 25 April 1925. For an obituary of Kilian, see *Annales de l'Université de Grenoble* (1925).

94. See "Notice historique sur les ressources botaniques dont dispose l'Université de Grenoble," 15 February 1925, in ADI, FdR, no. 196, 15 pp., on pp. 7–10.

95. Grenoble, Fac. Sci., PV (II), 15 November 1900, pp. 51–52; 9 February 1901, p. 59.

96. Including the Académie des Sciences. See "Notice historique sur les ressources botaniques," pp. 10–11.

97. In 1928–29, visitors came from Warsaw, Ghent, Holland, and Algiers. In *Grenoble, Univ. Conseil, 1928–1929*, p. 27.

98. Ibid., p. 7.

99. Grenoble, Fac. Sci., PV (II), 25 October 1904, pp. 143–144.

100. Ibid. (II), 17 May 1912, 292–293.

101. Ibid. (III), 27 June 1919.

102. See J. R. Partington, *A History of Chemistry*, Vol. IV (London: Macmillan, 1964), pp. 918–921; and George Kauffman, *Alfred Werner—Founder of Coordination Chemistry* (Berlin, 1966). And Nobel Prize Archives, Royal Swedish Academy of Sciences, courtesy of Elisabeth Crawford and Wilhelm Odelberg.

103. Pegoud, p. 153.

104. *Grenoble, Univ. Conseil, 1928–1929*, p. 25.

105. Grenoble, Fac. Sci., PV (III), 10 June 1922.

106. J. Collet, "Notice historique," p. 277.

107. Grenoble, Fac. Sci., PV (II), 26 April 1910, p. 257.

108. Ibid. (III), 3 November 1919.

109. In 1924, the University supplied funds for a new chair in general mathematics, hoping that the state would eventually take up the funding. The chair went to René Gosse. Grenoble, Fac. Sci., PV (IV), 2 July 1924.

110. Ibid., 2 February 1924 and 22 June 1924.

111. Ibid. (III), 23 May 1922. For a discussion of the curriculum of the Ecole Centrale des Arts et Manufactures in the first half of the nineteenth century, see John H. Weiss, *The Making of Technological Man: The Social Origins of French Engineering Education* (Cambridge: MIT Press, 1983), pp. 89–174.

112. Grenoble, Fac. Sci., PV (IV), 29 November 1930.

113. Jacques Hadamard, "L'Enseignement sous la IIIe république," in *L'Oeuvre de le troisième république*, Jean Benoit-Lévy, Gustave Cohen et al. (Montréal: L'Arbre, 1944), pp. 191–228. Fritz Ringer has discussed the strong commitment to Latin and classics in modern European education in his *Education and Society in Modern Europe* (Bloomington: University of Indiana Press, 1978). On mathematics and education, see Lewis Pyenson, *Neohumanism and the Persistence of Pure Mathematics in Wilhelmian Germany*, American Philosophical Society, Memoirs, vol. 150 (1984).

114. They were J. H. van't Hoff (Netherlands) in 1901; Svante A. Arrhenius (Sweden) in 1903; William Ramsay (Great Britain) in 1904; and Wilhelm Ostwald (Germany, originally from Latvia) in 1909. Henri Moissan, whose work was in electrochemistry, won the chemistry Nobel Prize in 1906.

115. J. H. van't Hoff, "Raoult Memorial Lecture," delivered 26 March 1902, *Jl. Chem. Soc.*, *81* (1902), 969–981, on p. 969.

116. Harry Clary Jones, *The Elements of Physical Chemistry* (London: Macmillan, 1902), pp. v–vi. Two of the best sources on the history of physical chemistry, its law and practitioners remain J. R. Partington, *History of Chemistry*, IV, 569–748; and Aaron J. Ihde, *The Development of Modern Chemistry* (New York: Harper and Row, 1964), pp. 391–417. Two excellent monographic sources are R. G. A. Dolby, "Debates over the Theory of Solution: A Study of Dissent in Physical Chemistry in the English-Speaking World in the Late Nineteenth and Early Twentieth Centuries," *HSPS*, 7 (1976), 297–404; and Robert S. Root-Bernstein, *The Ionists: Founding Physical Chemistry, 1872–1890* (Ph.D. diss., Princeton University, 1980). On discipline building, see Dolby's "The Case of Physical Chemistry," in *Perspectives on the Emergence of Scientific Disciplines*, Gerard Lemaine et al., eds. (Paris: Mouton, 1976), pp. 63–73.

117. François Raoult, "Causes des phénomènes d'endosmose électrique," *CR*, *36* (1853), 826–830, on p. 830. For Ramsay's remarks, William Ramsay, "Professor François-Marie Raoult," *Nature*, *64* (2 May 1901), 17–18. And van't Hoff, "Raoult Memorial Lecture," p. 970.

118. AN F$^{17}$ 21574. Cover sheet, NIs and RCs. A letter from Raoult to the minister, dated 18 September 1855, reports that Raoult has just received the licence at Paris.

119. Ibid., letter from Raoult to minister, dated 29 August 1862.

120. Ibid., RC for 1864.

121. Ibid., letter from the minister of public instruction to Raoult, dated 8 July 1864.

122. Ibid., letter from M. Monty, rector of the Académie de Dijon, to the minister of public instruction, dated 9 August 1863.

123. Ibid., NI for 1867.

124. Ibid., note by M. Faye on Raoult's nomination for Lory's replacement at Grenoble, in 1867.

125. Tragically, both died while still young, one in 1876 and the other in 1893, leaving Raoult one grandson.

126. AN F¹⁷ 21574, letter from Raoult to minister, dated 15 September 1869. He published two long papers during these years in the *Annales de Chimie et Physique:* "Forces électromotrices et quantités de chaleur dégagées dans les combinaisons chimiques," 2 (1864), 317–372; and "Recherches thermiques sur les voltamètres, et quantité de chaleur absorbée dans les décompositions électrochimiques," 4 (1865), 392–426. Also see "Sur les effets électriques produits par la dissolution des sels dans l'eau," *CR,* 69 (1869), 823–826.

127. Grenoble, Fac. Sci., PV (I), 18 June 1870, pp. 56*a*–56*b*.

128. AN F¹⁷ 21574. Notes from A. J. Balard to the minister of public instruction, undated; and from Jules Jamin, attached to the back of a letter from Raoult to the minister of public instruction, dated 21 February 1870, requesting his nomination to the chemistry chair at Grenoble.

129. Ibid., NI for 1875.

130. Ibid., cover sheet and exchange of letters between the Grenoble rector, M. Chappuis, and the Medical School's director, dated 21 June 1873 and 2 July 1873. Also, salaries in the NI for 1885.

131. Ibid., NIs for the Sciences Faculty and the Medical School in the 1880s. He also taught two hours of laboratory exercises at the Medical School in the 1870s.

132. Ibid.

133. Ibid., NI for 1890. And Frederick H. Getman, "François-Marie Raoult Master Cryoscopist," *Journal of Chemical Education,* 13 (1936), 153–155, on p. 155; and Boirac and Kilian, "Nécrologie. François Raoult," *Annales de l'Université de Grenoble,* 13 (1901), i–xvii, on p. iii.

134. Raoult set out this arrangement in a letter written to Marcellin Berthelot as inspector-general, dated 29 March 1881. Setting out the numbers of professors in different salaries and ranks in the old system, in comparison with the new system, he claimed that it was for those presently at the 8,000-franc level that competition would be stiffest for advancement into a new class and that advancement to the next class would take an average of seven years. In AN F¹⁷ 21574.

135. Ibid., RC of 1875, signed by C. Rollier.

136. Van't Hoff, "Raoult Memorial Lecture," p. 976. Raoult, "Sur les effets électriques produits par la dissociation des sels dans l'eau," *CR,* 69 (1869), 823–826. See the discussion in Partington, IV, pp. 693–694; and Harry C. Jones, *Elements of Physical Chemistry,* pp. 380–381.

137. On the interest of Catholic scientists in molecular processes, see Mary Jo Nye, "The Moral Freedom of Man and the Determinism of Nature: Catholic Science Viewed through the *Revue des Questions Scientifiques,*" *British Journal for*

*the History of Science*, 9 (1976), 274–292. Valson and Favre in *CR*, 75 (1872), 1000; and Valson in *Ann. Chim. Phys.*, 20 (1870), 361, and in *CR*, 74 (1872), 103; discussed in Wilhelm Ostwald, *Solutions*, trans. M. M. Pattison Muir from *Lehrbuch der allgemeinen Chemie*, 2d ed. (London: Longmans, Green and Co., 1891), p. 283.

138. See Sigalia Dostrovsky, "Jules Violle," in *DSB*, pp. 38–39. After he began teaching in Paris, Violle continued to spend some time both in Lyon and Grenoble. For about twenty years after 1884, Violle collaborated with the Lyonnais physicist Théophile Vautier in research on the propagation of sound. Some of their work was carried out along an underground cylindrical pipe built for the Grenoble water system which provided a path length of more than twelve kilometers.

139. Van't Hoff, "Raoult Memorial Lecture," p. 972.

140. See Raoult, "Absorption de l'ammoniaque par les dissolutions salines," *Ann. Chim. Phys.*, 1 (1874), 262–275; discussed in Ostwald, *Solutions*, pp. 26–28.

141. Ramsay, "Raoult," p. 17.

142. Wüllner, Poggendorff's *Annalen der Physik*, 103 (1858), 529; 105 (1858), 85; 110 (1860), 564; and Rudorff, *Annalen der Physik*, 114 (1861), 63; 145 (1871), 599.

143. Louis de Coppet, *Ann. Chim. Phys.*, 23 (1871), 366; 25 (1872), 502; and 26 (1872), 98. See Partington, IV, pp. 644–648, on the freezing point problem; pp. 648–655, on vapor pressure; and Ostwald, *Solutions*, pp. 156–245. Raoult's 1878 paper was "Sur la tension de vapeur et sur le point de congélation des solutions salines," *CR*, 87 (1878), 167–169.

144. See Berthelot, *Thermochimie, données et lois numériques*, 2 vols. (Paris: Gauthier-Villars, 1897), pp. 2, 162 in Vol. I; and see Dolby, "Debates over the Theory of Solution," pp. 301–303.

145. This is the suggestion of van't Hoff, "Raoult Memorial Lecture," on p. 973. Raoult, "Sur le point de congélation des liqueurs alcooliques," *CR*, 90 (1880), 865–868.

146. Raoult, "Loi de congélation des solutions aqueuses des matières organiques," *CR*, 94 (1882), 1517–1519.

147. Discussed in Partington, IV, p. 644, from de Coppet, *Ann. Chim. Phys.*, 23 (1871), 366; 25 (1872), 502; and 26 (1872), 539. See Ostwald, *Solutions*, p. 204.

148. Raoult, "Recherches sur la température de congélation des dissolutions," *Journal de Physique*, 3 (1882), 16–27.

149. Raoult, *CR*, 94 (1882), 1519.

150. Raoult, "Loi de congélation des solutions benzéniques des substances neutres," *CR*, 95 (1883), 187–189, on p. 189. For a general discussion of atomism and the Karlsruhe Congress, see the introductory essay in *The Question of the Atom: From the Karlsruhe Congress to the First Solvay Conference, 1860–1911. A Selection of Primary Sources*, Mary Jo Nye, ed. (Los Angeles: Erwin Tomash, 1984).

151. Van't Hoff, "Raoult Memorial Lecture," p. 975; Ostwald, *Solutions*, p. 168. And Raoult, "Loi générale de congélations des dissolvants," *CR*, 95 (1882), 1030–1033.

152. Raoult, ibid., pp. 1031–1032.

153. Ibid., p. 1032.

154. Ostwald, *Solutions*, p. 213. See Partington, IV, p. 646. In 1892, Raoult returned to what he regarded as the abnormal value of 18.5 for organic solutes in water, trying extreme dilutions to break down what he took to be dimerized molecules, but he found that the value remained 18.5. Partington, IV, p. 647.

155. See Marcellin Berthelot, "Réponse à la note de M. Wurtz," *CR, 84* (1877), 1189–1195, on pp. 1189, 1192; "Atomes et équivalents," *CR, 84* (1877), 1269–1276, on pp. 1273–1274. On Berthelot, see Mary Jo Nye, "Berthelot's Anti-Atomism: A 'Matter of Taste'?" *Annals of Science, 38* (1981), 585–590.

156. See Raoult, *CR, 95* (1882), p. 1030.

157. Raoult, *Journal de Physique, 3* (1882), p. 26.

158. Quoted by Partington, IV, p. 647, who cites Raoult, "Sur le point de congélation des dissolutions salines," *CR, 99* (1884), 324–326, on p. 326.

159. Raoult cites Favre and Valson, in *CR, 65,* p. 1000; and De Vries, *CR, 19* November 1883. In Raoult, *CR, 99* (1884), 324–326.

160. On de Coppet, see Raoult, "Influence de la dilution sur le coefficient d'abaissement du point de congélation des corps dissous dans l'eau," *CR, 100* (1885), 982–984, on p. 982.

161. Graph, p. 65, Raoult, "Recherches sur la température de congélation des dissolutions," *Journal de Physique, 5* (1886), 64–73; also, Raoult, *CR, 100,* 982–984.

162. Raoult, *CR, 100,* p. 984.

163. Ibid.; and the earlier paper, "Sur l'abaissement du point de congélation des dissolutions des sels alcalins," *CR, 98* (1884), 509–512, on p. 511 (also presented by Berthelot). The reference in *CR, 100,* 982–984, is to Berthelot's *Essai de mécanique chimique fondée sur la thermochimie,* chap. 8. Raoult emphasized that some of his results confirmed calorimeter studies by Berthelot, and he referred to Berthelot as the "illustre auteur de la mécanique chimique." In Raoult, "Les températures de congélations des dissolutions," conférence le 5 May 1886, in *Conférences faites à la Société Chimique de Paris en 1883–1884–1885–1886* (Paris: Bureaux des Deux Revues, 1886), pp. 139–163.

164. Van't Hoff, "Raoult Memorial Lecture," pp. 975–976.

165. Raoult, *Journal de Physique, 5,* pp. 72–73. And van't Hoff, "Raoult Memorial Lecture," p. 976.

166. Raoult, *Journal de Physique, 5,* pp. 72–73.

167. Raoult, in *Conférences faites à la Société Chimique,* pp. 158, 162.

168. His emphasis. Ibid., p. 154.

169. Quoted in Partington, IV, p. 647, from Raoult, "Tensions de vapeur des dissolutions," *Ann. Chim. Phys., 20* (1890), 295–369, on p. 355.

170. See Partington, IV, p. 658. Van't Hoff, who had studied in Wurtz's Paris laboratory in 1873, was already well known in France for his development simultaneously with Achille Le Bel of the notion of the tetrahedral valency distribution of carbon.

171. Cited in Root-Bernstein, *The Ionists,* p. 288.

172. See Dolby, "Debates over the Theory of Solution," pp. 304–305; and van't Hoff's paper, "The Function of Osmotic Pressure in the Analogy between Solutions and Gases," *Phil. Mag.*, 26 (1888), 81–105 (trans. William Ramsay). The 1885 paper, almost identical in content to the translation in the *Phil. Mag.*, was "Lois de l'équilibre chimique dans l'état dilué gazeux ou dissous," *Kongliga Svenska Vetenskaps-Academiens Handlingar [Stockholm]*, 21 (1884–85), no. 17, 1–58, presented 14 October 1885; also published in the *Zeit. phys. Chem.*, 1 (1887), 481–508. Also see Partington, IV, pp. 654–655.

173. Quoted in Partington, IV, p. 678. Van't Hoff's account is in *Phil. Mag.*, 26, pp. 98–99.

174. See Partington, IV, p. 675.

175. Svante Arrhenius, "The Theory of Electrolytic Dissociation," Faraday Lecture, delivered 25 May 1914, *Jl. Chem. Soc.*, 105 (1914), 1414–1426, on p. 1417.

176. On "Ionians," see ibid., pp. 1421–1422; and J. Vargas Eyre, *Henry Edward Armstrong* (London, 1958), pp. 220–221, quoted in Dolby, p. 397.

177. Quoted in van't Hoff, "Raoult Memorial Lecture," p. 977.

178. His publications in the journal were: Raoult, "Ueber die Dampfdrucke ätherischer Lösungen," *Zeit. phys. Chem.*, 2 (1888), 353–373; and "Ueber Präzisionskryoskopie; sowie einige Anwendung derselben auf wässerige auf Lösungen," ibid., 27 (1898), 617–661.

179. Cited in van't Hoff, "Raoult Memorial Lecture," p. 978.

180. For a description of his apparatus, and a comparison of his data with later data, see Getman, in *Journal of Chemical Education* (1936), pp. 153–155.

181. Regarding the newer ionist controversy, he said:

> *from the cryoscopic point of view*, all the dissolved particles (ions and whole molecules) are active; while from the viewpoint of electrical conductivity, the ions alone are active.

Raoult, "Les enseignements chimiques de la cryoscopie et de la tonométrie," *RS*, 14 (1900), 225–232, on p. 231. It should be noted that with the invention of apparatus by E. O. Beckmann for measuring elevations in boiling point of a solution, this became the preferred method over freezing point depressions after 1889. Beckmann, "Studien zur Praxis der Bestimmung des Molekulargewichts aus Dampfdruckerniedrigungen," *Zeit. phys. Chem.*, 4 (1889), 532–552.

182. On this point, see Ostwald, *Solutions*, p. 168.

183. See Dolby, "Debates over the Theory of Solution," pp. 393–404; and Mary Jo Nye, "The Nineteenth-Century Atomic Debates and the Dilemma of an 'Indifferent Hypothesis,' " *Studies in the History and Philosophy of Science*, 7 (1976), 245–268.

184. Armand Gautier, "Conférence faite à la demande du conseil de la Société," 17 May 1907, in *Centenaire de la Société Chimique de France (1857–1957)* (Paris: Masson, 1957), pp. 3–89, on p. 87.

185. On the relations between physical chemistry and the electrochemicals industry, see Martha Moore Trescott, *The Rise of the American Electrochemicals*

*Industry, 1880–1910: Studies in the American Technological Environment* (Westwood, Conn.: Greenwood Press, 1981); and Jeffrey L. Sturchio, *Chemists and Industry in Modern America: Studies in the Historical Application of Science Indicators* (Ph.D. diss., University of Pennsylvania, 1981), e.g., p. 204.

## 4: TOULOUSE

1. See "Toulouse," pp. 99–101, and "Universities," pp. 748–780, in *Encyclopedia Britannica*, 11th ed. (1910–11); and John Hine Mundy, *Liberty and Political Power in Toulouse 1050–1230* (New York: Columbia University Press, 1954), pp. 3–8. Parts of this chapter closely follow my previously published articles "The Scientific Periphery in France: A Study of the Faculty of Sciences at Toulouse (1880–1930)," *Minerva* 13 (1975), 374–403; and "Non-Conformity and Creativity," A Study of Paul Sabatier and the French Scientific Community," *Isis* 68 (1977), 375–391.

2. Paul Sabatier, "La Faculté des Sciences de Toulouse," in *Université de Toulouse. Bulletin*, 37 (1929), 190; and Camille-Georges Picavet, "Introduction," in *L'Université de Toulouse. Son passé—son présent* (Toulouse, 1929), pp. 3–26.

3. Sabatier, "La Faculté des Sciences de Toulouse," pp. 189–191.

4. See Matthew Arnold, *Works*, Vol. XII, *A French Eton. Higher School and Universities in France. Higher Schools in Germany* (London: Macmillan, 1904), pp. 7–12.

5. Antoine Prost, *Histoire de l'enseignement en France 1800–1967* (Paris: Armand Colin, 1968), map, p. 107.

6. Toulouse, *Univ. Conseil, 1887–1888*, pp. 9, 14; and Prost, p. 230.

7. See Louis Liard, *Universités et facultés* (Paris, 1890), p. 185; and Prost, p. 237.

8. See John M. Burney, *The University of Toulouse in the Nineteenth Century: Faculties and Students in Provincial France* (Ph.D. diss., University of Kansas, 1982), pp. 29–43.

9. See the front-page article by Jean Jaurès, *Dépêche, 21* (12 June 1890).

10. See Philippe Wolff, *Histoire de Toulouse* (Toulouse: Privat, 1958), p. 277; Burney, *University of Toulouse*, pp. 37, 42; and Madeleine Reberioux, "Jaurès et Toulouse (1890–1892)," *Annales du Midi*, 75 (1963), 295–310, on p. 300, and "Jean Jaurès élu municipal de Toulouse (1890–1893)," *Cahiers Internationaux*, no. 107 (1959), 69–75, on p. 70.

11. Harvey Goldberg, *The Life of Jean Jaurès* (Madison: University of Wisconsin Press, 1968), pp. 25–26; and Louis Rascol, *Claude-Marie Perroud* (Paris: H. Didier, 1941), pp. 149–155.

12. On the *Dépêche*, see Philippe Wolff, pp. 378–380, as well as Goldberg, p. 26; and Claude Bellanger, Pierre Albert et al., *Histoire générale de la presse française*, Vol. III, *De 1871 à 1940* (Paris, 1972), p. 400.

13. Goldberg, p. 62. According to Marcelle Auclair, *Jean Jaurès* (Paris, 1959), p. 82, Jaurès wrote Perroud after his electoral defeat in 1889 that he had seen

Liard, and "il est entendu qu j'aurai une chaire à Toulouse, j'en suis bien heureux."

14. Reberioux, in *Cahiers Internationaux*, pp. 72–73.

15. *Dépêche*, 21 (20 June 1890), p. 3; and Reberioux, in *Annales du Midi*, p. 306.

16. *Dépêche*, 21 (12 July 1890), p. 2; (21 July 1890), p. 2; and (2 August 1890), p. 2.

17. Ibid., 21 (21 July 1890), p. 2.

18. Ibid., 21 (12 June 1890), p. 1.

19. Ibid., 21 (4 June 1890), p. 1.

20. Prost, pp. 239–240.

21. Reberioux, in *Cahiers Internationaux*, pp. 72–73.

22. See Rascol, p. 245; and Robert Deltheil, Ernest Eslangon et al., *Benjamin Baillaud 1848–1934* (Toulouse: Privat, 1937), pp. 13–17, 27–28, 45. Edouard Privat, who headed the largest publishing house in Toulouse, was Baillaud's son-in-law. I have not been able to consult René Baillaud's autobiography, *Souvenirs d'un jeune Toulousain (1885–1907)* and *Souvenirs: parents et amis de Toulouse et de l'Aveyron, d'un siècle à l'autre* (Rodez: Carrere, 1976 and 1978).

23. Paul Sabatier, "La Faculté des Sciences de Toulouse" (1929), p. 200.

24. Baillaud's opinions of the Faculty were apparently confided to Perroud. See Rascol, pp. 246–247.

25. See Paul Sabatier, "La Faculté des Sciences de Toulouse" (1929), pp. 200–209; and *Toulouse, Univ. Conseil, 1907–1908*, p. 103. On Stieltjes, see Rascol, p. 247; and *Toulouse, Univ. Conseil, 1894–1895*, p. 4. Jules Drach (at Toulouse from 1908–1918) and a later addition to the Faculty, Sabatier's pupil Alphonse Mailhe (1922–1926), also completed their academic careers at Paris. Members of the Faculty remaining at Toulouse throughout their careers who became corresponding members of the Academy included Mathieu Leclerc du Sablon (who was dean between the tenures of Baillaud and Sabatier), Louis Roy, and Emile Mathias. Mathias was a candidate for a Sorbonne lectureship in the early 1890s and was judged at Paris to have "une grande habilité expérimentale" and "un sens critique très développé." See Paris, Fac. Sci., PV Conseil, Pièces Annexes (1883–1909), pp. 86, 104. In 1900, Mathias was considered for a post at the Ecole Normale Supérieure, but lost the competition to his Toulouse colleague Aimé Cotton. See Jean Rosmorduc, *Aimé Cotton (1869–1951). Le savant, l'homme, le citoyen* (Ecole Pratique des Hautes Etudes, thèse du doctorat de 3ᵉ cycle, 1971), p. 281. I would like to thank Spencer Weart for directing my attention to this source.

26. Letters from Benjamin Baillaud to Claude-Marie Perroud and to the ministry, 24 January 1882. AN F$^{17}$ 24106 (Paul Sabatier). Regarding the doleful state of provincial laboratories, see Robert Fox, "Scientific Enterprise and the Patronage of Research in France, 1800–1870," *Minerva*, 11 (1973), 442–473, on p. 462.

27. Toulouse, Fac. Sci., PV Assem., 10 January 1891; and Rascol, p. 246.

28. Baillaud's report is in *Toulouse, Univ. Conseil, 1892–1893*, pp. 31–32.

29. Rosmorduc, p. 115.

30. Toulouse, Fac. Sci., PV Assem., 27 May 1886.

31. *Toulouse, Univ. Conseil, 1905–1906*, pp. 117–118.

32. Prost (p. 234) reports 503 professors in 1880, 650 in 1890, and 1,048 in 1909. The Toulouse figures are from a report entitled "Faculté des Sciences de Toulouse, Etat comparatif du personnel en 1903 et en 1940," Toulouse, Fac. Sci., PV Assem., 19 July 1941. In *Benjamin Baillaud*, it is said that there were twenty chairs in 1893 (p. 15).

33. *Toulouse, Univ. Conseil, 1892–1893*, pp. 11–12.

34. Ibid., *1895–1896*, p. 6. State support for all universities, including personnel, increased only 10 percent from 1887–1900. Matériel funds decreased by 30 percent, and then further decreased 6 percent from 1900 to 1910. See Paul Forman, John Heilbron, and Spencer Weart, *Physics circa 1900: Personnel, Funding, and Productivity of the Academic Establishments*, HSPS, 5 (1975), p. 66.

35. *Toulouse, Univ. Conseil, 1895–1896*, p. 6.

36. *Toulouse, Livret Etud., 1894–1895*, p. 9, and ibid., *1899–1900*, p. 10; and *Toulouse, Univ. Conseil, 1900–1901*, p. 11. For a general discussion of gifts and legacies to French science through 1905, see François Picavet, "Dons, donations et legs," *RIE*, 49 (1905), 487–513, and 50 (1905), 22–48.

37. When not about the budget or the curriculum, the meetings at Toulouse were largely devoted to giving out or extending bourses, approving books for the library, or drawing up tables of juries for examination of candidates for diverse degrees.

38. Paul Sabatier, "La Faculté des Sciences et ses instituts techniques en 1910," pp. 166–193, in Vol. I, *Documents sur Toulouse et sa région*, 2 vols. (Toulouse: Privat, 1910), p. 167; and *Toulouse, Univ. Conseil, 1882–1883*, p. 55.

39. "Les Facultés ne sont plus seulement une juxtaposition de chaires savantes, elles deviennent, de plus en plus, des écoles normales secondaires." Quoted in Rascol, pp. 149–150.

40. *Toulouse, Univ. Conseil, 1882–1883*, pp. 54–55.

41. Ibid., *1887–1888*, pp. 75–76.

42. Ibid., *1886–1887*, p. 83; and *1887–1888*, p. 75. The dismay about poor performances in mathematics on the baccalaureate was one shared at Paris. See Toulouse, Fac. Sci., PV Assem., 28 January 1896; and Paris, Fac. Sci., PV Conseil, 2 July 1903, pp. 15–16.

43. Toulouse, Fac. Sci., PV Assem., 27 January 1894.

44. On Sabatier's entrétien, see ibid., 29 April 1893. This was probably not a formal seminar. In an interview with Thomas Kuhn and Théodore Kahan, Edmond Bauer commented that the lack of seminars in France was one of the major factors isolating students and professors from each other. Paul Langevin began a seminar in Paris around 1900, and according to Bauer, it was the only one in France for many years. Interview of T. S. Kuhn and Théodore Kahan with Edmond Bauer, 8 January 1963, no. 4 of 5 sessions, Archives for the History of Quantum Physics. Depository at the University of California at Berkeley.

45. Robert Gilpin comments on this practice, citing Jean Perrin, according to Ferdinand Lot, *Jean Perrin et les atomes* (Paris, 1963), pp. 162–167. In Robert Gilpin, *France in the Age of the Scientific State* (Princeton, N.J.: Princeton University Press, 1968), pp. 96–97.

46. The details included the sum of 300 francs for Sabatier's chemistry laboratory as expenses for students of the licence and agrégation, whereas Sabatier had already insisted upon 1,000–1,200 francs as an absolute minimum. See Toulouse, Fac. Sci., PV Conseil, 5 February 1892, 19 February 1892, and 26 February 1892.

47. Ibid., 26 February 1892. Ten people were present at the council meeting, including Baillaud, Berson (who objected to the report, but did not leave with Sabatier), Leclerc du Sablon, and Stieltjes.

48. The chargés de cours and lecturers received an additional 500 francs. New credits were to be given to faculty members when they had exhausted their original funds, and these funds were to be assigned by the council. See Toulouse, Fac. Sci., PV Conseil, 13 January 1894.

49. See Robert Deltheil, *L'Université de Toulouse et son rôle régional* (Toulouse: Privat, 1941), p. 12. On Baillaud, Rascol, p. 248. Sabatier himself said that the orientation of the Faculty changed because of the creation of the PCN and the turning of higher scientific education toward applied sciences (in "La Faculté des Sciences" [1910], p. 167).

50. To be sure, he did not mean to eulogize only applied science. He mentioned that it is the same men who are occupied "on the one hand with the most abstract parts of science, and on the other hand, with industrial and agricultural questions . . . if applications are the goal sought, the sciences are always the necessary means." *Toulouse, Univ. Conseil, 1903–1904*, pp. 122–123.

51. Gilpin, p. 357.

52. Lot, "De la situation faite à l'enseignement supérieur en France," *Cahiers de la Quinzaine*, 9 (1906), 1–107; and 11 (1906), 109–237, pp. 15, 55, 64.

53. According to Fox, the best-housed Sciences Faculties in the nineteenth century were at Caen, Nancy, and Clermont-Ferrand. Strasbourg was said to be the best staffed (pp. 461, 468). See Lot, p. 16; J. Gosselet, "L'Enseignement des sciences appliquées dans les universités," *RIE*, 37 (1899), 97–106, on p. 98; and Paul Appell, "L'Enseignement supérieur des sciences," *Rev. Gén. Sci.*, 15 (1904), 287–299, on pp. 293–294.

54. AN F[17] 24106, NIs and RCs for 1882 and 1887.

55. Sabatier, "La Faculté des Sciences" (1929), pp. 192–193; and Marie Edward Joseph Cathala, in *Centenaire. Paul Sabatier. Prix Nobel, Membre de l'Institut. 1854–1954* (Toulouse: Privat, 1954), unnumbered pages.

56. Sabatier had been Berthelot's aid at the Collège de France.

57. *Toulouse, Livret Etud., 1899–1900*, p. 94; and *Toulouse, Univ. Conseil, 1904–1905*, pp. 120–121.

58. Toulouse, Fac. Sci., PV Assem., 12 May 1886 and 6 January 1887.

59. Ibid., 24 April 1888.

60. Toulouse, Fac. Sci., PV Conseil, 16 March 1888; *Toulouse, Univ. Conseil, 1887–1888*, p. 77; and *1888–1889*, p. 80.

61. Toulouse, Fac. Sci., PV Conseil, 7 February 1900, 13 June 1900, and 10 February 1904.

62. Ibid., 18 July 1906; and Toulouse, Fac. Sci., PV Assem., 28 November 1906.

63. See articles or speeches in the *Journal d'Agriculture Pratique* in 1892, 1893, 1894, 1896, 1898, 1902, 1903, 1906, and 1910. Also, "La fixation de l'azote atmosphérique sur les plantes et sur le sol," *Rev. Gén. Sci.*, 4 (1893), 135–143; "Sur l'emploi des engrais," *Bulletin du Syndicat Agricole de la Haute Garonne*, 1896, p. 11ff.; *Leçons élémentaires de chimie agricole* (1889). In industrial chemistry, articles in *Revue de Chimie Pure et Appliquée*, 1905, 1914; *Zeitschrift für Eis-und Kälte Industrie*, 1913; *Le Gaz*, 1914; and *Chemistry and Industry*, 1927.

64. Lyon's Ecole de Chimie, founded in 1883, was the oldest of this type of institution. The institute at Nancy had been founded in 1889. (See chaps. 2 and 4.) Paris' Institut de Chimie Industrielle was new, although Paris also housed the Ecole Municipale de Physique et de Chimie, established by the city in 1882. See Toulouse, Fac. Sci., PV Assem., 23 May 1906.

65. See Paul Sabatier, "Institut de Chimie," in *L'Université de Toulouse. Son passé—son présent*, pp. 295–299. Also, Toulouse, Fac. Sci., PV Conseil, 13 December 1911.

66. P. Huc, "Le Professeur Paul Sabatier est mort . . . ," supplement technique de *La Halle aux Cuirs*. A photocopy of this article was given to me by Mlle. Parlange. And Toulouse, Fac. Sci., PV Conseil, 13 November 1912.

67. In 1904–05, 35 people enrolled in Bouasse's class. A poll of their intended vocations revealed telegraphists, civil engineering workers, and employees in electrical factories. *Toulouse, Univ. Conseil, 1904–1905*, p. 121.

68. See Deltheil, *L'Université de Toulouse* (1941), p. 17; and *Toulouse, Univ. Conseil, 1905–1906*, pp. 113–114. On the mayor's proposal, see Rascol, pp. 345–357; and Toulouse, Fac. Sci., PV Conseil, 18 July 1906, and PV Assem., 28 November 1906.

69. Toulouse, Fac. Sci., PV Assem., 6 January 1909.

70. *Toulouse, Univ. Conseil, 1905–1906*, p. 115.

71. Toulouse, Fac. Sci., PV Assem., 26 May 1909; Dupouy, in *Centenaire Paul Sabatier*, unnumbered pp.; and Paul Sabatier, "Institut Agricole," in *L'Université de Toulouse* (1929), p. 313. According to the minutes, the General Council of the Haute-Garonne was prodded by the minister of agriculture to vote an annual subvention of 2,000 francs. Other local groups voted a total annual subsidy of 1,000 francs.

72. Toulouse, Fac. Sci., PV Conseil, 25 November 1914, and PV Assem., 9 May 1924.

73. Toulouse, Fac. Sci., PV Assem., 8 December 1915. Some critics charged that the relationship of the Sciences Faculties to their institutes established a provincial cumul.

74. On enrollment, see *Toulouse, Univ. Conseil, 1928–1929*; on degrees and diplomas, see C. Richard, *L'Enseignement en France*, Vol. III (Paris, 1925), pp. 443–444.

75. According to Deltheil, in *L'Université de Toulouse* (1941), p. 4, the first chair of French law was established at Toulouse in 1679. See *Toulouse, Univ. Conseil, 1886–1887*, p. 7; ibid., *1905–1906*, p. 9; and *L'Université de Toulouse, Son passé—son présent*, p. 25. The figures for total enrollment and for enrollments in law, sciences, letters, and medicine and pharmacy are, respectively, 1886: 1,218, 777, 110, 110, 167; 1905–06: 2,856, 1,685, 340, 282, 549 (with some students apparent-

ly listed twice for science and medicine and pharmacy); and 1927–28: 3,524, 694, 1,401, 561, 868.

76. See table below, which has been compiled from data in *Statistique Générale, 16* (1895–96), p. 156; *30* (1910), p. 48; and *45* (1929), p. 32.

SCIENCES FACULTIES WITH LARGEST ENROLLMENTS AND FOREIGN
STUDENTS IN RELATION TO TOTAL ENROLLMENTS

| Year | | 1894–95 | 1901–02 | 1910–11 | 1928–29 |
|---|---|---|---|---|---|
| France | Total | 2,307 | 4,107 | 6,096 | 14,690 |
| | Foreign | 101 | 294 | 1,193 | 3,793 |
| Toulouse | Total | 137 | 236 | 648 | 1,479 |
| | Foreign | — | 4 | 197 | 816 |
| Bordeaux | Total | 200 | 262 | 301 | 598 |
| | Foreign | — | 6 | — | — |
| Grenoble | Total | 69 | 118 | 416 | 1,073 |
| | Foreign | — | 5 | 65 | 479 |
| Lille | Total | 129 | 155 | 335 | 792 |
| | Foreign | — | — | — | — |
| Lyon | Total | 209 | 515 | 530 | 925 |
| | Foreign | — | 20 | — | — |
| Marseilles | Total | — | 196 | 287 | — |
| | Foreign | — | 6 | — | — |
| Montpellier | Total | 96 | 209 | 234 | 815 |
| | Foreign | — | 46 | 71 | 341 |
| Nancy | Total | 166 | 329 | 751 | 1,140 |
| | Foreign | — | 38 | 331 | 542 |
| Paris | Total | 776 | 1,317 | 1,675 | 4,259 |
| | Foreign | 80 | 166 | 429 | 996 |

77. *Toulouse, Univ. Conseil, 1928–1928,* p. 80.

78. See *Statistique Générale, 18* (1898), p. 527; *32* (1912), p. 36; *44* (1928), p. 32; and *46* (1930), p. 42, for these figures. Also, *Toulouse, Univ. Conseil, 1928–1929,* pp. 78–79. In *Statistique Générale, 38* (1922), pp. 30–31, the science licences conferred throughout France from 1880 to 1922 are cataloged.

79. *Statistique Générale*, *45* (1929), pp. 32–33, and *46* (1930), p. 42.

80. See table to n. 76. According to Spencer Weart, the distinguished physicist Lew Kowarski noted that many Eastern Europeans were unsure of their ability to compete successfully in Paris for academic laurels.

81. See *Toulouse, Univ. Conseil, 1909–1910*, pp. 15–18, 19–20.

82. On the Institut de Chimie, see Paul Sabatier, "Institut de Chimie," in *L'Université de Toulouse: Son passé—son présent* (1929), p. 299. In his dean's report (*Toulouse, Univ. Conseil, 1928–1929*, pp. 72–73), Sabatier records 194 Polish, 143 Rumanian, 131 Bulgarian, 85 Russian, 41 Lithuanian, 36 Spanish, and other foreign students.

83. Toulouse, Fac. Sci., PV Assem., 26 March 1919, and PV Conseil, 16 March 1917. According to *Toulouse, Univ. Conseil, 1929–1930*, p. 16, Walter B. Cannon of Harvard University visited Toulouse in 1930 to lecture on his work on the nervous system.

84. Quoted in Charles Camichel, "Eloge de M. Paul Sabatier lu en séance publique le 30 mai 1943," *Académie des Jeux Floraux* (Toulouse, 1943), pp. 103–129. A copy of this paper was given to me by Mlle. Parlange.

85. According to Reberioux, *Annales du Midi*, p. 300, the population in 1886 was 147,617, and that of 1894, 149,963. Sabatier comments on the agricultural situation in "Allocution prononcée à la séance publique du dimanche 30 décembre 1906," *Journal d'Agriculture Pratique . . . du Midi*, *102* (1906), 585–591, on p. 591. Quotation on p. 587.

86. On the Fifth Plan and planning, see Gilpin, pp. 237, 314–315.

87. Ibid., p. 351. The four-page pamphlet "Université Paul Sabatier" (Toulouse, 1972) is a useful brief source on the modern Sciences Faculty. Also *Toulouse, Univ. Conseil, 1966–1967*, pp. 739–740, for a complete listing of the Faculties and institutes associated with the university.

88. Toulouse, Fac. Sci., PV Assem., 6 November 1918.

89. Jean-Baptiste Nadal, in *Centenaire: Paul Sabatier*, unnumbered pp.

90. See Toulouse, Fac. Sci., PV Conseil, 8 January 1908.

91. Ibid., 8 December 1915, quoted 28 June 1916; also 16 March 1917, 31 October 1917, and 11 December 1918. The Cie. Française de Métaux contributed 3,000 francs for research, the Tréfileries et Laminoirs du Hauvre 10,000 francs for studies in hydrodynamics, and the Compagnie des Chemin de Fer du Midi monies for scholarships and an annual subvention of 3,000 francs.

92. For example, see Paris, Fac. Sci., PV Conseil, 17 February 1897, p. 226, and 21 January 1902, p. 348. According to Henri Maillart, Gabriel Lippmann's budget for his physics laboratory was 13,000 francs before the war and 13,000 francs after the war, although the costs of heating had increased from 1,800 francs to 6,000 francs. See Henri Maillart, *L'Enseignement supérieure* (Paris, 1925), p. 73.

93. This probably could not have taken place at Paris. In the early 1890s, the ministry designated the dean as manager of expenses, but Dean Darboux proposed that the Sorbonne Faculty elect a commission of finances. He assured the faculty that he would take part in the work of the commission. See Paris, Fac. Sci., PV Conseil, 11 December 1890, pp. 157–158.

94. Toulouse, Fac. Sci., PV Conseil, 24 November 1926, 30 March 1927, and 7 October 1927.

95. After placing first in the concours of 1902, Jacob studied Alpine geology at Grenoble under the direction of Wilfrid Kilian. He became professor at Toulouse in 1912, where he worked on the geology of the Pyrenees. Jacob took a leave of absence from 1918–1922 for research in Indochina. He was elected to the Academy of Sciences in 1931 and later became director of the CNRS. As a member of the Academy he initiated reforms to abolish differences between its members. See Jean Sarrailh, Jean Cabannes, Louis de Broglie et al., "Allocutions prononcées à l'occasion du jubilé scientifique de Charles Jacob" (Paris, 1950).

96. Toulouse, Fac. Sci., PV Assem., 8 June 1921, and PV Conseil, 30 March 1927.

97. According to Prost, p. 234, the total faculty in France increased only from 1,048 in 1909 to 1,145 in 1930. In 1909, the Sciences Faculty assembly at Toulouse included 18 members, whereas in 1929 there were 24.

98. "Faculté des Sciences de Toulouse—état comparatif du personnel en 1903 et en 1940," Toulouse, Fac. Sci., PV Assem., 19 July 1941.

99. This information is from the Academy's *Annuaire pour 1930* and *J. C. Poggendorfs biographisch-literarisches Handwörterbuch*, 1904 and 1931.

100. See the remarks of Jean Langevin, Léon Brillouin, and Edmond Bauer in interview with T. S. Kuhn and others. Interview of T. S. Kuhn and Théodore Kahan with Jean Langevin, January 1963, no. 1, pp. 7a and 8. Interview of P. P. Ewald, G. Uhlenbeck, T. S. Kuhn, and Mrs. Ewald with Léon Brillouin, 29 March 1962, no. 1 of 2 sessions, p. 8; and interview of T. S. Kuhn and Théodore Kahan with Edmond Bauer, 14 January 1963, no. 4 of 5 sessions, p. 2. Archives for the History of Quantum Physics, University of California at Berkeley depository.

In my June 14, 1978, conversation with Professor Aubry of Nancy, he mentioned that there were very hard feelings between Sabatier and Bouasse, and that Bouasse expressed contempt for chemists by comparison to physicists, claiming that chemists were only making la cuisine. Insulting comments about Baillaud and Sabatier are to be found (pp. 12, 57) in Bouasse's published critique of the baccalaureate system, *Bachot et bachotage, étude sur l'enseignement en France* (Toulouse, 1910). Bouasse attacked technical institutes as purveyors of pay supplements to faculty salaries, predicting correctly that the institutes were the wave of the future (pp. 190, 204–205). For the remarks about the Academy and the 1911 Toulouse meeting, see Toulouse, Fac. Sci., PV Conseil, 1 February 1911.

101. Mary Jo Nye, "Science and Socialism: The Case of Jean Perrin in the Third Republic," *French Historical Studies*, 9 (1975), 141–169.

102. According to Lucien Babonneau, "Paul Sabatier," in *Génies occitans de la science* (Toulouse: Privat, 1947), pp. 167–189, p. 186, Sabatier was a "Christian and a believer who considered man but a reflection of God."

103. See the NIs and the observations in the RCs by the inspector, dean, and rector for 1884, 1885, and esp. 1886. AN F$^{17}$ 24106 (Paul Sabatier).

104. Letter from Rector Claude-Marie Perroud to M. le Directeur [perhaps of the Paris Observatory], dated 15 October 1907, AN F$^{17}$ 23735 (Benjamin Baillaud).

105. See the rectors' observations in the NIs for 1910, 1911 (Rector Jeanmarie), and 1913 (Rector Lapie). AN F$^{17}$ 24106.

106. On this pattern of careers, see Prost, p. 76. Théodore Sabatier became a physics professor at the Carcassonne lycée in the 1870s. AN F$^{17}$ 24106, and Cathala, in Centenaire: Paul Sabatier, unnumbered pp.

107. For general biographical information about Paul Sabatier, see Charles Camichel's speeches in Centenaire: Paul Sabatier and in Gabriel Bertrand, Charles Camichel, and Gaston Dupouy, Ceremonies de centenaire de la naissance de Paul Sabatier à Toulouse, les 4 et 5 novembre 1954 (Hendaye, for the Académie des Sciences, 1954). Also, Camichel's "Eloge de M. Paul Sabatier lu en séance publique le 30 mai 1943," Académie des Jeux Floraux (Toulouse, 1943), pp. 103–129; and Hyacinthe Vincent, "Nécrologie," CR, 213 (1941), 281–283, on p. 281.

108. Letter from Marcellin Berthelot to the minister of higher education, 12 April 1878, and letter from Bersot to the minister of higher education, 14 April 1878, AN F$^{17}$ 24106. Also François Tresserre, "Réponse au remerciement de M. Paul Sabatier, nommé mainteneur, le 14 février 1909," Académie des Jeux Floraux (Toulouse, 1909), p. 8.

109. University personnel could vote only after 1848. Prost, pp. 80–81; and Robert J. Smith, "L'atmosphère politique à l'Ecole Normale Supérieure à la fin du xix$^e$ siècle," Revue d'Histoire Moderne et Contemporaine, 20 (1973), 484–268, on p. 249, and The Ecole Normale Supérieure and the Third Republic (Albany: SUNY Press, 1982), pp. 104–131.

110. See the letter from Berthelot to Ernest Renan, dated 25 September 1879 in their Correspondance, 1847–1892 (Paris, 1898), p. 483. On bias against Catholic universitaires at the Ecole Normale, see Prost, p. 80, or Smith, "L'atmosphère politique," p. 250.

111. Tresserre, "Réponse," p. 9. And Sabatier, "Pasteur. Histoire d'une esprit," Rev. Gén. Sci., 8 (1897), 76–77; and also Camichel, "Eloge," pp. 121–122.

112. Cathala, who had been Sabatier's student at Toulouse, said at the centenary celebration that he personally heard the word "disgrace" from Sabatier's lips. In Cathala, Centenaire: Paul Sabatier. On the notation of equivalents, see Alan J. Rocke, "Atoms and Equivalents: The Early Development of the Chemical Atomic Theory," HSPS, 9 (1978), 225–263.

113. Paul Sabatier, Recherches thermiques sur les sulfures (Doctoral diss., University of Paris, 1880), e.g., pp. 44, 46, and table, pp. 92–94. Also in Ann. Chim. Phys., 22 (1881), 5–99.

114. Letters from Henri Sainte-Claire Deville (and Milne Edwards) to the rector, Académie de Paris, 21 July 1886, AN F$^{17}$ 24106.

115. Letter from Marcellin Berthelot to the minister of higher education, 27 July 1880; two letters from Dean Benjamin Baillaud to minister of higher education, both dated 24 January 1882; letter from Rector Claude-Marie Perroud to the minister of higher education, 25 January 1882; telegram from Sabatier to

Berthelot, 26 January 1882; and letter from Sabatier to Berthelot, 4 November 1883, thanking Berthelot for his "bienveillant appui." AN F¹⁷ 24106.

116. NIs and RCs, with comments by the inspector, dean, and rector for 1884, 1885, and 1886. AN F¹⁷ 24106.

117. "ne fut-il pas plus tard, il y a environ cinquante ans, blâmé officiellement par son recteur?" See Cathala, *Centenaire: Paul Sabatier*, unnumbered pp.

118. Rectors Jeanmarie (1910 and 1911) and Lapie (1913) commented on "germs of discord" within the Faculty and recommended that Sabatier should "broaden his manner" as dean. In the NIs for 1910, 1911, and 1913. AN F¹⁷ 24106.

119. After his retirement in 1929, he remained active for many years at the Faculty, although he was not officially teaching or directing the Institute of Chemistry. He died in 1941.

120. See J. R. Partington, *A History of Chemistry*, Vol. IV (London: Macmillan, 1964), p. 897, and n. 6 on p. 897. Lecoq de Boisbaudran, who verified Mendeleev's prediction of eka-aluminum by his 1875 discovery of gallium, gave Mendeleev's principle little help in France when he wrote that if he had sought gallium in the precipitates suggested by Mendeleev's theory, he would not have found gallium at all. In *CR*, 81 (1875), 1100–1105, on p. 1105. Also see Daniel Q. Posin, *Mendeleev: The Story of a Great Scientist* (New York: Whittlesey House, 1948), p. 202.

121. Adolphe Wurtz, *The Atomic Theory*, trans. L. Cleminshaw, 6th ed. (London, 1892), pp. 154–155, 175.

122. Marcellin Berthelot, *Les origines de l'alchimie* (Paris, 1885), pp. 302–304, 312.

123. Henri Moissan, ed., *Traité de chimie minérale*, 5 vols. (Paris: Masson, 1904), Vol. I, pp. 37, 28–29. Moissan discusses the French chemical textbooks of the 1880s and 1890s on p. 24 and in notes 1–4, p. 24. Cathala and Lucien Babonneau have said that Sabatier was probably the first provincial French professor to take the atomic notation and the periodic classification as the basis of his teaching of inorganic chemistry. In Cathala, *Centenaire: Paul Sabatier*, unnumbered pp.; and Babonneau, "Paul Sabatier," pp. 172–173.

124. See Sabatier, "Sur la classification des corps simples par la loi périodique," *Annales de la Faculté des Sciences de Toulouse*, 4 (1890), B.1–B.14, esp. pp. B.1, 3, 13–14; and "Sur les relations de la couleur des corps avec leur nature chimique," ibid., 6 (1892), E.1–E.38.

125. One former student wrote that lycée chemistry taught through formulas written in equivalents had been "grim" and "unprofitable" but that in listening to Sabatier's lectures, "chemistry became a logical science, a deductive science." Paul Dop, "Rapport adressé à M. le Ministre de l'Instruction Publique par le conseil de l'Université de Toulouse," in *Toulouse, Univ. Conseil, 1929–1930*, pp. 5–18, p. 9.

126. On Mendeleev, see Posin, pp. 245, 278. And Sabatier, in "Dinner to Professor Sabatier," *Journal of the Royal Society of Arts*, 74 (1925–26), 890–891; and Sabatier, "Sir William Ramsay et son oeuvre," *RS*, 54 (1916), 609–616.

127. H. Tisnès, 'Le Prix Nobel à Toulouse," *Dépêche*, 14 November 1912; and "Le Professeur Sabatier consacrera le montant du prix Nobel à l'Institut de Chimie," *Le Matin*, 14 November 1912. Sabatier also mentioned that Moissan told him in 1906, the year before Moissan's death, that he would support Sabatier for the Prize. Newspaper microfilm, Nobel Prize Archives, Royal Swedish Academy of Sciences, courtesy of Elisabeth Crawford and Wilhelm Odelberg. In a letter in the Archives of the Académie des Sciences, dated 10 January 1928, Sabatier credits Fischer and von Baeyer with supporting him.

128. Edmond Bauer commented on Langevin's course in the interview cited earlier with T. S. Kuhn and Théodore Kahan.

129. A. W. Tilden, "Mendeleef Memorial Lecture," *Jl. Chem. Soc.*, 95 (1909), 2077–2105, on p. 2094.

130. On the Prix La Caze, see *CR*, *126* (1898), 86–87. There is no published bibliography of Sabatier's complete scientific and other writings. His granddaughter, Mlle. Parlange, graciously gave me a copy of a list compiled in 1932; it includes 343 entries.

131. He used as an indicator for end points of neutralization not heat but the change in coloration of chromates and dichromates. Paul Sabatier, "Spectre d'absorption des chromates alcalins et de l'acide chromique," *CR*, *103* (1886), 49–52, and "Partage d'une base entre deux acides, cas particulier des chromates alcalins," ibid., 138–141.

132. Sabatier, "Essai critique sur les principes de la thermochimie," pp. 291–292, 297–300. And Lothar Meyer, "Les théories modernes de l'affinité," *RS, 40* (1887), 1–5.

133. M. Le Chanoine Palfray, "Notice sur la vie et les travaux de M. Le Chanoine Senderens," *Bull. Soc. Chim.*, 6 (1939), 1–29, on p. 2. In 1897, Senderens presided over the Commission du Black Rot de la Haute-Garonne, of which Sabatier was a member. Senderens's thesis was *Action du soufre sur les oxydes et les sels en présence de l'eau* (1892). Also see *Universelle de 1900: Institut Catholique de Toulouse Exposition* (Toulouse: Edouard Privat, 1900).

134. Sabatier and Senderens, "Sur une nouvelle classe de combinaisons: les métaux nitrés," *Bull. Soc. Chim.*, 9 (1893), 669–674, esp. p. 670; Sabatier, "Les hydrogénations directes sur le nickel," *Bull. Soc. Chim.*, 6 (1939), 1261–1268, on p. 1261.

135. Henri Moissan and Charles Moureu, "Action de l'acétylène sur le fer, le nickel et le cobalt reduits par l'hydrogène," *CR*, 122 (1896), 1240–1243.

136. Sabatier, "La méthode d'hydrogénation directe par catalyse," 1912 Nobel Prize Lecture (Stockholm, 1913), 12 pp., on p. 4. Reprinted in *RS*, 51 (1913), 289; and in English translation, in *Nobel Lectures: Chemistry, 1901–1921* (New York: American Elsevier, 1966), pp. 221–231. Also, Sabatier's "How I Have Been Led to the Direct Hydrogenation Method by Metallic Catalysts," *Industrial and Engineering Chemistry*, 18 (1926), 1005–1008, on p. 1006.

137. Sabatier and Senderens, "Action du nickel sur l'éthylène," *CR*, 124 (1897), 616–618.

138. Sabatier and Senderens, "Action du nickel sur l'éthylène. Synthèse de

l'éthane," *CR, 124* (1897), 1358–1361. In accounting for his discovery, Sabatier was in the habit of saying "Ce fut ma chance!" according to Cathala, in *Centenaire. Paul Sabatier*, unnumbered pp.

139. On this, see Alphonse Mailhe, "Paul Sabatier," *Chemiker-Zeitung, 149* (1912), 1451 (in a 4-page reprint, p. 3).

140. Sabatier and Senderens, "Hydrogénations directes réalisées en présence du nickel réduit; préparation de l'hexahydrobenzene," *CR, 132* (1901), 210–212; and Sabatier, "Les hydrogénations directes sur le nickel" (1939), p. 1263. Berthelot's experiments with hydrogen iodide are detailed in *Les carbures d'hydrogène* (Paris, 1910), Vol. III, as summarized in a note in *CR, 138* (1904), 248–249.

141. For brief descriptions of Sabatier's and Senderens's experiments, see Sabatier's *La catalyse en chimie organique* (Paris: C. Béranger, 1913), trans. E. Emmet Reid, *Catalysis in Organic Chemistry*, 2d printing (New York: Van Nostrand, 1923); also "Catalysis," *Journal of the Society of Chemical Industry, 33* (1914), reprt. of 11 pp., on p. 9. On the petroleums, see Sabatier, "Synthèse de divers pétroles—contribution à la théorie de formation des pétroles naturels," *CR, 134* (1902), 1185–1188; and Babonneau's account, "Paul Sabatier," pp. 177–179.

142. Palfray, pp. 9, 21; Eric K. Rideal, "Obituary Notice. J.-B. Senderens," *Nature, 141* (1928), 148; and Senderens's "Aperçu historique sur le procédé Sabatier-Senderens," 6 pp., typescript in Nobel Prize Archives, Royal Swedish Academy of Sciences, courtesy of Elisabeth Crawford and Wilhelm Odelberg.

143. Mailhe took his doctorate at Toulouse in 1902, and in 1922, became the first titular of Toulouse's Chaire d'Etat de Chimie Industrielle. In 1926, he left Toulouse for Paris, where he held a Chair of Combustibles. Other students who worked closely with Paul Sabatier included Marcel Murat (1912–1914), Léo Espil (1914), Georges Gaudion (1918–19), Bonasuke Kubota, and Antonio Fernandez (in the 1920s). In addition to Kubota, the Japanese chemist Itizo Kasiwagi studied with Sabatier briefly in 1918, before going to Harvard University to study with Arthur Michael. I am grateful to Kasiwagi's son, Hazime Kasiwagi, for this information.

144. E.g., Sabatier and Alphonse Mailhe, "Réduction directe des désirés halogènes aromatiques par le nickel divisé et l'hydrogène," *CR, 138* (1904), 245–248.

145. Sabatier and Senderens, "Décomposition catalytique de l'alcool éthylique par les métaux divisé; formation régulière d'aldehyde," *CR, 136* (1903), 738–741; and "Transformation des aldehydes et des cétones en alcools par hydrogénation catalytique," *CR, 137* (1903), 301–303. A summary of the results is Sabatier and Mailhe, "Actions des oxydes métalliques sur les alcools primaires," *Ann. Chim. Phys., 20* (1910), 289–352. See also Sabatier, "How I Have Been Led to the Direct Hydrogenation Method," p. 1008.

146. See Vincent, "Nécrologie," pp. 282–283. Manipulating the conditions under which catalysis occurred involved varying the catalyst itself, the temperature, and the presence or absence of impurities. See Sabatier, *Catalysis*, pp. 7–10, 31–35; and Sabatier and Senderens, "Nouvelles méthodes générales d'hydrogénation et de dédoublement moléculaire, basés sur l'emploi des métaux divisés," *Ann. Chim. Phys., 4* (1905), 334.

147. See Hugh S. Taylor, "Paul Sabatier," *Jl. Amer. Chem. Soc.*, 66 (1944), 1615–1617, esp. p. 1615; and B. B. Corson, "Industrial Catalysis," *Journal of Chemical Education*, 24 (1947), 99–103, esp. p. 100. Sabatier took out eight patents, one of which was for the transformation of liquid fatty acids into solid fatty acids (1907, Brevet no. 394957).

148. Finely divided platinum was used in some reactions to effect oxidation, and there were a few instances (including that of de Wilde in 1874) of reductions. See Sabatier, "Catalysis," p. 9; and for a short history of catalysis, Wilhelm Ostwald's 1909 Nobel Lecture "On Catalysis," in *Nobel Lectures: Chemistry, 1901–1921*, pp. 150–169, on p. 154–163. See Berzelius on catalysis in *Traité de chimie*, 2d ed. (Paris, 1845), Vol. I, pp. 110–112. His views are discussed in Wilhelm Ostwald, "La force catalytique et ses applications," *RS, 17* (1902), 641–650, on p. 641.

149. Moissan, *Traité*, p. 13; and Jacques Duclaux, "Applications de la théorie cinétique à l'étude des phénomènes de catalyse," *CR, 152* (1911), 1176–1179.

150. Ostwald, Nobel Lecture, pp. 161–162; and "La force catalytique et ses applications," pp. 646–647. The latter was a translation of a lecture made by Ostwald at the Hamburg Natural Scientists meeting of 1901.

151. Sabatier, *Catalysis*, p. 41.

152. Senderens's biographer writes of their collaboration: "It is thus that in the course of ten years of collaboration between the two scientists, so intimate was their work that one does not know how to try to distinguish the part of each, without presumption and without injustice." In Palfray, p. 4.

153. Cited in ibid., p. 48. According to Corson, "Industrial Catalysis," p. 100, Michael Faraday adopted a chemical theory of catalysis, attributing the catalytic effect of platinum on the combustion of hydrogen to the *ad*sorption of hydrogen and oxygen on the platinum surface, where the end reaction took place through successive intermediate forms.

154. According to Sabatier, James Dewar assigned the formula $Pd_3H_2$ to a hypothetic intermediary in the hydrogenation of hydrocarbons upon palladium. In *Catalysis*, pp. 52–53.

155. Ibid., p. 48.

156. Ibid., pp. 42–43.

157. In the 1920s, Wilhelm Schlenk and Theodor Weichselfelder successfully demonstrated the existence of an unstable brown, solid nickel hydride on the surface of grains of reduced nickel. They used x-ray diffusion. Cited in Sabatier, "Les hydrogénations," p. 1264.

158. Sabatier and Mailhe, "Sur la décomposition catalytique de l'acide formique," *CR, 152* (1911), 1212–1215. They conclude the paper by saying that their results are irreconcilable with the idea that the catalyst's action is one of lowering the temperature (p. 1215). See discussion in the Nobel Lecture, p. 11.

159. It is still cited in chemical literature today. J. E. Germain writes in *Catalytic Conversion of Hydrocarbons* (New York: Academic Press, 1969), p. vii: "most of the advances in organic catalysis have been made so far by a purely empirical and analogical approach, typical of the work of the great pioneers, such as Sabatier and Ipatieff."

160. The Faraday Society meetings of 1922 and 1932 are discussed in Hugh S. Taylor, "Catalysis: Retrospect and Prospect," Fifth Spiers Memorial Lecture, *Discussion of the Faraday Society, 8* (1950), 9–18, on p. 9. See also I. Langmuir, *J. Am. Chem. Soc., 37* (1915), 1139, and *38* (1916), 1145, cited in Wilder D. Bancroft, "Theories of Contact Catalysis," in Reid's translation of Sabatier's *Catalysis*, pp. 58–69, p. 59.

161. Sabatier, "Sur l'inversion du rôle des catalyseurs," *CR, 185* (1927), 17–19.

162. Sabatier, "Les hydrogénations," p. 1268.

163. Marcel Brillouin, "Pour la matière," *Rev. Gén. Sci., 6* (1895), 1032–1034.

164. Emile Picard, *La science moderne et son état actuel* (Paris: Flammarion, 1905), esp. pp. 117–122.

165. In his book on catalysis (p. 47 of the 1923 translation), Sabatier cites Job's work on salts as excitators of oxidation in homogeneous milieux. See, e.g., André Job, "Activité de quelques sels de terres rares, comme excitateurs d'oxydation," *CR, 136* (1903), 45–47, and "L'Acétate de nickel modifié, nouveau type d'excitateur d'oxydation pour l'hydroquinon," *CR, 144* (1907), 1266–1267.

166. Sabatier, Nobel Lecture, pp. 11–12. The acceptance speech is discussed in Camichel, "Eloge," p. 117; and Naves, in *Centenaire: Paul Sabatier*, unnumbered pp.

167. Toulouse's literary Académie de Jeux Floraux elected Sabatier to its membership in 1909. In his "Remerciement," p. 5, he spoke of his election as a demonstration of the linkages between arts and science. Camichel speaks of Tannery's influence on Sabatier in "Eloge," pp. 118–119.

168. On this unfortunate matter, see Palfray, p. 10; Heinrich Rheinboldt, "Fifty Years of the Grignard Reaction," *Journal of Chemical Education, 27* (1950), 476–488, on p. 485, n. 39. Sabatier's lecture was "Hydrogénations et deshydrogénations par catalyse," *Ber. Deut. Chem. Gesell., 44* (1911), 1984–2001. The remark that Senderens was his student is on p. 1984. Sabatier's retraction is "Bemerkung zu meinem Vortrag vom 13 Mai 1911 über 'Hydrogénations et deshydrogénations par catalyse,'" ibid., *44* (1911), 3180. Senderens became a corresponding member of the Academy of Sciences only in 1922. Among his good friends in Paris were Jean Perrin and Georges Urbain (Palfray, p. 22).

169. The information on nominations, courtesy of Elisabeth Crawford. Also, letters, from Moissan, dated 26 January 1907; Haller, dated 8 February 1908; and Lemoine, dated 25 January 1909. And Senderens, "Aperçu historique sur le procédé Sabatier-Senderens," in Nobel Prize Archives, Royal Swedish Academy of Sciences, microfilm courtesy of Elisabeth Crawford and Wilhelm Odelberg. On procedures for soliciting Nobel Prize nominations, see essays by Elisabeth Crawford and by Elisabeth Crawford and Robert M. Friedman, in C. G. Bernhard et al., eds., *Science, Technology and Society in the Time of Alfred Nobel* (London: Pergamon, 1982).

170. Charles Moureu and Raymond Poincaré, "Jubilé de M. Paul Sabatier," *Institut de France: Académie des Sciences* (Paris, 1913), p. 8; and *Dépêche, 44* (18 September 1913), p. 2. The Academy votes are reported in *CR, 156* (1913), 1196, 1293, 1344; *157* (1913), 662, 891, 1104.

171. Sabatier's example is by no means the only one of the election of a scientist to his nation's scientific academy subsequent to the Nobel award. See,

e.g., D. S. Greenberg, "The National Academy of Sciences—Profile of an Institution," *Science*, 156 (1967), 362, discussed in Stuart Blume, *Toward a Political Sociology of Science* (New York: Free Press, 1974), p. 83.

172. On the Sorbonne chair, see Paris, Fac. Sci., PV Conseil, 18 July 1900, p. 321, and 15 May 1907, p. 62. At the Paris Faculty's first council meeting after Moissan's death, the dean informed the group that he had asked Sabatier to occupy the chair left vacant by Moissan. "The unanimity with which Sabatier has been designated as the most capable candidate to succeed M. Moissan" has honored him greatly, the dean said to the council, but Sabatier had refused to leave Toulouse (p. 62). There is no record of the discussion or vote taken designating Sabatier for the chair.

173. In a conversation with Mlle. Parlange, 4 June 1972, at Paris.

174. Quoted in Babonneau, "Paul Sabatier," p. 186. The rule change involved the creation in 1913 of a division for six nonresident members. The number was increased to twelve in 1945. The Academy more recently suppressed completely the clause of Paris residency for regular members, while leaving to the provinces their twelve nonresident members. In Lucien Plantefol, "L'Académie des Sciences durant les trois premiers siècles de son existence," in *Académie des Sciences: Troisième Centenaire, 1666–1966* (Paris: Gauthier-Villars, 1967), pp. 53–139, p. 124.

175. Duhem's first essays in the philosophy of science were published in the 1890s in the *Revue des Questions Scientifiques* as a refutation of the journal's earlier atomic views. See chap. 6.

## 5: LYON

1. W. Lexis, *Die neuen Französischen Universitäten. Denkschrift aus der Pariser Weltausstellung von 1900* (Munich, 1901), p. 56.

2. *Lyon en 1906: Lyon et la région lyonnaise en 1906* (Lyon: A. Rey, 1906), Vol. I, pp. 94–96. All references in this chapter are to Vol. I of the two-volume work.

3. *Lyon en 1906*, p. 145; and *Lyon, 1906–1926* (Lyon: A. Rey, 1926), p. 35.

4. *Lyon en 1906*, p. 99; and *Université de Lyon, 1900* (Lyon: A. Storck, 1900), p. iv.

5. Auguste Ehrard, *L'Université de Lyon* (Lyon: A. Rey, 1919), pp. 31–32; André Latreille, ed., *Histoire de Lyon et du lyonnais* (Toulouse: Privat, 1975), p. 492; and Theodore Zeldin, *France, 1848–1945*, Vol. II, *Intellect, Taste, and Anxiety* (Oxford: Oxford University Press, 1977).

6. Louis M.Greenberg, *Sisters of Liberty: Marseille, Lyon, Paris and the Reaction to a Centralized State, 1868–1871* (Cambridge: Harvard University Press, 1971), pp. 223, 218–220, 226–237, 246–256.

7. Edouard Herriot, preface to *Lyon en 1906*, xv–xvii, on p. xvii.

8. Jean Bouvier, *Naissance d'un banque: Le Crédit Lyonnais* (Paris: 1968), pp. 69, 74.

9. One of the Lumière brothers taught a cours libre in photography at the Sciences Faculty in 1889. See Lyon, Fac. Sci., PV, 19 June 1889. In 1895, the Lumière brothers contributed 2,500 francs to the Faculty in support of Koehler's

expenses for a zoological expedition in the Atlantic Ocean. Lyon, Fac. Sci., PV, 31 May 1895.

10. See Louis F. Haber, *The Chemical Industry during the Nineteenth Century* (Oxford: Clarendon Press, 1958), pp. 201–202; and Paul M. Hohenberg, *Chemicals in Western Europe: 1850–1914. An Economic Study of Technical Change* (Chicago: Rand McNally, 1967), pp. 35–36.

11. Hohenberg, p. 41.

12. *Lyon en 1906*, pp. 143–144. The chairs were in analysis and mathematics (later called "pure mathematics"), astronomy, physics, chemistry, zoology, botany, mineralogy and geology (later, two separate chairs), physical astronomy (established when the old astronomy chair was transformed into "applied mathematics"), physiology, and chemistry applied to industry and agriculture.

13. *Lyon, 1906–1926*, pp. 37–38. The other two mathematics chairs were renamed "general mathematics" and "applied and rational mechanics." The applied chemistry chairs were designed expressly for Louis Meunier and Couturier, when established. See Lyon, Fac. Sci., PV, 16 June 1922.

14. *Université de Lyon: 1900*, p. xxiii.

15. AN F$^{17}$ 23767 (Georges Gouy). Letter from Victor Grignard to the rector of the Académie de Lyon, 29 June 1925.

16. Emile Picard, "La vie et l'oeuvre de Paul Villard et Georges Gouy," *Institut de France: Académie des Sciences* (Paris, 1937), 29 pp.

17. Ibid., p. 2. AN F$^{17}$ 23767; see RCs for 1885, 1886, 1889, 1890, 1893, 1899.

18. See Picard, "La vie et l'oeuvre," pp. 19–24; and J. B. Gough, "Gouy," in *DSB*, pp. 483–484.

19. See Georges Gouy, "Note sur le mouvement brownien," *Journal de Physique*, 7 (1888), 561–564, on p. 564; and "Sur la mouvement brownien," *CR*, 109 (1889), 102–105. Gouy's remarks at Lyon in 1894: "Le mouvement brownien et les mouvements moléculaires," (Lyon, 1895), 23 pp., published in *Rev. Gén. Sci.*, 6 (1895), 1–7. On Jean Perrin, see Mary Jo Nye, *Molecular Reality: A Perspective on the Scientific Work of Jean Perrin* (London: Macdonald, 1972).

20. Paris, Fac. Sci., PV Conseil, 3 March 1898, p. 266.

21. See Franck Bourdier, "Charles Depéret," in *DSB*, Vol. IV, p. 39.

22. See Lucienne Félix, "Ernest Vessiot," in *DSB*, Vol. XIV, pp. 13–14; and Jean Dieudonné, "Elie Cartan," in *DSB*, Vol. III, pp. 95–96. Also, J. H. C. Whitehead, in the *Obituary Notices of the Royal Society of London*, on Cartan.

23. See Ehrard, *L'Université de Lyon*, p. 134.

24. *Université de Lyon. 1900*, p. xxiii.

25. Ibid., p. xxiv.

26. Ehrard, pp. 212–213.

27. *Lyon en 1906*, pp. 245–246.

28. Ibid., pp. 238–247.

29. See J.B. Carles, "Jules Raulin," in *DSB*, Vol. IV, pp. 310–311. And, *Lyon, 1906–1926*, p. 451.

30. Lyon, Fac. Sci., PV, 9 April 1886.

31. Ibid.

32. Ibid., 16 March 1889.

33. Ibid., 24 December 1892.

34. By this time, he had published fifty notes or memoirs in mineral and organic chemistry, a dozen of them on industrial topics, and he had received the first Prix Nicolas LeBlanc, given by the Société Chimique de Paris, in 1891. See ibid. Also, ibid., 17 July 1896. As holder of the chair, he directed the Ecole de Chimie Industrielle, the Ecole Française de Tannerie, and, for a while, the Station Agronomique. See *Lyon en 1906*, pp. 155–156; and *Lyon, Livret Etud., 1910–1911*, p. xxvi.

35. *Lyon en 1906*, p. 147.

36. Ehrard, pp. 144–145.

37. See Victor Grignard, "Sur les instituts universitaires de sciences appliquées," *RS*, 1919, pp. 13–15.

38. Ehrard, p. 209.

39. *Lyon, 1906–1926*, p. 459.

40. Ibid., pp. 38–39, 92–93.

41. See Roger Grignard, *Centenaire de la naissance de Victor Grignard, 1871–1971* (Lyon: Audin, 1972), pp. 25–26.

42. Ibid., p. 26.

43. This is the main thrust of the argument in Michel Bauer and Elie Cohen, "Politiques d'enseignement et coalitions industrialo-universitaires. L'exemple de deux 'grandes écoles' de chimie, 1882–1976," *Rev. Fr. Soc.*, 22 (1981), 183–203. They contrast the Lyon Chemical Institute with that at Nancy, where students received more physics in their training than did chemistry students at Lyon.

44. Lyon, Fac. Sci., PV, 25 March 1898.

45. Ibid., 20 October 1928.

46. Ibid.; and *Lyon, Livret Etud., 1929–1930*, p. 38.

47. Lyon, Fac. Sci., PV, 17 May 1906.

48. See Bauer and Cohen, pp. 188–189, n. 7. They mention the foundation in the 1960s of the GECO (Groupe d'Etude de Chimie Organique) for studying the application of quantum and wave mechanics in organic reaction mechanisms.

49. On firing, Lyon, Fac. Sci., PV, 16 June 1922. On "nominal" chairs, ibid., 23 February 1923. Like Grenoble, Lyon also responded to the ministerial inquiry of 7 December 1921, which asked whether all the provincial universities should be maintained, by saying that nothing should be changed at their Faculty, except that more chairs were needed. Ibid., 23 December 1921.

50. Ibid., 12 May 1911.

51. Ibid., 15 October 1925. Another instance of a commitment by local industrialists to long-term funding for a position within the Faculty was the arrangement worked out for Adjoint Professor Vauthier in 1892. A business group in gas and electricity offered Vauthier an annual credit of 8,000 francs for five years to study technical questions in photometry and electricity for them. Vauthier's Laboratory of Photometry flourished for decades as an annex to the Faculty's service of general physics. Lyon, Fac. Sci., PV, 23 July 1892 and 16 November 1896.

52. See Gouy's protest of this practice, Lyon, Fac. Sci., PV, December 1910.

53. The Sciences Faculty council routinely began discussing deficits in 1924 and 1925. As happened at Toulouse and elsewhere, council members, in this case led by Victor Grignard, openly noted that the formal budget was a "false" budget that did not show the real deficits. See Lyon, Fac. Sci., PV, 3 May 1924, 15 May 1925, 28 April 1927, 26 March 1928.

54. The university's annual report in 1914 said "L'influence économique de la France et de la région lyonnaise compte en Orient un bon instrument de plus." M. Huvelin, "Rapport annuel du conseil de l'Université de Lyon pour l'année scolaire 1912–1913," in *Enquêtes Enseign.*, no. 108 (Paris, 1914), pp. 277–317, p. 281.

55. See *Lyon, 1906–1926*, pp. 14, 33, 72.

56. *Lyon, Livret Etud.*, 1929–1930, pp. 108–116. Enrollments increased and fees were raised, but the offerings at Lyon changed very little from 1920 to 1930; only one new certificate of higher studies was introduced in ten years.

57. Lyon, Fac. Sci., PV, 11 May 1923.

58. But then, Lyon, unlike Grenoble, was unlikely to lose status in such a specialized system. See Lyon, Fac. Sci., PV, 15 December 1903, 15 May 1908, 23 December 1921.

59. Ibid., 26 December 1919 and 10 January 1920.

60. *Lyon. 1906–1926*, p. 35.

61. Lyon, Fac. Sci., PV, 28 April 1927.

62. Ibid., 10 January 1930.

63. Bauer and Cohen, p. 186.

64. Clement Vautel, "Propos d'un Parisien," quoted in Roger Grignard, *Centenaire*, pl. 38; also Roger Grignard, "Un savant ami de la Belgique, Victor Grignard," *Académie Royale de Belgique: Bulletin de Classes des Sciences*, 59 (1973–1978), 750–762, séance of 28 July 1973; and AN F$^{17}$ 26752 (Victor Grignard).

65. One of his teachers at Cherbourg, Paul Petit, later was among Grignard's colleagues at the Sciences Faculty at Nancy. See Roger Grignard, "Un savant ami de la Belgique," pp. 754–755; Heinrich Rheinboldt, "Fifty Years of the Grignard Reaction," *Journal of Chemical Education*, 27 (1950), 476–488, on p. 488; Charles Courtot, "Notice sur la vie et les travaux de Victor Grignard (1871–1935)," *Bull. Soc. Chim.*, 3 (1936), 1433–1472, on p. 1433; and Victor Grignard, "Discours prononcé par Victor Grignard à l'occasion du cinquantenaire de l'Ecole de Chimie Industrielle de Lyon, le 12 mai 1933," in Charles Adam et al., *Victor Grignard* (Lyon, 1936), pp. 33–38, on pp. 34–35.

66. Courtot, p. 1434.

67. V. Grignard, "Discours," in Charles Adam et al., *Victor Grignard*, p. 35.

68. See J. B. Piobetta, *Le baccalauréat* (Paris: J. B. Baillière, 1937), p. 949.

69. Rheinboldt, p. 477.

70. Commonly, they are benzene rings where some of the CH groups are reduced to $CH_2$ or substituted $CH_2$ groups. See Aaron Ihde, *The Development of Modern Chemistry* (New York: Harper and Row, 1964), pp. 463–464; and J. R. Partington, *A History of Chemistry*, Vol. IV (London: Macmillan, 1964), p. 867.

71. On Bouveault, see A. Béhal, in *Bull. Soc. Chim.*, 7 (1910), i–xxii.

72. In Paris, Fac. Sci., PV Conseil, Pièces-Annexes, pp. 237–238, presented at the council meeting of 2 July 1901.

73. According to Rheinboldt, p. 485, n. 37.

74. Paris, Fac. Sci., PV Conseil, Pièces-Annexes, p. 218.

75. On Barbier, see M. Blaise, "Extrait des procès-verbaux des séances. Séance du 10 November 1922, présidence de M. Blaise," *Bull. Soc. Chim.*, 31 (1922), 1242–1248; Jean Colonge, "Historique et aspects particuliers de la réaction de Victor Grignard," conférence du 26 mai 1950, *Bull. Soc. Chim.*, 1950, 910–918, on p. 912; and René Locquin, "Genèse et évolution de la découverte des composés organo-magnesiens mixtes," pp. 13–45, in "Journée Victor Grignard, 13 mai 1950," extract from *Annales de l'Université de Lyon*, pp. 27–29. Rheinboldt notes that Barbier destroyed his curriculum vitae shortly before his death; n. 37, p. 485.

76. Quoted in Rheinboldt, p. 486.

77. Gouy, "Le mouvement brownien et les mouvements moléculaires" (Lyon, 1895), p. 14.

78. "Discours" (1933); perhaps he also heard Arthur Hannequin, whose book skeptically examining the experimental and logical arguments for real atoms and molecules was published in the *Annales de l'Université de Lyon* and then as *Essai critique sur l'hypothèse des atomes* (Paris: Alcan, 1895).

79. See Charles Depéret's arguments in 1891 for the establishment of a chair of mineralogy; the chair was instituted in 1895. Lyon, Fac. Sci., PV, 25 April 1891.

80. In 1902, Offret was a second-line candidate for the Sorbonne chair of mineralogy behind Wallerant, who was lecturer at the Ecole Normale. See Paris, Fac. Sci., PV Conseil, Pièces-Annexes, pp. 244–246. His qualifications for adjoint professor at Lyon are discussed in Lyon, Fac. Sci., PV, 24 December 1892. In 1886, he shared the Prix Vaillant.

81. Lyon, Fac. Sci., PV, 18 December 1899.

82. Locquin, pp. 28, 32.

83. Miquey's account, in R. Grignard, *Centenaire*, p. 33.

84. Barbier and V. Grignard, "Sur l'acétylbutyrate d'ethyle B-isopropylé et les acides diisopropylhexène dioïques stéréoisomères," *CR*, 126 (1898), 251; V. Grignard, "Sur un nouvel hydrocarbure hexavalent, la méthyl-2-hexènine-3.5," *Bull. Soc. Chim.*, 21 (1899), 574; and V. Grignard, "Sur le méthyl-2 heptatriène-4.5.6," *Bull. Soc. Chim.*, 21 (1899), 576.

85. Philippe Barbier, "Synthèse du dimethylhéptenol," *CR*, 128 (1899), 110–111, on p. 111.

86. Locquin, pp. 30–31; and Henry Gilman, "Victor Grignard," *Jl. Amer. Chem. Soc.*, 59 (1937), 17–19, on p. 17.

87. Locquin, p. 31.

88. Ibid., p. 33.

89. Ibid., pp. 14–18. Frankland, *Ann. Chem.*, 111 (1859), 63, and *Phil. Trans.*, 149 (1859), 412; and Wanklyn, *Jl. Chem. Soc.*, 13 (1861), 125.

90. Locquin, p. 19.

91. Locquin, p. 21; G. Wagner and A. Saytzeff, *Ann. Chem.*, 175 (1875), 363.

92. Locquin, p. 22.

93. Ibid., p. 23. W. Hallwachs and A. Schafarik, *Ann. Chem.*, 109 (1859), 206; and A. Cahours, *Ann. Chem.*, 114 (1860), 240.

94. Löhr, *Ann. Chem.*, 261 (1891), 48, 72.

95. Fleck was an American from Philadelphia. See Fleck, *Ann. Chem.* (1893), 129. Fleck set out to prepare phenylmagnesium bromide, but did not succeed. Apparently he destroyed the compound by adding an excess of bromine. See Gilman, "Victor Grignard," p. 18.

96. *Ann. Chem.* (1894), 320. Their inaugural dissertations at Tübingen were on these topics, in 1889, 1892, and 1894, respectively. See Locquin, pp. 23–26; and V. Grignard, *Sur les combinaisons organo-magnésiens mixtes et leur application à des synthèses d'acides, d'alcools et d'hydrocarbures* (Doctoral thesis [Thèse d'Etat], University of Lyon, 1901) in *Annales de l'Université de Lyon, 6* (1901), 118 pp., on pp. 2–7.

97. See V. Grignard, *Notice sur les titres et travaux scientifiques* (Lyon: Imprimerie J. Marlhens, 1926), pp. 6–7; also Jean Colonge, who studied Victor Grignard's laboratory notebooks, p. 913. Henry Gilman claimed in an obituary notice for Grignard that Barbier had smiled at Grignard's suggestion that it should be possible to prepare organomagnesium compounds and would not let him try it (p. 17).

98. Colonge, p. 913.

99. See the very good discussion in C. E. Waters, "Report: The Grignard Reaction," *Amer. Chem. Jl.*, 33 (1905), 304–326.

100. V. Grignard, "Sur quelques nouvelles combinaisons organométalliques du magnésium et leur application à des synthèses d'alcools et d'hydrocarbons," *CR, 130* (1900), 1322–1324, on pp. 1322–1323.

101. *Ibid.*, p. 1323.

102. V. Grignard, *Thèse*, p. 10.

103. Standard procedures now include:

1. The formation of a hydrocarbon by treatment of the "Grignard reagent" with water, an alcohol, or a phenol.

$$Mg{\overset{\displaystyle C_6H_5}{\underset{\displaystyle Br}{}}} + C_2H_5O\,H \rightarrow Mg{\overset{\displaystyle OC_2H_5}{\underset{\displaystyle Br}{}}} + C_6H_6$$

$$Mg{\overset{\displaystyle CH_3}{\underset{\displaystyle I}{}}} + HO\,H \rightarrow Mg{\overset{\displaystyle OH}{\underset{\displaystyle I}{}}} + CH_4$$

2. Preparation of alcohols by direct addition to the carbonyl linkage of an aldehyde or ketone, as we have seen above.

3. Preparation of an acid by the addition of $CO_2$ to the Grignard reagent.

$$C \begin{matrix} \diagup\!\!\!\diagup O \\ \diagdown\!\!\!\diagdown O \end{matrix} + Mg \begin{matrix} \diagup C_2H_5 \\ \diagdown I \end{matrix} \rightarrow C_2H_5 - C \begin{matrix} \diagup\!\!\!\diagup O \\ - O \end{matrix} - MgI$$

$$C_2H_5 - \overset{O}{\underset{\|}{C}} - O - MgI + HCl \rightarrow C_2H_5\overset{O}{\underset{\|}{C}} - OH + MgICl$$

104. See the 1905 bibliography in Waters, "Report: The Grignard Reaction," pp. 318–326.

105. Ihde, p. 342.

106. *Victor Grignard: Prix Nobel de Chimie, 1912.* Commentary on a film put together in honor of the centenary celebration, in 1971, of Grignard's birth. Centre regional de recherche et de documentation pedagogique, 47–49 rue Philippe de Lassalle, 69004 Lyon. On p. 11.

107. For example, in water, alcohol, ammonia, and hydrochloric acid. See Ihde, p. 632.

108. See Georges Urbain, "L'oeuvre de Victor Grignard. Son *Précis de chimie organique,*" xvi pp., on p. xiv. Also, see prefaces to Béhal's textbooks.

109. On Blaise, see Henry Gault, "Notice sur la vie et les travaux de Emile Edmond Blaise (1872–1939)," *Bull. Soc. Chim.,* 8 (1941), 269–346.

110. E. E. Blaise, "Nouvelles réactions des dérivés organométalliques," *CR,* 132 (1901), 38–41.

111. For the letters, see Jean Colonge, p. 913.

112. V. Grignard, "Action des éthers d'acides gras monobasiques sur les combinaisons organomagnésiens mixtes," *CR,* 132 (1901), 336–338, on p. 338.

113. Locquin, p. 38.

114. V. Grignard, Thèse, p. 15. The two March papers are V. Grignard, "Sur les combinaisons organo-magnésiens mixtes," *CR,* 132 (1901), 558–561; and V. Grignard and Tissier, "Action des chlorures d'acides et des anhydrides d'acides sur les composés organométalliques du magnésium," *CR,* 132 (1901), 683–685.

115. Armand Valeur, "Action des éthers d'acides bibasiques sur les composés organométalliques," *CR,* 132 (1901), 833–834; Charles Moureu, "Nouvelles réactions des composés organomagnésiens," ibid., 837–839; V. Grignard and Tissier, "Sur les composés organo-métalliques du magnésium," ibid., 835–836; and Blaise, "Sur les dérivés éthéro-organomagnésiens," ibid., 839–841.

116. Blaise, *CR,* 132 (1901), 840.

117. *Bull. Soc. Chim.,* 25 (1901), p. 497.

118. *CR,* 132 (1901), 480–482.

119. *Bull. Soc. Chim.,* 25 (1901), 531. This is in the minutes for the meeting of

26 April 1901. The Grignard-Béhal correspondence is reprinted in Colonge, p. 914.

120. Letter reprinted in Colonge, p. 914.

121. Letter of 21 April 1901 from Berthelot, and letter of 22 April 1901 from Moissan; reprinted in Colonge, pp. 914–915; and also reprinted in R. Grignard, *Centenaire*, pp. 36–40.

122. See P. Jolibois, *CR*, 155 (1912), 353, and 156 (1913), 712; Job and Dubien, *Bull. Soc. Chim.*, 39 (1926), 583; V. Grignard, *CR*, 132 (1901), 560, and Thèse, pp. 11 and 20–22, as well as *Notice sur les titres*, p. 26; W. Schlenk and W. Schlenk, Jr., *Ber. Deut. Chem. Gesell.*, 62 (1929), 920; and W. Schlenk, Jr., ibid., 64 (1931), 734.

123. See Lauder W. Jones, "Review: Some Recent Work in Organic Chemistry," *Jl. Amer. Chem. Soc.*, 27 (1905), 1553–1568, on p. 1553.

124. See Partington, Vol. IV, pp. 787–788, citing A. von Baeyer and V. Villiger, *Ber. Deut. Chem. Gesell.*, 35 (1902), 1202.

125. Blaise, *CR*, 134 (1902), 551, and *Bull. Soc. Chim.*, 35 (1906), 90; and *Bull. Soc. Chim.* (new series), 1 (1907), 610.

126. V. Grignard, "Sur le mode de scission des combinaisons organomagnésiens mixtes. Action de l'oxyde d'éthylène," *CR*, 136 (1903), 1260; and same title, *Bull. Soc. Chim.*, 29 (1903), 944; also, *Bull. Soc. Chim.* (new series), 1 (1907), 256. And V. Grignard, "Les combinaisons organomagnésiens mixtes et la synthèse organique," *Rev. Gén. Sci.*, 14 (1903), 1040–1050, on p. 1044.

127. See V. Grignard, *Rev. Gén. Sci.* (1903), pp. 1044–1045.

128. Courtot, p. 1440.

129. See Louis F. Fieser and Mary Fieser, *Introduction to Organic Chemistry* (Boston: D. C. Heath, 1957), p. 109; and Jean Colonge, p. 916.

130. See Ihde, pp. 536–541. Also, on the application of electron theory to chemistry, Anthony N. Stranges, *Electrons and Valence: Development of the Theory, 1900–1925* (College Station: Texas A&M University Press, 1982), and Robert Kohler, Jr., "The Lewis-Langmuir Theory of Valence and the Chemical Community, 1920–1928," *HSPS*, 6 (1975), 431–468.

131. For quotation, V. Grignard, ed., *Traité de chimie organique*, Vol. I (Paris: Masson, 1935), pp. x–xi.

132. Kohler discusses the reaction of organic chemists to the paired-electron valence theory. He notes that following initial interest in the new theory, the American organic chemist James B. Conant recommended in 1928 that the safest course for organic chemists was to think of valence electrons as "merely the number of negative charges which the atom can gain or lose in making up the mystic number of eight" and to leave atomic structure to atomic physicists. Quoted in Kohler, "The Lewis-Langmuir Theory of Valence," p. 440.

133. He previously had been teaching two conférences in mineral chemistry, as well as three or four sessions of laboratory work for a total of 32 students in two conférences of one hour each, as well as a three-hour laboratory section in mineral chemistry for students at the licence level.

134. AN F[17] 26752.

135. Waters, "Report: The Grignard Reaction," 304–326; Klages, *Chemische Zeitung, 29* (1905), 19.

136. Jones, "Review," p. 1553.

137. This lecture received wide attention through publication in the popular *Revue Générale des Sciences, 14* (1903), 1040–1050. It was reprinted in Albin Haller, ed., *Les récents progrès de la chimie* (Paris: Gauthier-Villars, 1904), pp. 121–157.

138. R. Grignard, *Centenaire*, pp. 46–47.

139. V. Grignard, "Remarques sur la nomenclature," *Chemische Zeitung* (1910), 1239; and "Remarques et nouvelles remarques sur les rapports du Comité du travail pour la réforme de nomenclature en chimie organique," Union Internationale de la Chimie Pure et Appliquée, Conférence de Varsovie (1927), de La Haye (1928).

140. Barbier and Depéret both talked to the Lyon rector in October to ensure that this understanding was a strong one, stressing the value of the research collaboration between Barbier and Grignard. See F$^{17}$ 26752, letters from Depéret, Grignard, and Lyon rector, and the deputy Cazeneuve to the Ministry of Public Instruction, during the period 2 June 1902 to 12 October 1905. Grignard enlisted the aid of Cazeneuve in a letter of 21 July 1905. There also was some discussion of Grignard filling a vacancy at Clermont, in letters between the ministry and the Lyon rector, 17 March 1904 and 19 March 1904.

141. Rheinboldt, p. 483.

142. Haller's letter is quoted in R. Grignard, *Centenaire*, pp. 47–48.

143. Henri Moissan, "Rapport sur les travaux de M. Grignard," *CR, 143* (1906), 1023–1026.

144. Rheinboldt, p. 483.

145. Armand Gautier, "Conférence faite à la demande du conseil de la Société," 17 May 1907, in *Centenaire de la Société Chimique de France (1857–1957)* (Paris: Masson, 1957), pp. 3–89, on p. 73.

146. Philippe Barbier, "Sur l'origine de l'introduction du magnésium dans la synthèse organique," *Bull. Soc. Chim., 7* (1910), 206–208, quotation on p. 208.

147. V. Grignard, "Sur l'emploi du magnésium en chimie organique," *Bull. Soc. Chim., 7* (1910), 453–454.

148. Quoted in R. Grignard, *Centenaire*, p. 50.

149. Ibid., p. 51; and Paris, Fac. Sci., PV Conseil, 25 November 1909, p. 107.

150. Quoted in R. Grignard, *Centenaire*, p. 50.

151. In 1910, he married the recently widowed Augustine-Marie Boulant Paindestre, whom he had known from school days in Cherbourg. She brought to their marriage her son Robert Paindestre, and in 1911, a second son, Roger Grignard, was born. He eventually was to study chemistry at Paris under one of his father's best friends, Georges Urbain, and to work in chemical industry. Roger Grignard's doctoral research was interrupted by the Second World War and by Urbain's death, and he decided to work in chemical industry rather than continue with the doctoral thesis. R. Grignard, *Centenaire*, pp. 52–53; AN F$^{17}$ 26752; and my conversation with Roger Grignard, at Lyon, 26 July 1978.

152. See M. Floquet, "Rapport de M. Floquet, doyen de la Faculté des

Sciences, sur la situation et les travaux de la Faculté pendant l'année scolaire 1911–1912," in *Nancy, Rentrée Facs., 1912–1913*, pp. 125–167.

153. See E. M. Bellet, *Action du cyanogène et du chlorure de cyanogène sur les combinaisons organomagnésiens mixtes: Synthèse de nitriles et de cétones* (Doctoral thesis [Thèse d'Université], University of Nancy, 1913); and Charles Courtot, *Etudes dans la série des fulvènes* (Doctoral thesis [Thèse d'Etat], University of Nancy, 1915). V. Grignard and Bellet in *CR*, 155 (1912), 44; 158 (1914), 457. V. Grignard, Bellet, and Courtot, in *Ann. Chim. Phys.*, I, 4, (1915), 28; and II, 12 (1919), 364. V. Grignard and Courtot, in *CR*, 152 (1911), 272; 152 (1911), 1493; 154 (1912), 361; 158 (1914), 1763; 158 (1914), 1763 (two different papers); 160 (1915), 500; and *Bull. Soc. Chim.* (1915), 228. See Rheinboldt, p. 484; and Courtot, pp. 1445–1447.

154. Courtot, p. 1445.

155. Photograph of letter, dated 13 November 1912, in R. Grignard, *Centenaire*, pl. ix.

156. Ibid., p. 56.

157. Letter of 8 January 1913, reprinted in R. Grignard, *Centenaire*, p. 57; and article reprinted in R. Grignard, *Centenaire*, pl. xxxviii.

158. Nominations, courtesy of Elisabeth Crawford. According to Crawford, the first joint award in chemistry was justified not only by the similarity of Grignard's and Sabatier's methods but also by the consideration that if one were rewarded, several years would go by before the committee could recommend the other, since both their fields and their nationalities were the same. On the 1909 prize, see Elisabeth Crawford and Robert M. Friedman's essay in the Nobel Symposium volume.

159. Letter of 25 November 1912, quoted in R. Grignard, *Centenaire*, p. 56.

160. R. Grignard, *Centenaire*, p. 61.

161. V. Grignard, "Discours" (1933), p. 36; R. Grignard, *Centenaire*, p. 103; and Urbain, "L'oeuvre de Victor Grignard," p. i.

162. AN F17 26752. He now had a yearly salaried income of 22,000 francs.

163. V. Grignard, *Notice sur les titres*, pp. 22–23.

164. See Rheinboldt, p. 486; and Grignard, *Notice sur les titres*, pp. 9–10.

165. Among organic chemists of the next generation who directed some of the most important research with organomagnesium compounds were Charles Paul Prevost, Albert Kirrmann, and Henri Marie Normant. I am grateful to Pierre Laszlo of the University of Liège for discussing with me more recent work in organic chemistry.

166. Rheinboldt, p. 487.

167. R. Grignard, *Centenaire*, p. 109.

168. Reprinted in ibid., pl. xlviii.

169. Reprinted in ibid., pl. xlvii. On Perrin and the Popular Front, see Mary Jo Nye, "Science and Socialism: The Case of Jean Perrin in the Third Republic," *French Historical Studies*, 9 (1975), 141–169.

170. See A. Rosier, "Du chômage intellectuel" (31 March 1933), MS, in ADI, FdR, 78 pp.

171. R. Grignard, *Centenaire*, pp. 66–72.

172. Rheinboldt, p. 486; and V. Grignard, *Notice sur les titres*, p. 9.

173. See letter from Victor Grignard to W. A. Noyes, dated 2 June 1923, and reprinted in R. Grignard, *Centenaire*, pp. 96–99. According to Frederick Brown in his study of chemical warfare, the English arrived at a similar decision not to use gases in warfare (1) out of respect for the Hague Conventions of 1899 and 1907 prohibiting the "diffusion of asphyxiating or deleterious gases"; (2) out of fear of opening up an unlimited gas warfare; and (3) out of appreciation for the weakness of the British chemical industry. See Frederick J. Brown, *Chemical Warfare: A Study in Restraints* (Princeton, N.J.: Princeton University Press, 1968), pp. 7–9.

174. Letter from Victor Grignard to W. A. Noyes, dated 4 August 1923, and reprinted in translation in W. A. Noyes, *Building for Peace*, Vol. II, *International Letters* (New York: Chemical Catalogue Co., 1924), pp. 7–9.

175. See account of Cyrille Toussaint, one of these collaborators, in R. Grignard, *Centenaire*, pp. 74–80. After the war, Toussaint went to the Rhineland to direct the German factories of Farbwerke and Badische Anilin. Georges Rivat returned to Lyon, where he made a good deal of money in a dyeing factory. Jean Gérard built the "Maison de la Chimie." Ibid., p. 80.

176. Brown, n. 89, p. 43.

177. See account in G. E. Gibson and W. J. Pope, "Victor Grignard, 1871–1935," *Jl. Chem. Soc.* (1937), 171–179, on p. 176. Victor Grignard and Edouard Urbain, "Sur la préparation du phosgène au moyen du tétrachlorure de carbone et de l'oléum ou de l'acide sulfurique ordinaire," *CR, 169* (1919), 17; and "Sur la préparation du phosgène par la tétrachlorure de carbone et l'oléum ou l'acide sulfurique ordinaire et sur l'utilisation des résidus," by Victor Grignard, Bardet, Gérard, D. Simon, and Edouard Urbain, *Bull. Soc. Chim.* (1920), 322.

178. Brown, p. 11.

179. Guthrie discovered mustard gas by absorbing ethylene in sulfur chloride; *Jl. Chem. Soc.*, 12 (1860), 109. See Toussaint, in R. Grignard, *Centenaire*, pp. 76–77.

180. Gibson and Pope, p. 176; Victor Grignard, Georges Rivat, and G. Scatchard, "Sur le sulfure d'éthyle BB'-biiodé et son application à la détection et au dosage de l'ypérite," *CR, 15* (1921), 5; and "Sur la detection de l'ypérite et du thiodiglycol," *Bulletin, Association Française pour l'Avancement des Sciences* (1920), 125.

181. V. Grignard, *Notice sur les titres*, p. 72; Brown, pp. 3, 12.

182. See Gordon Wright, *France in Modern Times: 1760 to the Present* (Chicago: Rand McNally, 1966), p. 396.

183. See Jean Mayer, "Science," in Julian Park, ed., *The Culture of France in our Time* (Ithaca: Cornell University Press, 1954), pp. 266–336; André Langevin, *Paul Langevin, mon père* (Paris: Les Editeurs Français Réunis, 1971), 77–79; Robert Gilpin, *France in the Age of the Scientific State* (Princeton, N.J.: Princeton University Press, 1968), p. 155.

184. During the First World War, these committees were in mechanics, including aviation, wireless telegraphy, radiography; chemistry; medicine, surgery, and hygiene; and food supply. See Paul Appell, *Souvenirs d'un alsacien,*

*1858–1922* (Paris: Payot, 1923), pp. 259–261. On the Franco-Prussian War of 1870–71, see Maurice Crosland, "Science and the Franco-Prussian War," *Social Studies of Science, 6* (1976), 185–214.

185. R. M. Gattefossé, "L'industrie chimique à Lyon," in *Lyon. 1906–1926,* pp. 448–464, p. 448.

186. The United States did not ratify the agreement until 1974. Many chemists warned that other nations would secretly gain a lead both in chemical armaments and in related industrial know-how. See Daniel P. Jones, "American Chemists and the Geneva Protocol," *Isis, 71* (1980), 426–440.

187. Brigitte Schroeder-Gudehus, *Les scientifiques et la paix: La communauté scientifique internationale au cours des années 20* (Montreal: Presses de l'Université de Montréal, 1978).

188. My conversation with Aubry at Nancy, 12 June 1977.

189. Appell, *Souvenirs,* pp. 192–205.

190. See Schroeder-Gudehus, *Les scientifiques et la paix,* and Daniel J. Kevles, "Into Hostile Political Camps: The Reorganization of International Science in World War I," *Isis, 62* (1971), 47–60.

191. See Lyon, Fac. Sci., PV, 23 February 1923.

192. Ibid., 20 March 1926.

193. W. A. Noyes, *Building for Peace, Vol. I, A Chemist's Summer in Europe* (New York: Chemical Catalogue Co., 1923), esp. pp. 3–15.

194. See Kevles, p. 48. For the wording and names of the signers of the manifesto, see Georg F. Nicolai, *The Biology of War* (New York, 1918), pp. xi–xiii.

195. Letter from V. Grignard to Noyes, 2 June 1923, in R. Grignard, *Centenaire,* p. 99; and letter from V. Grignard to Noyes, 4 August 1923, in Noyes, Vol. II, p. 9.

196. Letter from V. Grignard to Noyes, 2 June 1923, in R. Grignard, *Centenaire,* p. 99.

197. Letter from V. Grignard to Noyes, 7 September 1923, in Noyes, Vol. II, pp. 13–15.

198. Julien Benda, *La trahison des clercs* (Paris: Grasset, 1927).

## 6: BORDEAUX

1. See the figures in chap. 4, pp. 133–134, and table in n. 76, chap. 4.

2. Camille Jullian, *Histoire de Bordeaux depuis les origines jusqu'en 1895* (Bordeaux: Feret et Fils, 1895), p. 741.

3. See "Bordeaux," in *Encyclopedia Britannica,* 11th ed. (London: 1910–11), pp. 244–245.

4. Ibid., and Jullian, pp. 756–758.

5. See Joseph Lajugie, "La population et la vie professionelle," and "L'essouflement de la vie économique," in Lajugie, ed., *Bordeaux au xxᵉ siècle* (Bordeaux, 1972), pp. 73–114 and 115–163, on pp. 74 and 141–144. (This is Vol. VII of Charles Higounet, ed., *Histoire de Bordeaux.*) Also, Jullian, p. 738.

6. Louis Desgraves and Charles Higounet, "Le mouvement intellectuel et musical," in L. Desgraves and Georges Depeux, eds., *Bordeaux au xixᵉ siècle*

(Bordeaux, 1969), pp. 455–520, on p. 489. (This is Vol. VI of *Histoire de Bordeaux.*)

7. In 1839, the society opened primary and professional courses to working-class male adults, and in 1866, to women. See ibid., pp. 455, 463–464, 488.

8. See George Weisz, *The Emergence of Modern Universities in France, 1863–1914* (Princeton, N.J.: Princeton University Press, 1983), pp. 98–107, 120–121. Also, see Harry W. Paul, "Science and the Catholic Institutes in Nineteenth-Century France," *Societas*, 1 (1971), 271–285; and for a more thorough treatment of Catholicism and science in France, Harry W. Paul, *The Edge of Contingency: French Catholic Reaction to Scientific Change from Darwin to Duhem* (Gainesville: University of Florida Press, 1979).

9. John McManners, *Church and State in France, 1870–1914* (New York, 1973), pp. 4–12, 39–52. Also Joseph N. Moody, "French Anti-Clericalism: Image and Reality," *Catholic Historical Review*, 56 (1970), 630–648, on p. 642.

10. R. Darricau, "Puissance du catholicisme," in *Bordeaux au xix$^e$ siècle*, pp. 297–323.

11. R. Darricau, "L'épiscopat et la république," in ibid., pp. 347–374, esp. pp. 347–352. On the *cercles catholiques*, see John M. Burney, *The University of Toulouse in the Nineteenth Century: Faculties and Students in Provincial France* (Ph.D. diss., University of Kansas, 1982), pp. 280–284.

12. Darricau, pp. 353–355.

13. Ibid., pp. 368–370.

14. Georges Rayet, *Histoire de la Faculté des Sciences de Bordeaux (1838–1894)* (Bordeaux: Gounouilhou, 1898), pp. 16–18.

15. Bordeaux, *Livret Etud., 1899–1900*, pp. 155–156.

16. Rayet, pp. 26–29.

17. Letter of 30 August 1838, cited in Rayet, pp. 20–21.

18. Bordeaux, Fac. Sci., PV (I), 16 December 1839, 30 December 1839, and 28 January 1841. The ministry furnished some journals free of charge to the Faculties. See Bordeaux, Fac. Sci., PV Assem. (I), 2 December 1898.

19. Bordeaux, Fac. Sci., PV (I), 23 January 1842.

20. Rayet, pp. 35–36.

21. *Bordeaux, Rentrée Facs., 1869–1870*, pp. 16–17.

22. Quoted in Rayet, p. 41.

23. Ibid., pp. 40–41.

24. G. Dupeux, "Les luttes politiques," in *Bordeaux au xix$^e$ siècle*, pp. 323–346, on pp. 327–328.

25. Rayet, pp. 47–50.

26. Bordeaux, Fac. Sci., PV Conseil (I), 22 February 1886.

27. Rayet, p. 58.

28. On this point, see Weisz, *Emergence of Modern Universities*, pp. 210–211.

29. Bordeaux, Fac. Sci., PV (I), 6 February 1841 and 20 April 1841.

30. Ibid., 7 February 1855.

31. Ibid., 21 April 1864 and 7 July 1870.

32. There were ten candidates for the licence in 1869, of whom three (of eight) passed in mathematics and two (of two) in physics. See *Bordeaux. Rentrée Facs., 1869–1870*, p. 20.

33. Albert Maire, *Catalogue des thèses de sciences soutenues en France de 1810 à 1890 inclusivement* (Paris: H. Welter, 1892).

34. Bordeaux, Fac. Sci., PV (I), 4 December 1873, 18 October 1873, and 15 January 1874.

35. See *Statistique Générale*, 3 (1880), 250; 15 (1892–1894), 266; and 20 (1900), 88.

36. M. Vèzes, "Rapport annuel du conseil de l'Université de Bordeaux. 1912–1913," in *Enquêtes Enseignt.*, no. 108 (Paris, 1914), pp. 150–161, on pp. 156–157. In a 1907 report, the Sciences Faculty dean complained bitterly that a *cabinet de physique*, "the object of pride of any Sciences Faculty, does not exist at Bordeaux," that PCN manipulations had to be done in the corridors of the physics and botany sections. In 1908, the complaint was that money had been given anonymously to the Medical Faculty, but the Sciences Faculty was in a grave condition. M. Marandot, "Rapport au conseil de l'Université de Bordeaux pour l'année scolaire, 1906–1907," in *Enquêtes Enseignt.*, no. 95 (Paris, 1908), pp. 77–96, p. 94; and M. P. Camena d'Almeida, "Rapport annuel du conseil de l'Université de Bordeaux pour l'année scolaire 1907–1908," in ibid., no. 97 (Paris, 1909), pp. 149–184, p. 183.

37. *Bordeaux, Univ. Conseil, 1920–1921*, pp. 15, 105; *Statistique Générale*, 46 (1930), 39–40. In the 1920s the students were mainly from Spain, Greece, and Serbia.

38. Desgraves and Higounet, pp. 476–478.

39. The Philomatic Society and Bordeaux Chamber of Commerce jointly established in 1873 an Ecole Supérieure de Commerce et d'Industrie, to which an electricity division and colonial section were added in 1902. This school provided two-year programs of instruction. In ibid., pp. 464–465.

40. *Bordeaux, Univ. Conseil, 1920–1921*, pp. 8–9.

41. Bordeaux, Fac. Sci., PV Conseil (IV), 9 March 1924 and 24 June 1920.

42. Ibid., 20 February 1923, 1 March 1923, 13 March 1923, 27 April 1923, and 24 May 1927. The sources at my disposal have made it difficult to estimate accurately the number of students enrolled in industrial physics and other engineering programs. In 1930, according to *Statistique Générale*, 46 (1930), 39–40, there were 674 students enrolled in the Bordeaux Sciences Faculty, of whom 102 were licence students, 133 PCN, 16 agrégation, and 1 doctorate. Two hundred forty-five were enrolled for "other diplomas" and the rest were in none of these categories.

43. *Annuaire Général de l'Université et de l'Enseignement Française 1929–1930* (Paris, 1929), pp. 261–262; and *Bordeaux, Livret Etud., 1930–1931*, p. 236.

44. *Catalogue des thèses et écrits académiques*, 1869–1897, 1905–06, 1909–10, 1914, 1925, 1930.

45. Maire, *Catalogue des thèses* (1892).

46. The latter included Charles Hugot in mineral (and physical) chemistry, Charles Henri Chevallier in physics, F. Caubet in physics, and Octave Manville in mineralogy.

47. Bordeaux, Fac. Sci., PV Conseil (III), 21 December 1909. His "polemical" articles were in the *Rev. Gén. Sci.* (1906) and *Lumière Electrique* (1909). The 1906

article seems mild in its call for more effective decentralization, in its criticism of the recent reorganization of the Ecole Normale Supérieure, and in its well-placed claim that the two-year education given by the Ecole Polytechnique was a superficial education in its coverage of a wide range of subjects. See Albert Turpain, "Les réformes de l'enseignement supérieur," *Rev. Gén. Sci.*, 17 (1906), 166–177, esp. p. 177. Turpain was one of the few young provincial scientists to dispute René Blondlot's claims for N-rays; see chap. 2.

48. Bordeaux, Fac. Sci., PV Conseil (I), 30 November 1897. This kind of move remained typical, as we have seen elsewhere. In 1929, Foch, professor of experimental physics at Bordeaux, left for a lectureship at the Sorbonne. Bordeaux, Fac. Sci., PV Conseil (IV), 20 December 1929.

49. Among the more eminent Faculty members in the 1920s was the chemist Georges Dupont, who completed his career in Paris. The organic chemist Laurent Richard and the astronomer Luc Picart remained at Bordeaux. Alfred Kastler, Nobel Prize winner in physics in 1966, finished his doctoral thesis at Bordeaux in 1936 and taught there from 1938 to 1941. He moved to Clermont-Ferrand to become a lecturer. Kastler, who died in 1984, completed his career at the Ecole Normale Supérieure. Charles Higounet and L. G. Planes, "La vie scientifique et littéraire," in *Bordeaux au xx<sup>e</sup> siècle*, pp. 605–646, on p. 618.

50. On Auguste Laurent, see J. R. Partington, *A History of Chemistry*, Vol. IV (London: Macmillan, 1964), pp. 376–393.

51. A. Baudrimont, "Observations sur les équivalents chimiques comparés aux éléments corpusculaires," 12-p. extract from *Moniteur Scientifique Quesneville* (1877) (Paris: Renou, Maulde, and Cocke, 1877), p. 10. Also Rayet, pp. 259–288.

52. Baudrimont, p. 4.

53. Ibid., pp. 7–9. See also, Partington, Vol. IV, pp. 393–395.

54. On Gayon, see the remarks after his death in 1929 by Richard and P. Cousin in *Bordeaux, Univ. Conseil, 1928–1929*, pp. 5–7, 171–172; also, Rayet, pp. 289–296. At the death of Baudrimont in 1880, the Bordeaux Sciences Faculty bought many of his books. Eighty of the 500 volumes came from the library of Lavoisier, carrying Lavoisier's signature and ex libris and comprising a series of chemical treatises published from 1688 to 1785, (Rayet, pp. 73–74). The Baudrimont family remained in Bordeaux for some generations. Albert Baudrimont was on the staff at the Medical Faculty in the 1920s and defended a thesis in medicine in 1929, as did Edouard-Jules M. A. A. Baudrimont in 1938–39. Rolard Baudrimont received a doctorate in the natural sciences at Bordeaux in 1973.

55. Bordeaux, Fac. Sci., PV Conseil (I), 12 November 1894.

56. AN F<sup>17</sup> 23295 (Pierre Duhem). Donald G. Miller, "Duhem," in *DSB*, Vol. IV, pp. 225–233; Emile Picard, *La vie et l'oeuvre de Pierre Duhem* (Paris: Gauthier-Villars, 1922); and *Mémoires de la Société des Sciences Physiques et Naturelles de Bordeaux*, 1 (1927): *L'Oeuvre scientifique de Pierre Duhem*, with articles by Octave Manville, Jacques Hadamard, and A. Darbon, and a bibliography.

57. See Duhem, *Physique de croyant*, extract from *Annales de Philosophie Chrétienne* (Paris: Bloud et Cie, 1905), p. 6.

58. See Miller, "Duhem," p. 228; and Picard, p. 3.

59. Miller, "Duhem," p. 229; and Picard, p. 6.

60. Reports by Darboux and Bouty, dated September 1888, in AN F¹⁷ 23295. The role of Gabriel Lippmann in the rejection of Duhem's first doctoral thesis is discussed in Stanley L. Jaki, *Uneasy Genius: The Life and Work of Pierre Duhem* (The Hague: M. Nijhoff, 1984), pp. 51–52.

61. Duhem placed first in the physics agrégation concours. See letter from Georges Perrot to the minister, dated 16 October 1886, in AN F¹⁷ 23295. Duhem succeeded Pionchon as an aide at the Ecole Normale, and later succeeded him once more, in a chair at Bordeaux.

62. Picard, pp. 15ff.

63. Quoted in Picard, p. 13. See Miller, "Duhem," p. 229; and Duhem, "Commentaires aux principes de la thermodynamique," *Journal de Mathématiques Pures et Appliquées, 8* (1892), 269–330; *9* (1893), 293–359; and *10* (1894), 207–285.

64. Duhem, *Traité élémentaire de mécanique chimique fondée sur la thermodynamique,* 2 vols. (Paris: Hermann, 1897), in I, vi.

65. See his general monograph, the *Traité d'energétique* (1911). Also during this period (1896–1910), he became involved in a polemical discussion regarding "false equilibria," where "false equilibria" are instances of extremely slow reaction rates. See Miller, "Duhem," pp. 229–230.

66. Picard, p. 23.

67. For his comments on kinetic theory, see Duhem, "La loi des phases à propos d'un livre récent de M. Wilder D. Bancroft," *RdQS, 44* (1898), 54–82, on pp. 56–57.

68. See Duhem, "Les théories électriques de James Clerk Maxwell. Etude historique et critique," *RdQS, 49* (1901), 5–21, on p. 19.

69. "To be positivist is to affirm that there is no other logical method than the method of the positive sciences; that which is unknowable to the positive sciences is in itself and absolutely unknowable." In "Physique et métaphysique," *RdQS, 34* (1893), 55–83, on p. 70.

70. Duhem, book review of the French translation by Emile Bertrand of Mach's *La mécanique,* trans., 4th German ed. (Paris: Hermann, 1904), in *RdQS, 55* (1904), 198–217, on p. 198.

71. See Mary Jo Nye, "The Boutroux Circle and Poincaré's Conventionalism," *Journal of the History of Ideas, 40* (1979), 107–120.

72. For Duhem's debts to Jules Tannery, see *Physique de croyant,* pp. 7–8. Duhem wrote an obituary notice for Jules Tannery's brother, Paul Tannery, who shared with Duhem a fervent and open Catholic faith and an avocation in the history of science. On Paul Tannery and the anticlerical, political factors influencing his career, see Harry W. Paul, "Scholarship and Ideology: The Chair of the General History of Science at the Collège de France, 1892–1913," *Isis, 67* (1976), 376–397. Also, Duhem, "Paul Tannery. 1843–1904," extract from *Revue de Philosophie* (Montligeon, 1905), 15 pp.

73. On this point, see the very fine chapter on Duhem in Harry W. Paul, *The Edge of Contingency,* pp. 137–178, as well as the different emphasis (Pascal, rather than neo-Thomism) of Niall Martin in his review of Paul's book, "Darwin and Duhem," *History of Science, 20* (1982), pp. 64–72.

74. See Duhem, *Physique de croyant*, pp. 8–9.

75. Ibid., and p. 23. On the impossibility of a "crucial experiment," see Duhem, "Quelques réflexions au sujet de la physique expérimentale," *RdQS*, 36 (1894), 179–229.

76. RC for 1893–94, dated June 1894 by the Rennes rector. AN F$^{17}$ 23295.

77. Rayet, pp. 228–234; and Bordeaux, Fac. Sci., PV Conseil (I), 12 November 1894.

78. Ibid., 27 November 1894.

79. Letter from Rayet to Louis Liard, director of higher education, and former Bordeaux Letters Faculty member, dated 4 December 1894. AN F$^{17}$ 23295.

80. On this, see Paul Forman, John Heilbron, and Spencer Weart, *Physics ca. 1900, HSPS*, 5 (1975), pp. 30–32; and Bordeaux, Fac. Sci., PV Conseil (I), 22 January 1895 and 30 April 1895.

81. Ibid., 22 January 1897.

82. Ibid., 9 October 1895.

83. Bordeaux, Fac. Sci., PV Assem. (I), 26 February 1898.

84. Ibid., 16 May 1900; and Bordeaux, Fac. Sci., PV Conseil (II), 3 July 1906.

85. See Bordeaux, Fac. Sci., PV Assem. (II), 17 July 1906; and (I), 18 November 1902.

86. Ibid. (I), 16 May 1900.

87. Bordeaux, Fac. Sci., PV Conseil (II), 19 December 1902 and 13 November 1903.

88. For Duhem's remarks, ibid., 3 July 1906.

89. Ibid., 29 June 1906.

90. Ibid., 3 July 1906.

91. In May 1907, Sauvageau complained that tied votes were giving the new dean, Padé, increased power because he could vote to break a deadlock. Bordeaux, Fac. Sci., PV Assem. (II), 31 May 1907.

92. The new physics professors were Joseph Guinchant, who had been adjoint professor at Caen, and Bénard, who had been lecturer at Lyon. See Bordeaux, Fac. Sci., PV Conseil (II), 30 October 1909; ibid. (III), 5 November 1909, 3 December 1909, 21 December 1909, 18 February 1910, 4 March 1910, 13 October 1918.

93. Duhem, "Usines et laboratoires," *Revue Philomathique de Bordeaux et du Sud-Ouest*, 2 (1899), 385–400.

94. One of his students was Paul Saurel, who taught mathematics at the City University of New York and came to Bordeaux to study with Duhem. Sidney Ratner, professor in the history department at Rutgers University, initially called Saurel to my attention.

95. Bordeaux, Fac. Sci., PV Assem. (I), 23 November 1900.

96. Ibid., 27 November 1900. In 1906, the ratio of state-supported to regionally supported "complementary" courses and conferences, or small lecture courses, was 3:7. See Bordeaux, Fac. Sci., PV Conseil (II), 25 May 1906.

97. It was not until 1920 that "physical chemistry" was added to the title. See Bordeaux, Fac. Sci., PV Conseil (I), 2 March 1900, 15 March 1901, 17 May 1901; and ibid. (III), 20 February 1920.

98. *Bordeaux, Livret Etud., 1909–1910,* p. 77.

99. Bordeaux, Fac. Sci., PV Conseil (II), 14 January 1902.

100. Ibid.

101. RC for 1899, AN F$^{17}$ 23295.

102. RC for 1892. Letter from Dean Demartres to the Lille rector, dated 10 July 1893; letter from the rector to the ministry, dated 12 July 1893. AN F$^{17}$ 23295. Donald Miller discusses this incident in "Ignored Intellect: Pierre Duhem" *Physics Today, 19* (1966), 47–53, on p. 50.

103. RC for 1900, dated 5 May 1900. AN F$^{17}$ 23295.

104. Duhem's speech was published in the antirepublican, clerical *Nouvelliste,* according to Bizos. See letter from Rector Bizos to minister, dated 28 June 1899. AN F$^{17}$ 23295.

105. According to Miller, "Ignored Intellect," p. 52. RCs for 1903 and 1899, AN F$^{17}$ 23295. By 1913, a new university rector had a different opinion of Duhem, which may have had some influence in Duhem's election in late 1913 as one of the first six nonresident members of the Academy of Sciences. See Letter from rector to Ministry of Public Instruction, dated 31 March 1913, in which the rector expressed his confidence in Duhem's loyalty, in AN F$^{17}$ 23295. One of Duhem's biographers had denied the validity of the political allegations against Duhem, stating that "Duhem never engaged in politics, and even acted as if he had no opinions. However, his secret preference pushed him from the side of the democrats [*du côté des démocrates*], among whom he counted many friends." In Pierre Humbert, *Pierre Duhem* (Paris: Librairie Bloud et Gay, 1932), n. 1, p. 126.

106. Paul Saurel did his dissertation with Duhem on systems equilibria. Both Duhem and Saurel were regular contributors to the *American Journal of Physical Chemistry* in its early years. Duhem's student Lucien Marchis wrote his thesis on the deformation of glass (1898), Fernand Caubet on liquefaction of gas mixtures (1901), and Octave Manville on the finite deformation of a continuous milieu (1903). They all taught at Bordeaux, and, like Duhem, they contributed articles to the *Zeitschrift für physikalische Chemie* or the *Journal de Chimie Physique* in the early 1900s. Duhem's other doctoral students included E. Monnet (1897), H. Pélabon (1898), Albert Turpain (1899), E. Lenoble (1900), and C. H. Chevallier (1901). On these students, see Jaki, *Uneasy Genius,* pp. 131–143; also p. 153.

107. According to Humbert, *Pierre Duhem,* pp. 18–19. Also, see Harry Paul, "Scholarship and Ideology," p. 397, for the suppression of the chair after Wyrouboff's death in 1913, until its reestablishment for Pierre Boutroux in 1920.

108. *Bordeaux, Livret Etud., 1909–1910,* pp. 73–74.

109. NIs and RCs for 1909, 1910, AN F$^{17}$ 23295.

110. See Niall Martin, "The Genesis of a Medieval Historian: Pierre Duhem and the Origins of Statics," *Annals of Science, 33* (1976), 119–129.

111. Duhem, *Origines de la statique,* Vol. I (Paris, 1905), p. ii. Niall Martin cites Duhem's statement that during winter 1903–04 he found "there was no way out of analyzing every manuscript relating to statics in the Bibliothèque Nationale and Bibliothèque Mazarine" in order to discover what Renaissance mechanics owed Jordanus de Nemore and his disciples. Duhem's boyhood training in Greek and Latin was very useful. See Martin, p. 126.

112. See Mary Jo Nye, "The Moral Freedom of Man and the Determinism of Nature: The Catholic Synthesis of Science and History in the *Revue des Questions Scientifiques*," *British Journal for the History of Science, 9* (1976), 274–292.

113. *La théorie physique; son objet et sa structure* (Paris, 1906), first published in *Revue de Philosophie*; and *Essai sur la notion de théorie physique de Platon à Galilée* (Paris: Hermann, 1908); first published in the *Annales de Philosophie Chrétienne*. Also see his "Les théories éléctriques de James Clerk Maxwell," where he credits Maxwell with the genius of imagination but identifies the highest kind of scientific work with "the model of order we find in Gauss" (pp. 19–20).

114. Duhem, "L'Evolution des théories physiques de xviii$^e$ siècle jusqu'à nos jours," *RdQS, 40* (1896), 463–499, on pp. 497–498.

115. Abel Rey, "La philosophie scientifique de M. Duhem," *Revue de Métaphysique et de Morale, 12* (1904), 699–744, on pp. 740–741, 744.

116. *Physique de croyant*, pp. 31–34; and Paul, *Edge of Contingency*, pp. 137–178.

117. Two of his most ambitious historical works were to come after 1905, the three volumes of the *Etudes sur Léonard da Vinci, ceux qu'il a lus et ceux qui l'ont lu* (Paris: 1906–1913) and *Le système du monde. Histoire des doctrines cosmologiques de Platon à Copernic*, 10 vols. (Paris, 1913–1959), the first five volumes of which were published from 1913 to 1917.

118. See *La théorie physique*, pp. 85–149; *La science allemande* (Paris: A. Hermann, 1915) and *La chimie est-elle une science française?* (Paris: A. Hermann, 1916). Also, see Harry Paul's discussion, "Pierre Duhem as Propagandist: A Subtle Revision," in Paul, *The Sorcerer's Apprentice. The French Scientist's Image of German Science 1840–1919* (Gainesville: University of Florida Press, 1972), pp. 54–76.

119. Picard, p. 49.

## 7: CONCLUSION

1. Joseph Ben-David, *The Scientist's Role in Society: A Comparative Study* (Englewood Cliffs, N.J.: Prentice-Hall, 1971); Terry Shinn, "The French Science Faculty System 1808–1914: Institutional Change and Research Potential," *HSPS, 10,* (1979), 271–332; and Robert Fox and George Weisz, "Introduction: The Institutional Basis of French Science in the Nineteenth Century," in Fox and Weisz, eds., *The Organization of Science and Technology in France 1808–1914* (Cambridge, England: Cambridge University Press, 1980), pp. 1–28, on p. 14.

2. Ben-David, *Scientist's Role*, pp. 133, 176–179.

3. Charles C. Gillispie, *Science and Polity in France at the End of the Old Regime* (Princeton, N.J.: Princeton University Press, 1980); Gerald Geison, ed., *Professions and the French State, 1700–1900* (Philadelphia: University of Pennsylvania Press, 1984).

4. Ted Feldman has analyzed data demonstrating this to be the case. The data is housed by the History of Physics Project at the Office for History of Science and Technology at the University of California at Berkeley.

5. Paul Forman, John Heilbron, and Spencer Weart, in *Physics circa 1900*, special issue of *HSPS, 5* (1975), claim that the number of students taking

courses beyond elementary physics decreased in German universities in the late 1880s and early 1890s, and that a similar decline in physics in Italian universities was blamed on industrial employers' preference for graduates of technical schools (on p. 29).

6. See *Institut de France: Index biographique de l'Académie des Sciences, 1666–1978* (Paris: Gauthier-Villars, 1978), pp. 11–12.

7. On German scientific imperialism, see Lewis Pyenson, "Cultural Imperialism and Exact Sciences: German Expansion Overseas, 1900–1930," *History of Science, 20,* (1980), 1–43. My impression about foreign students in German universities after 1900 has been confirmed by Steven Turner, who has extensively studied science in the German universities.

8. Spencer R. Weart, "The Physics Business in America, 1919–1940: A Statistical Reconnaissance," in Nathan Reingold, ed., *The Sciences in the American Context: New Perspectives* (Washington, D.C.: Smithsonian Institution, 1979), pp. 295–358, on p. 300. On foreign scientists in France, see Dominique Pestre, *Physique et physiciens en France, 1918–1940* (Paris: Editions des archives contemporaines, 1984), pp. 155–156, and esp. n. 10, pp. 166–167.

9. See Paul Courteault, "L'Université de Bordeaux et la guerre," Conference at Athens, 4 March 1918, extracted from *Revue Philomathique de Bordeaux et du Sud-Ouest, 21* (1918), 4–10.

10. On this point, see Jean-Jacques Salomon, *Science and Politics,* Noël Lindsay, trans. (Cambridge: M.I.T. Press, 1973), p. 28.

11. See Spencer Weart, *Scientists in Power* (Cambridge: Harvard University Press, 1979), pp. 16–18, 23–25; and Jean Perrin, *L'Organisation de la recherche en France* (Paris, 1938).

12. Weart, *Scientists in Power,* pp. 30–31; and Mary Jo Nye, "Science and Socialism: The Case of Jean Perrin in the Third Republic," *French Historical Studies, 9* (1975), 141–169, on pp. 161–163.

13. See Piaganiol and Villecourt, *Pour une politique scientifique* (Paris, 1963), p. 138.

14. Perrin, *L'Organisation de la recherche.* Also, Pestre, *Physique et physiciens,* p. 220.

15. Shinn, "French Science Faculty System," p. 323.

16. Harry W. Paul, "Apollo Courts the Vulcans: The Applied Science Institutes in Nineteenth-Century French Science Faculties," in Fox and Weisz, eds., *Organization,* pp. 155–182, on p. 155.

17. John L. Heilbron, *"Fin-de-siècle* Physics," in C. G. Bernhard et al., eds., *Science, Technology and Society in the Time of Alfred Nobel* (Oxford: Pergamon Press, 1982), pp. 51–73, on p. 67.

18. Michael Sanderson, *The Universities and British Industry, 1850–1970* (London: Routledge and Kegan Paul, 1972), pp. 29, 80–81.

19. Peter Lundgreen, "The Organization of Science and Technology in France: A German Perspective," in Fox and Weisz, eds., *Organization,* pp. 322–332, on pp. 313, 320.

20. Heilbron, *"Fin-de-siècle* Physics," p. 68.

21. See, e.g., Peter Borscheid, *Naturwissenschaft, Staat und Industrie in Baden*

*(1848–1914)* (Stuttgart, 1976); Karl-Heinz Manegold, *Universität, Technische Hoch-schule und Industrie* (Berlin, 1970); and the review essay by David Cassidy, "Recent German Perspectives on German Technical Education," *HSPS*, 14 (1983), 187–200.

22. Craig Zwerling, "The Emergence of the Ecole Normale Supérieure as Centre of Scientific Education in the Nineteenth Century," pp. 31–60, and Victor Karady, "Educational Qualifications and University Careers in Science in Nineteenth-Century France," pp. 95–124, in Fox and Weisz, eds., *Organization*.

23. H. Paul, p. 161; and Sanderson, p. 10.

24. See T. J. Markovitch, "L'Evolution industrielle de la France," *Revue d'Histoire Economique et Social*, 33 (1975), 266–288; and Robert A. Nye, *Crime, Madness and Politics in Modern France: The Medical Model of National Decline* (Princeton, N.J.: Princeton University Press, 1984).

25. Antoine Prost, *Histoire de l'enseignement en France, 1800–1967* (Paris: Armand Colin, 1968), p. 234.

26. See pp. 67–68 in A. Rosier, "Du chômage intellectuel," MS, 78 pp. Completed 31 March 1933 and deposited in the ADI, FdR. (This manuscript was published in *L'enseignement public* (Paris: Delagrave, 1948).

27. Ibid., p. 19, and tables, pp. 32, 36.

28. Sanderson, pp. 4–5. Also see Dominique Pestre, *Physique et physiciens.*

29. See Mary Jo Nye, "Recent Sources and Problems in the History of French Science," *HSPS*, 13 (1983), 401–415.

30. John H. Weiss, *The Making of Technological Man: The Social Origins of French Engineering Education* (Cambridge: M.I.T. Press, 1981), pp. 73–75; and Zwerling, p. 31.

31. Cassidy, p. 197.

32. Spencer Weart has made a strong case for the invigorating effects of industrial support on American physics after the First World War: "the possibilities for inquiry were broadened all around" in a scientific community in which industrial physicists made up 25 percent of the membership of the American Physical Society in 1920. In Weart, "The Physics Business," pp. 302, 304–305.

33. See Maurice Levy-Leboyer, "The Contribution of French Scientists and Engineers to the Development of Modern Managerial Structures in the Early Part of the Twentieth Century," in *Science, Technology and Society in the Time of Alfred Nobel*, pp. 283–297, esp. pp. 288–289. Also Vital Chomel, ed., *Histoire de Grenoble* (Toulouse: Privat, 1976), pp. 299–301.

34. See "Techniciens en quête de promotion," "Electronique-Electrotechnique," "Informatique," and "Chimie," in *Le Monde de Education*, no. 85 (1982), 29–35.

35. Sanderson, p. 22.

36. Quoted in Erwin N. Hiebert, "Nernst and Electrochemistry," in George Dubpernell and J. H. Westbrook, eds., *Selected Topics in the History of Electrochemistry* (Princeton, N.J.: The Electrochemical Society, 1978), pp. 180–200.

37. *Physics circa 1900*, table 1, p. 6; and pp. 115–118.

38. Ibid., p. 128. And Mary Jo Nye, "Scientific Decline: Is Quantitative Evalu-

ation Enough?" *Isis*, 76 (1985), 697–708. For a recent evaluation of French physics, see Pestre, esp. pp. 84–97, which includes comparisons with American physics.

39. Robert Nye, *Crime, Madness and Politics in Modern France: The Medical Concept of National Decline* (Princeton, N.J.: Princeton University Press, 1984), pp. 134–135.

40. Victor Grignard and Georges Rivat, "Sur une modification à apporter à la loi sur les brevets d'invention," *Chimie et Industrie*, 1 February 1919, repr., 5 pp.

41. See n. 4.

42. On this, see Harry W. Paul, "The Debate over the Bankruptcy of Science in 1895," *French Historical Studies*, 5 (1968), 299–327.

43. Gerald Geison,"Scientific Change, Emerging Specialties, and Research Schools," *History of Science*, 19 (1981), 20–40, esp. chart 2, p. 24.

44. For a review of some of this literature, see Robert K. Merton, "The Sociology of Science: An Episodic Memoir," pp. 3–141 in Merton and Jerry Gaston, eds., *The Sociology of Science in Europe* (Carbondale, Ill.: Southern Illinois University Press, 1977).

45. Pestre, *Physique et physiciens*, p. 221, where he suggests, "Le résultat est qu'en province les cours de faculté sont plus modernes, plus à jour nous l'avons vu, tandis qu'à Paris une tendance à la sclérose marque certains d'entre eux."

46. For theories of innovation and growth, see Michael Mulkay, *The Social Process of Innovation: A Study in the Sociology of Science* (London: Macmillan, 1972). Also, Robert K. Merton, "The Perspectives of Insiders and Outsiders," in Merton's *The Sociology of Science* (Chicago: The University of Chicago Press, 1973), pp. 99–136.

# GLOSSARY

academy: geographical division within the national system of education organized under Napoleon

Academy: an organization of scientists or scholars, artists, writers, such as the
Academy of Sciences in Paris

*agrégation:* competitive national qualifying examination for teachers in state
lycées and usually expected of university Faculty members

baccalaureate *(baccalauréat):* competitive national examination at the end of secondary (lycée) education, expected of students entering traditional programs in the Faculties and *grandes écoles*

*chargé de conférences:* "deputy" lecturer

*chargé de cours:* "deputy" professor, a teacher of a regular university course,
with the expectation of promotion to professor

*chargé des travaux:* laboratory assistant, second in rank to the laboratory director

*chef des travaux:* laboratory supervisor, without professorial rank

*concours:* competitive examination used to select a predetermined number of
qualifiers on the basis of the highest examination scores

*cours libres:* open lectures often taught by lecturers who were not Faculty members, but approved by them

*département:* geographical division within the administrative system of the
French government

*grandes écoles:* schools of higher education with admission by special national
competitive examinations which prepare students for professions, traditionally in state service; many of these schools, like the Ecole Polytechnique, Ecole des Mines, and Ecole des Ponts et Chaussées, were controlled by government ministries other than the Ministry of Public
Instruction

*licence:* a degree requiring two years of university-level study

lycée: state-supported secondary-level school, preparing students for the baccalaureate examination and certificate

*maître de conférences:* lecturer in the university

*normalien:* graduate of the Ecole Normale Supérieure, the Parisian grande école
preparing students for the agrégation

*palais:* a large building or group of buildings

*préparateur:* aide or assistant

*professeur adjoint:* "adjunct" or associate professor

*suppléant:* temporary replacement

NOTE: In the university hierarchy, titles included, in order of descending rank: professeur, professeur sans chaire, professeur adjoint, chargé de cours, maître de conférences, chargé de conférences, and maître adjoint de conférences.

# INDEX

318      INDEX

Becquerel, Henri (Becquerel rays or radioactivity), 57, 58–59, 60, 71, 72, 75–76, 99, 237
Becquerel, Jean, 66, 69, 71, 73–74
Béhal, Auguste, 174, 176, 177, 182
Beilstein, Friedrich, 173, 181, 192
Bellet, E. M., 184
Ben-David, Joseph, 224–225, 226, 227
Bernard, Claude, 5
Bernheim, Hippolyte, 71
Berson, G., 125
Bert, Paul, 18, 197
Berthelot, Marcellin, 18, 115, 127, 140, 151, 177, 182, 189, 219; as thesis director, 43, 139, 159, 167; chemical work and epistemological views, 106, 110, 116, 139, 141, 143–146 passim, 152, 166, 209, 210, 222; influence in recommending appointments, 76, 138–139
Berzelius, J. J., 110, 115, 143, 146
Besançon, 3, 12–13, 126, 181, 182
Bichat, Ernest, 38, 40, 41, 44, 69, 76–77, 239, 240
Bilon, F. M. H., 82
Bineau, A., 159
Bizos, Gaston, 218, 219
Blagden, Charles, 106
Blaise, Emile Edmond, 174–184 passim
Bloch, Felix, 227
Blondlot, René, 4, 8, 44, 47, 53–77, 173, 236–237, 239, 240; early researches on electric waves, x-rays, 54–58; Le Conte Prize, 75–76, 236; N-ray research, 56–69; social networks of, 57, 73, 76–77.
Bohr, Niels, 180
Boirac, E., 85, 103
Boltzmann, Ludwig, 112, 210
Bordeaux, 4, 7–8, 27, 31, 38, 119, 122, 139, 151, 194, 195–223, 241; applied science at, 126, 128, 202, 203–205, 217–218, 223, 230; construction of new facilities, 21, 201; doctorates, 202, 205, 215; Ecole de Chimie, 203, 217–218, 231; Ecole de Radiotélégraphie, 204; faculty positions increased, 200, 204; foreign students, 29 table,

203–204; Institut du Pin, 204, 217; Sciences Faculty established, 197, 199; Société des Sciences Physiques et Naturelles, 97, 215; Société Philomathique, 197, 306 n. 39; Station Agronomique et Oenologique, 204; student enrollments, 130, 195, 203, 216, 306 n. 42
Bordier, H., 63–64
Bouasse, Henri, 127–128, 129, 132, 136–137, 286 n. 100
Bouglé, Celestin, 31
Bouin, Pol, 45
Bourgeois, Léon, 120
Bourion, F., 165, 183
Boussinesq, J., 213
Boussingault, J. B., 157, 159
Boutroux, Emile, 15, 16, 50, 181, 211, 212, 221
Boutroux, Léon, 181
Bouty, Edmond, 26, 66, 72, 76, 209
Bouveault, Louis, 166–167, 174, 181, 183, 240
Bréal, Michel, 27
Brenier, Casimir, 88, 239
Bret, H., 82
Brillouin, Marcel, 66, 73, 121, 149, 152
Broca, André, 73–74
Brown, Frederick J., 303 n. 173
Bruhat, Georges, 241
Bruneau, Charles, 32
Brunel, Georges, 218
Büchner, Ernst, 150
Buhl, Adolphe, 134, 135
Burke, John, 62, 63
Burney, John, 3
Butlerov, Alexander, 170

Cabannes, Jean, 241
Caen, 12–13, 126
Cahan, David, 23
Cahours, Auguste, 171
Cailletet, L. P., 66, 72
Camichel, Charles, 66, 121, 129, 131, 132, 135, 136, 151
Cannizzaro, Stanislao, 108, 109
Carbonelle, Ignace, 220
Cardwell, David, 25

Designer:     U.C. Press Staff
Compositor:   Auto-Graphics, Inc.


Text:     10/13 Palatino
Display:  Palatino

Milton Keynes UK
Ingram Content Group UK Ltd.
UKHW012009180823
427123UK00001B/2